T0363911

CARBON NANOTUBES

Elsevier Journals of Related Interest

Applied Superconductivity
Carbon
Journal of Physics and Chemistry of Solids
Nanostructured Materials
Polyhedron
Solid State Communications
Tetrahedron
Tetrahedron Letters

CARBON NANOTUBES

Edited by

MORINUBO ENDO
Shinshu University, Japan

SUMIO IIJIMA
NEC, Japan

MILDRED S. DRESSELHAUS
Massachusetts Institute of Technology, USA

PERGAMON

U.K. Elsevier Science Ltd, The Boulevard, Langford Lane, Kidlington, Oxford OX5 1GB, U.K.

U.S.A. Elsevier Science Inc., 660 White Plains Road, Tarrytown, New York 10591-5153, U.S.A.

JAPAN Elsevier Science Japan, Tsunashima Building Annex, 3-20-12 Yushima Bunko-ku, Tokyo 113, Japan

---

First edition 1996

Library of Congress Cataloging in Pulication Data

A catalog record for this book is available from the Library of Congress

British Library Cataloguing in Publication Data

A catalogue record for this book is available in the British Library

ISBN 008 0426824

Reprinted from:

Carbon, Vol. 33, Nos 1, 2, 7, 12

Transferred to digital printing 2005
Printed and bound by Antony Rowe Ltd, Eastbourne

CONTENTS

EDITORIAL

Carbon nanotubes have been studied extensively in relation to fullerenes, and together with fullerenes have opened a new science and technology field on nano scale materials. This book aims to cover recent research and development in this area, and so provide a convenient reference tool for all researchers in this field. It is also hoped that this book can serve to stimulate future work on carbon nanotubes.

Carbon nanotubes have the same range of diameters as fullerenes, and are expected to show various kinds of size effects in their structures and properties. Carbon nanotubes are one-dimensional materials and fullerenes are zero-dimensional, which brings different effects to bear on their structures as well as on their properties. A whole range of issues from the preparation, structure, properties and observation of quantum effects in carbon nanotubes in comparison with *O*-D fullerenes are discussed in this book.

In order to review the wide research area of carbon nanotubes this book focuses on recent intensive work published in *Carbon*. The papers are written from the viewpoint that carbon nanotubes, as well as fullerenes, are the most interesting new carbon allotropes. Readers can then understand the fascination of graphene sheets when they are rolled into a nanometer size tubular form from a flat network corresponding to conventional graphite. This book also contains complementary reviews on carbon nanoparticles such as carbon nano-capsules, onion-like graphite particles and metal-coated fullerenes.

We hope this book will contribute to the dissemination of present understanding of the subject and to future developments in the science and technology of carbon nanotubes and fullerenes, and of carbon science more generally.

The editors thank all authors who contributed so many excellent papers covering all aspects of carbon nanotubes and the related fields. We are indebted to the Editor-in-Chief of *Carbon*, Professor Peter A. Thrower, for his suggestion and kind efforts, and also to Dr V. Kiruvanayagam for her kind cooperation related to this book.

Morinobu Endo
Sumio Iijima
Mildred S. Dresselhaus
Editors

PREFACE

Since the start of this decade (the 1990's), fullerene research has blossomed in many different directions, and has attracted a great deal of attention to Carbon Science. It was therefore natural to assemble, under the guest editorship of Professor Harry Kroto, one of the earliest books on the subject of fullerenes [1], a book that has had a significant impact on the subsequent developments of the fullerene field. Stemming from the success of the first volume, it is now appropriate to assemble a follow-on volume on Carbon Nanotubes. It is furthermore fitting that Dr Sumio Iijima and Professor Morinobu Endo serve as the Guest Editors of this volume, because they are the researchers who are most responsible for opening up the field of carbon nanotubes. Though the field is still young and rapidly developing, this is a very appropriate time to publish a book on the very active topic of carbon nanotubes.

The goal of this book is thus to assess progress in the field, to identify fruitful new research directions, to summarize the substantial progress that has thus far been made with theoretical studies, and to clarify some unusual features of carbon-based materials that are relevant to the interpretation of experiments on carbon nanotubes that are now being so actively pursued. A second goal of this book is thus to stimulate further progress in research on carbon nanotubes and related materials.

The birth of the field of carbon nanotubes is marked by the publication by Iijima of the observation of multi-walled nanotubes with outer diameters as small as 55 Å, and inner diameters as small as 23 Å, and a nanotube consisting of only two coaxial cylinders [2]. This paper was important in making the connection between carbon fullerenes, which are quantum dots, with carbon nanotubes, which are quantum wires. Furthermore this seminal paper [2] has stimulated extensive theoretical and experimental research for the past five years and has led to the creation of a rapidly developing research field.

The direct linking of carbon nanotubes to graphite and the continuity in synthesis, structure and properties between carbon nanotubes and vapor grown carbon fibers is reviewed by the present leaders of this area, Professor M. Endo, H. Kroto, and co-workers. Further insight into the growth mechanism is presented in the article by Colbert and Smalley. New synthesis methods leading to enhanced production efficiency and smaller nanotubes are discussed in the article by Ivanov and coworkers. The quantum aspects of carbon nanotubes, stemming from their small diameters, which contain only a small number of carbon atoms ($<10^2$), lead to remarkable symmetries and electronic structure, as described in the articles by Dresselhaus, Dresselhaus and Saito and by Mintmire and White. Because of the simplicity of the single-wall nanotube, theoretical work has focussed almost exclusively on single-wall nanotubes. The remarkable electronic properties predicted for carbon nanotubes are their ability to be either conducting or to have semiconductor behavior, depending on purely geometrical factors, namely the diameter and chirality of the nanotubes. The existence of conducting nanotubes thus relates directly to the symmetry-imposed band degeneracy and zero-gap semiconductor behavior for electrons in a two-dimensional single layer of graphite (called a graphene sheet). The existence of finite gap semiconducting behavior arises from quantum effects connected with the small number of wavevectors associated with the circumferential direction of the nanotubes. The article by Kiang *et al.* reviews the present status of the synthesis of single-wall nanotubes and the theoretical implications of these single-wall nanotubes. The geometrical considerations governing the closure, helicity and interlayer distance of successive layers in multilayer carbon nanotubes are discussed in the paper by Setton.

Study of the structure of carbon nanotubes and their common defects is well summarized in the review by Sattler, who was able to obtain scanning tunneling microscopy (STM) images of carbon nanotube surfaces with atomic resolution. A discussion of common defects found in carbon nanotubes, including topological, rehybridization and bonding defects is presented by Ebbesen and Takada. The review by Ihara and Itoh of the many helical and toroidal forms of carbon nanostructures that may be realized provides insight into the potential breadth of this field. The joining of two dissimilar nanotubes is considered in the article by Fonseca *et al.*, where these concepts are also applied to more complex structures such as tori and coiled nanotubes. The role of semi-toroidal networks in linking the inner and outer walls of a double-walled carbon nanotube is discussed in the paper by Sarkar *et al.*

From an experimental point of view, definitive measurements on the properties of individual carbon nanotubes, characterized with regard to diameter and chiral angle, have proven to be very difficult to carry out. Thus, most of the experimental data available thus far relate to multi-wall carbon nanotubes and to bundles of nanotubes. Thus, limited experimental information is available regarding quantum effects for carbon nanotubes in the one-dimensional limit. A review of structural, transport, and susceptibility measurements on carbon nanotubes and related materials is given by Wang *et al.*, where the interrelation between structure and properties is emphasized. Special attention is drawn in the article by Issi *et al.* to quantum effects in carbon nanotubes, as observed in scanning tunneling spectroscopy, transport studies and magnetic susceptibility measurements. The vibrational modes of carbon nanotubes is reviewed in the article by Eklund *et al.* from both a theoretical standpoint and a summary of spectroscopy studies, while the mechanical that thermal properties of carbon nanotubes are reviewed in the article by Ruoff and Lorents. The brief report by Despres *et al.* provides further evidence for the flexibility of graphene layers in carbon nanotubes.

The final section of the volume contains three complementary review articles on carbon nanoparticles. The first by Y. Saito reviews the state of knowledge about carbon cages encapsulating metal and carbide phases. The structure of onion-like graphite particles, the spherical analog of the cylindrical carbon nanotubes, is reviewed by D. Ugarte, the dominant researcher in this area. The volume concludes with a review of metal-coated fullerenes by T. P. Martin and co-workers, who pioneered studies on this topic.

The guest editors have assembled an excellent set of reviews and research articles covering all aspects of the field of carbon nanotubes. The reviews are presented in a clear and concise form by many of the leading researchers in the field. It is hoped that this collection of review articles provides a convenient reference for the present status of research on carbon nanotubes, and serves to stimulate future work in the field.

M. S. Dresselhaus

REFERENCES

1. H. W. Kroto, *Carbon* **30,** 1139 (1992).
2. S. Iijima, *Nature* (London) **354,** 56 (1991).

PYROLYTIC CARBON NANOTUBES FROM VAPOR-GROWN CARBON FIBERS

MORINOBU ENDO,[1] KENJI TAKEUCHI,[1] KIYOHARU KOBORI,[1] KATSUSHI TAKAHASHI,[1] HAROLD W. KROTO,[2] and A. SARKAR[2]
[1]Faculty of Engineering, Shinshu University, 500 Wakasato, Nagano 380, Japan
[2]School of Chemistry and Molecular Sciences, University of Sussex, Brighton BN1 9QJ, U.K.

(*Received* 21 *November* 1994; *accepted* 10 *February* 1995)

Abstract—The structure of as-grown and heat-treated pyrolytic carbon nanotubes (PCNTs) produced by hydrocarbon pyrolysis are discussed on the basis of a possible growth process. The structures are compared with those of nanotubes obtained by the arc method (ACNT; arc-formed carbon nanotubes). PCNTs, with and without secondary pyrolytic deposition (which results in diameter increase) are found to form during pyrolysis of benzene at temperatures ca. 1060°C under hydrogen. PCNTs after heat treatment at above 2800°C under argon exhibit have improved stability and can be studied by high-resolution transmission electron microscopy (HRTEM). The microstructures of PCNTs closely resemble those of vapor-grown carbon fibers (VGCFs). Some VGCFs that have micro-sized diameters appear to have nanotube inner cross-sections that have different mechanical properties from those of the outer pyrolytic sections. PCNTs initially appear to grow as ultra-thin graphene tubes with central hollow cores (diameter ca. 2 nm or more) and catalytic particles are not observed at the tip of these tubes. The secondary pyrolytic deposition, which results in characteristic thickening by addition of extra cylindrical carbon layers, appears to occur simultaneously with nanotube lengthening growth. After heat treatment, HRTEM studies indicate clearly that the hollow cores are closed at the ends of polygonized hemi-spherical carbon caps. The most commonly observed cone angle at the tip is generally ca. 20°, which implies the presence of five pentagonal disclinations clustered near the tip of the hexagonal network. A structural model is proposed for PCNTs observed to have spindle-like shape and conical caps at both ends. Evidence is presented for the formation, during heat treatment, of hemi-toroidal rims linking adjacent concentric walls in PCNTs. A possible growth mechanism for PCNTs, in which the tip of the tube is the active reaction site, is proposed.

Key Words—Carbon nanotubes, vapor-grown carbon fibers, high-resolution transmission electron microscope, graphite structure, nanotube growth mechanism, toroidal network.

1. INTRODUCTION

Since Iijima's original report[1], carbon nanotubes have been recognized as fascinating materials with nanometer dimensions promising exciting new areas of carbon chemistry and physics. From the viewpoint of fullerene science they also are interesting because they are forms of giant fullerenes[2]. The nanotubes prepared in a dc arc discharge using graphite electrodes at temperatures greater than 3000°C under helium were first reported by Iijima[1] and later by Ebbesen and Ajyayan[3]. Similar tubes, which we call pyrolytic carbon nanotubes (PCNTs), are produced by pyrolyzing hydrocarbons (e.g., benzene at ca. 1100°C)[4–9]. PCNTs can also be prepared using the same equipment as that used for the production of so called vapor-grown carbon fibers (VGCFs)[10]. The VGCFs are micron diameter fibers with circular cross-sections and central hollow cores with diameters ca. a few tens of nanometers. The graphitic networks are arranged in concentric cylinders. The intrinsic structures are rather like that of the annual growth of trees. The structure of VGCFs, especially those with hollow cores, are very similar to the structure of arc-formed carbon nanotubes (ACNTs). Both types of nanotubes, the ACNTs and the present PCNTs, appear to be essentially Russian Doll-like sets of elongated giant fullerenes[11,12]. Possible growth processes have been proposed involving both open-ended[13] and closed-cap[11,12] mechanisms for the primary tubules. Whether either of these mechanisms or some other occurs remains to be determined.

It is interesting to compare the formation process of fibrous forms of carbon with larger micron diameters and carbon nanotubes with nanometer diameters from the viewpoint of "one-dimensional" carbon structures as shown in Fig. 1. The first class consists of graphite whiskers and ACNTs produced by arc methods, whereas the second encompasses vapor-grown carbon fibers and PCNTs produced by pyrolytic processes. A third possible class would be polymer-based nanotubes and fibers such as PAN-based carbon fibers, which have yet to be formed with nanometer dimensions. In the present paper we compare and discuss the structures of PCNTs and VGCFs.

2. VAPOR-GROWN CARBON FIBERS AND PYROLYTIC CARBON NANOTUBES

Vapor-grown carbon fibers have been prepared by catalyzed carbonization of aromatic carbon species using ultra-fine metal particles, such as iron. The particles, with diameters less than 10 nm may be dispersed on a substrate (substrate method), or allowed to float in the reaction chamber (fluidized method). Both

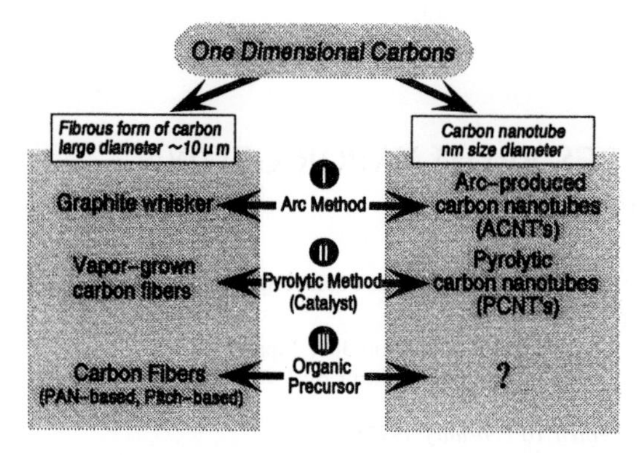

Fig. 1. Comparative preparation methods for micrometer size fibrous carbon and carbon nanotubes as one-dimensional forms of carbon.

methods give similar structures, in which ultra-fine catalytic particles are encapsulated in the tubule tips (Fig. 2). Continued pyrolytic deposition occurs on the initially formed thin carbon fibers causing thickening (ca. 10 μm diameter, Fig. 3a). Substrate catalyzed fibers tend to be thicker and the floating technique produces thinner fibers (ca. 1 μm diameter). This is due to the shorter reaction time that occurs in the fluidized method (Fig. 3b). Later floating catalytic methods are useful for large-scale fiber production and, thus, VGCFs should offer a most cost-effective means of producing discontinuous carbon fibers. These VGCFs offer great promise as valuable functional carbon filler materials and should also be useful in carbon fiber-reinforced plastic (CFRP) production. As seen in Fig. 3b even in the "as-grown" state, carbon particles are eliminated by controlling the reaction conditions. This promises the possibility of producing pure ACNTs without the need for separating spheroidal carbon particles. Hitherto, large amounts of carbon particles have always been a byproduct of nanotube production and, so far, they have only been eliminated by selective oxidation[14]. This has led to the loss of significant amounts of nanotubes—ca. 99%.

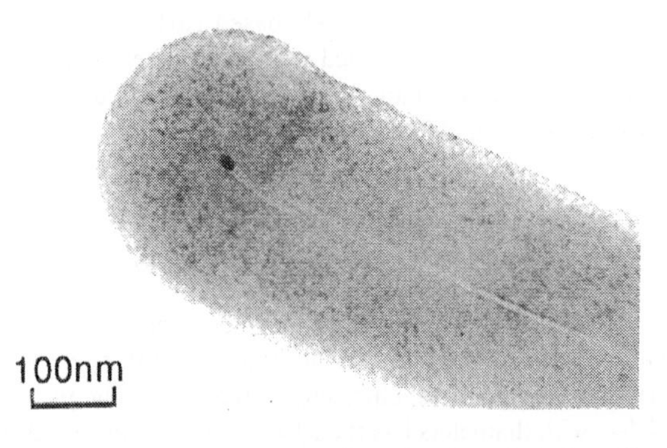

Fig. 2. Vapour-grown carbon fiber showing relatively early stage of growth; at the tip the seeded Fe catalytic particle is encapsulated.

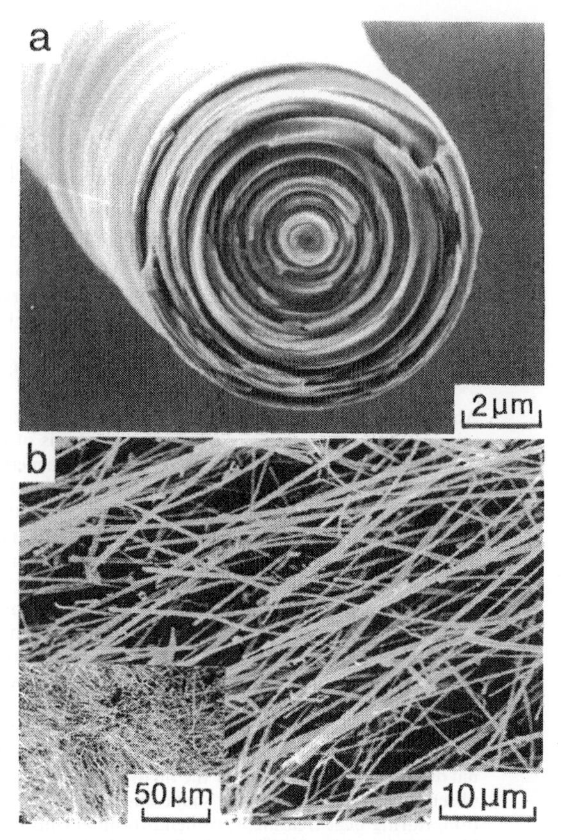

Fig. 3. Vapor-grown carbon fibers obtained by substrate method with diameter ca. 10 μm (a) and those by floating catalyst method (b) (inserted, low magnification).

3. PREPARATION OF VGCFs AND PCNTs

The PCNTs in this study were prepared using the same apparatus[9] as that employed to produce VGCFs by the substrate method[10,15]. Benzene vapor was introduced, together with hydrogen, into a ceramic reaction tube in which the substrate consisted of a centrally placed artificial graphite rod. The temperature of the furnace was maintained in the 1000°C range. The partial pressure of benzene was adjusted to be much lower than that generally used for the preparation of VGCFs[10,15] and, after one hour decomposition, the furnace was allowed to attain room temperature and the hydrogen was replaced by argon. After taking out the substrate, its surface was scratched with a toothpick to collect the minute fibers. Subsequently, the nanotubes and nanoscale fibers were heat treated in a carbon resistance furnace under argon at temperatures in the range 2500–3000°C for ca. 10–15 minutes. These as-grown and sequentially heat-treated PCNTs were set on an electron microscope grid for observation directly by HRTEM at 400kV acceleration voltage.

It has been observed that occasionally nanometer scale VGCFs and PCNTs coexist during the early stages of VGCF processing (Fig. 4). The former tend to have rather large hollow cores, thick tube walls and well-organized graphite layers. On the other hand,

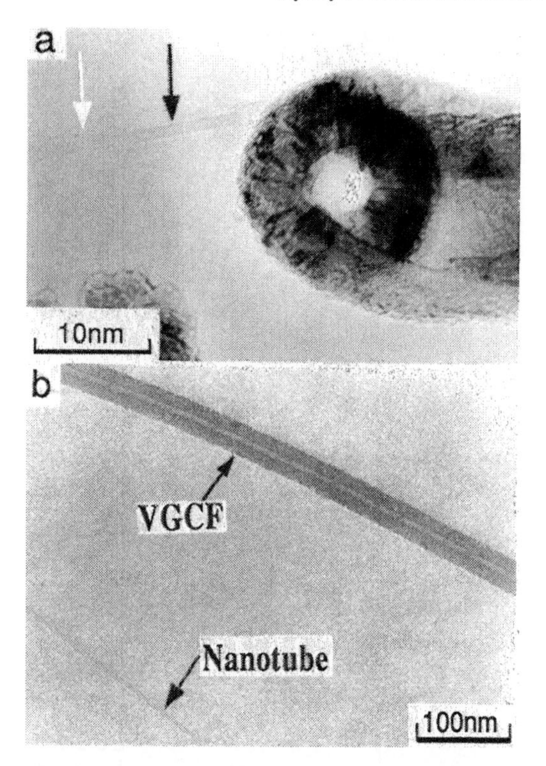

Fig. 4. Coexisting vapour-grown carbon fiber, with thicker diameter and hollow core, and carbon nanotubes, with thinner hollow core, (as-grown samples).

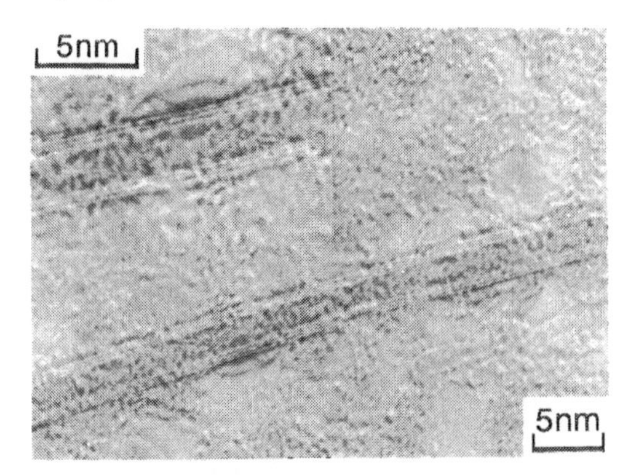

Fig. 5. Heat-treated pyrolytic carbon nanotube and enlarged one (inserted), without deposited carbon.

PCNTs tend to have very thin walls consisting of only a few graphitic cylinders. Some sections of the outer surfaces of the thin PCNTs are bare, whereas other sections are covered with amorphous carbon deposits (as is arrowed region in Fig. 4a). TEM images of the tips of the PCNTs show no evidence of electron beam opaque metal particles as is generally observed for VGCF tips[10,15]. The large size of the cores and the presence of opaque particles at the tip of VGCFs suggests possible differences between the growth mechanism for PCNTs and standard VGCFs[7–9]. The yield of PCNTs increases as the temperature and the benzene partial pressure are reduced below the optimum for VGCF production (i.e., temperature ca. 1000°–1150°C). The latter conditions could be effective in the prevention or the minimization of carbon deposition on the primary formed nanotubules.

4. STRUCTURES OF PCNTs

Part of a typical PCNT (ca. 2.4 nm diameter) after heat treatment at 2800°C for 15 minutes is shown in Fig. 5. It consists of a long concentric graphite tube with interlayer spacings ca. 0.34 nm—very similar in morphology to ACNTs[1,3]. These tubes may be very long, as long as 100 nm or more. It would, thus, appear that PCNTs, after heat treatment at high temperatures, become graphitic nanotubes similar to ACNTs. The heat treatment has the effect of crystallizing the secondary deposited layers, which are usually composed of rather poorly organized turbostratic carbon. This results in well-organized multi-walled concentric graphite tubules. The interlayer spacing (0.34 nm) is slightly wider on average than in the case of thick VGCFs treated at similar temperatures. This small increase might be due to the high degree of curvature of the narrow diameter nanotubes which appears to prevent perfect 3-dimensional stacking of the graphitic layers[16,17]. PCNTs and VGCFs are distinguishable by the sizes of the well-graphitized domains; cross-sections indicate that the former are characterized by single domains, whereas the latter tend to exhibit multiple domain areas that are small relative to this cross-sectional area. However, the innermost part of some VGCFs (e.g., the example shown in Fig. 5) may often consist of a few well-structured concentric nanotubes. Theoretical studies suggest that this "single grain" aspect of the cross-sections of nanotubes might give rise to quantum effects. Thus, if large scale real-space super-cell concepts are relevant, then Brillouin zone-folding techniques may be applied to the description of dispersion relations for electron and phonon dynamics in these pseudo one-dimensional systems.

A primary nanotube at a very early stage of thickening by pyrolytic carbon deposition is depicted in Figs. 6a–c; these samples were: (a) as-grown and (b), (c) heat treated at 2500°C. The pyrolytic coatings shown are characteristic features of PCNTs produced by the present method. The deposition of extra carbon layers appears to occur more or less simultaneously with nanotube longitudinal growth, resulting in spindle-shaped morphologies. Extended periods of pyrolysis result in tubes that can attain diameters in the micron range (e.g., similar to conventional (thick) VGCFs[10]. Fig. 6c depicts a 002 dark-field image, showing the highly ordered central core and the outer inhomogeneously deposited polycrystalline material (bright spots). It is worthwhile to note that even the very thin walls consisting of several layers are thick enough to register 002 diffraction images though they are weaker than images from deposited crystallites on the tube.

Fig. 7a,b depicts PCNTs with relatively large diameters (ca. 10 nm) that appear to be sufficiently tough

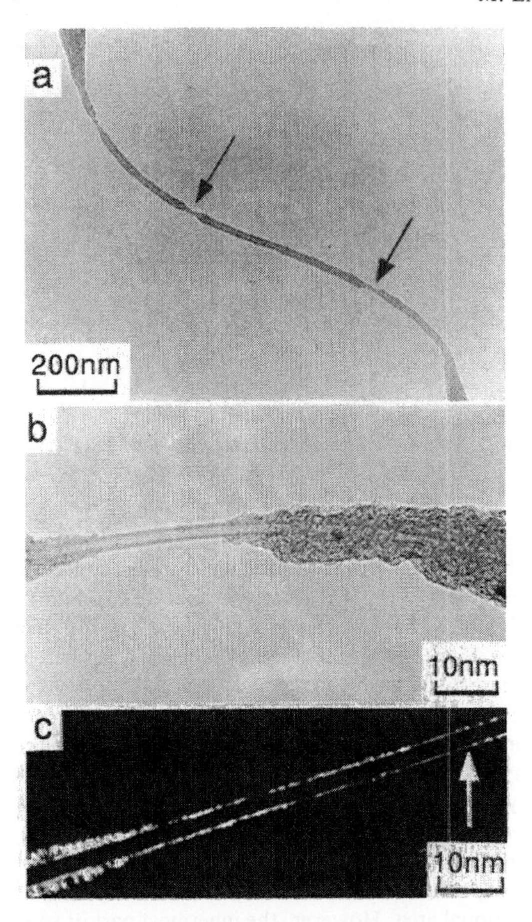

Fig. 6. PCNTs with partially deposited carbon layers (arrow indicates the bare PCNT), (a) as-grown, (b) partially exposed nanotube and (c) 002 dark-field image showing small crystallites on the tube and wall of the tube heat treated at 2500°C.

and flexible to bend, twist, or kink without fracturing. The basic structural features and the associated mechanical behavior of the PCNTs are, thus, very different from those of conventional PAN-based fibers as well as VGCFs, which tend to be fragile and easily broken when bent or twisted. The bendings may occur at propitious points in the graphene tube network[18].

Fig. 8a,b shows two typical types of PCNT tip morphologies. The caps and also intercompartment diaphragms occur at the tips. In general, these consist of 2–3 concentric layers with average interlayer spacing of ca. 0.38 nm. This spacing is somewhat larger than that of the stackings along the radial direction, presumably (as discussed previously) because of sharp curvature effects. As indicated in Fig. 9, the conical shapes have rather symmetric cone-like shells. The angle, ca. 20°, is in good agreement with that expected for a cone constructed from hexagonal graphene sheets containing pentagonal disclinations—as is Fig. 9e. Ge and Sattler[19] have reported nanoscale conical carbon materials with infrastructure explainable on the basis of fullerene concepts. STM measurements show that nanocones, made by deposition of very hot carbon on HOPG surfaces, often tend to

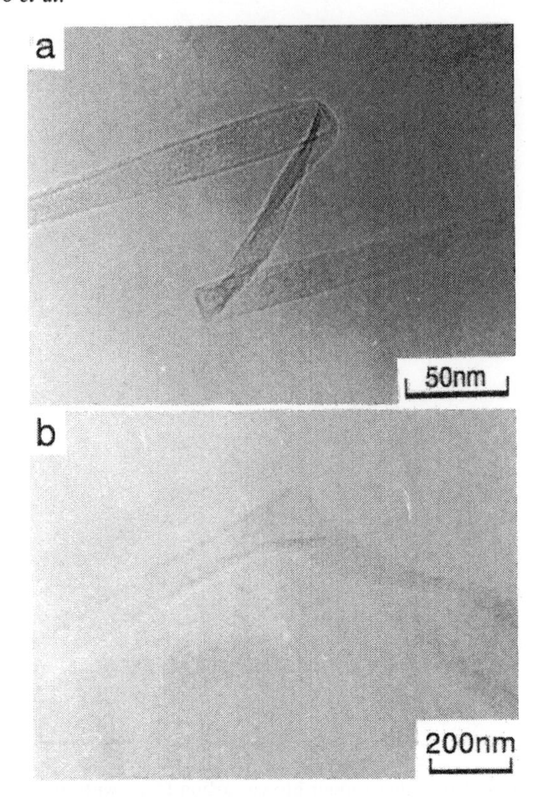

Fig. 7. Bent and twisted PCNT (heat treated at 2500°C).

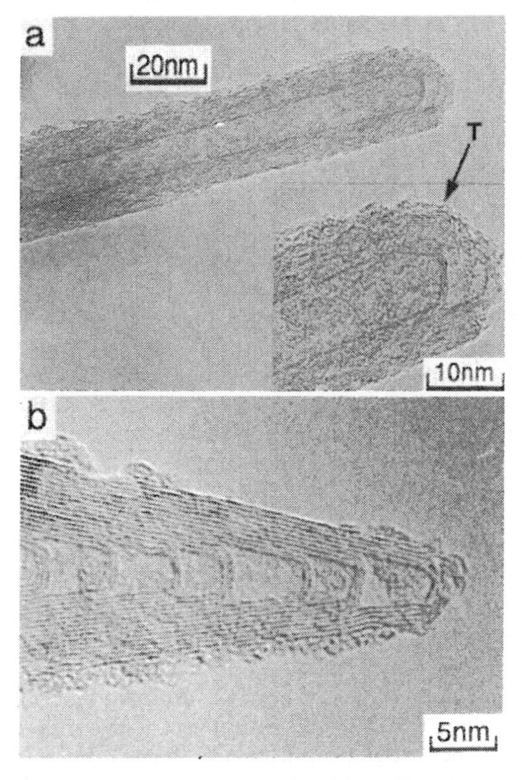

Fig. 8. The tip of PCNTs with continuous hollow core (a) and the cone-like shape (b) (T indicates the toroidal structure shown in detail in Fig. 11).

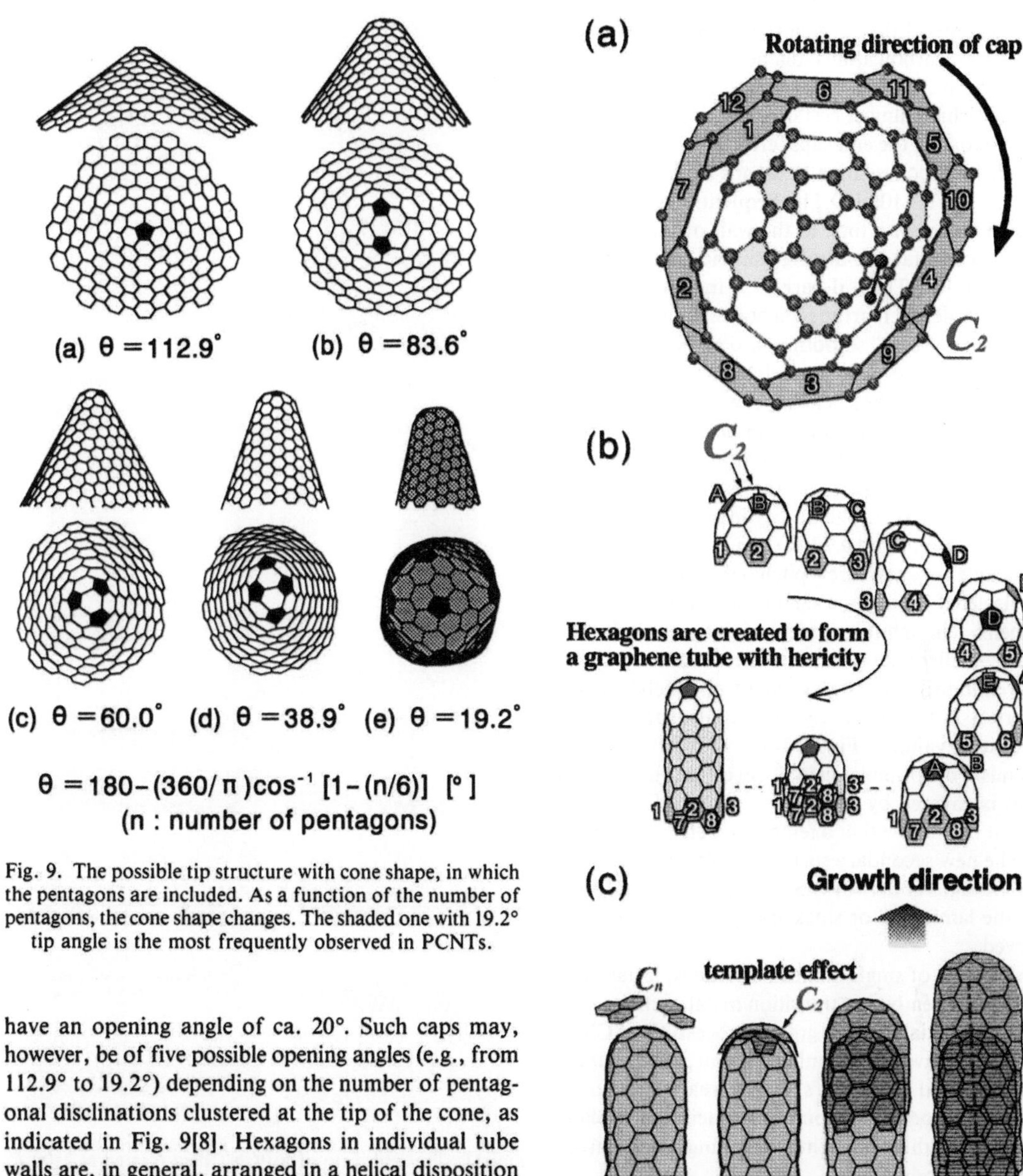

Fig. 9. The possible tip structure with cone shape, in which the pentagons are included. As a function of the number of pentagons, the cone shape changes. The shaded one with 19.2° tip angle is the most frequently observed in PCNTs.

Fig. 10. Growth mechanism proposed for the helical nanotubes (a) and helicity (b), and the model that gives the bridge and laminated tip structure (c).

have an opening angle of ca. 20°. Such caps may, however, be of five possible opening angles (e.g., from 112.9° to 19.2°) depending on the number of pentagonal disclinations clustered at the tip of the cone, as indicated in Fig. 9[8]. Hexagons in individual tube walls are, in general, arranged in a helical disposition with variable pitches. It is worth noting that the smallest angle (19.2°) that can involve five pentagons is most frequently observed in such samples. It is frequently observed that PCNTs exhibit a spindle-shaped structure at the tube head, as shown in Fig. 8b.

5. GROWTH MODEL OF PCNTs

In the case of the PCNTs considered here, the growth temperature is much lower than that for ACNTs, and no electric fields, which might influence the growth of ACNTs, are present. It is possible that different growth mechanisms apply to PCNT and ACNT growth and this should be taken into consideration. As mentioned previously, one plausible mechanism for nanotube growth involves the insertion of small carbon species C_n ($n = 1,2,3 \ldots$) into a closed fullerene cap (Fig. 10a–c)[11]. Such a mechanism is related to the processes that Ulmer *et al.*[20] and McElvaney *et al.*[21] have discovered for the growth of small closed cage fullerenes. Based on the observation of open-ended tubes, Iijima *et al.*[13] have discussed a plausible alternative way in which such tubules might possibly grow. The closed cap growth mechanism effectively involves the addition of extended chains of sp carbon atoms to the periphery of the asymmetric 6-pentagon cap, of the kind whose Schlegel diagram is depicted in Fig. 10a, and results in a hexagonal graphene cylinder wall in which the added atoms are arranged in a helical disposition[9,11] similar to that observed first by Iijima[1].

It is proposed that during the growth of primary tubule cores, carbon atoms, diameters, and longer linear clusters are continuously incorporated into the active sites, which almost certainly lie in the vicinity of the pentagons in the end caps, effectively creating helical arrays of consecutive hexagons in the tube wall as shown in Fig. 10a,b[9,11]. Sequential addition of 2 carbon atoms at a time to the wall of the helix results in a cap that is indistinguishable other than by rotation[11,12]. Thus, if carbon is ingested into the cap and wholesale rearrangement occurs to allow the new atoms to "knit" smoothly into the wall, the cap can be considered as effectively fluid and to move in a screw-like motion leaving the base of the wall stationary—though growing by insertion of an essentially uniform thread of carbon atoms to generate a helical array of hexagons in the wall. The example shown in Fig. 10a results in a cylinder that has a diameter (ca. 1 nm) and a 22-carbon atom repeat cycle and a single hexagon screw pitch—the smallest archetypal (isolated pentagon) example of a graphene nanotube helix. Though this model generates a tubule that is rather smaller than is usually the case for the PCNTs observed in this study (the simplest of which have diameters > 2–3 nm), the results are of general semiquantitative validity. Figure 10b,c shows the growth mechanism diagrammatically from a side view. When the tip is covered by further deposition of aromatic layers, it is possible that a templating effect occurs to form the new secondary surface involving pentagons in the hexagonal network. Such a process would explain the laminated or stacked-cup-like morphology observed.

In the case of single-walled nanotubes, it has been recognized recently that transition metal particles play a role in the initial filament growth process[23]. ACNTs and PCNTs have many similarities but, as the vapor-growth method for PCNTs allows greater control of the growth process, it promises to facilitate applications more readily and is thus becoming the preferred method of production.

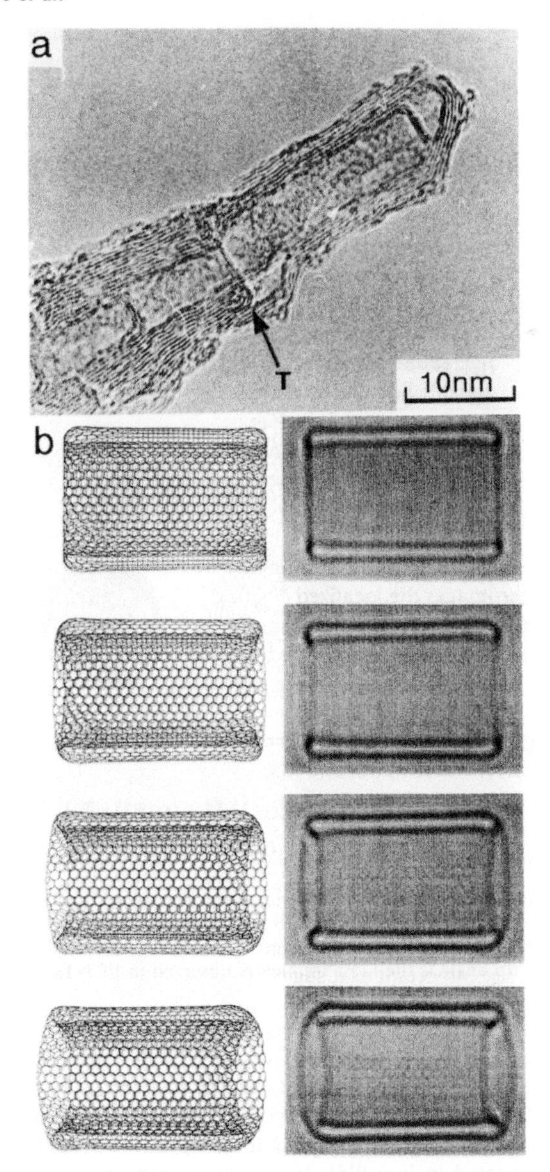

Fig. 11. The sealed tip of a PCNT heat treated at 2800°C with a toroidal structure (T) and, (b) molecular graphics images of archetypal flattened toroidal model at different orientations and the corresponding simulated TEM images.

6. CHARACTERISTIC TOROIDAL AND SPINDLE-LIKE STRUCTURES OF PCNTs

In Fig. 11a is shown an HRTEM image of part of the end of a PCNTs. The initial material consisted of a single-walled nanotube upon which bi-conical spindle-like growth can be seen at the tip. Originally, this tip showed no apparent structure in the HRTEM image at the as-grown state, suggesting that it might consist largely of some form of "amorphous" carbon. After a second stage of heat treatment at 2800°C, the amorphous sheaths graphitize to a very large degree, producing multi-walled graphite nanotubes that tend to be sealed off with caps at points where the spindle-like formations are the thinnest. The sealed-off end region of one such PCNT with a hemi-toroidal shape is shown in Fig. 11a.

In Fig. 11b are depicted sets of molecular graphics images of flattened toroidal structures which are the basis of archetypal double-walled nanotubes[24]. As the orientation changes, we note that the HRTEM interference pattern associated with the rim changes from a line to an ellipse and the loop structures at the apices remain relatively distinct. The oval patterns in the observed and simulated HRTEM image (Fig. 11b) are consistent with one another. For this preliminary investigation a symmetric (rather than helical) wall configuration was used for simplicity. Hemi-toroidal connection of the inner and outer tubes with helical structured walls requires somewhat more complicated dispositions of the 5/6/7 rings in the lip region. The general validity of the conclusions drawn here are, however, not affected. Initial studies of the problem indicated that linking between the inner and outer walls is not, in general, a hindered process.

The toroidal structures show interesting changes in morphology as they become larger—at least at the lip. The hypothetical small toroidal structure shown in Fig. 11b is actually quite smooth and has an essentially rounded structure[24]. As the structures become larger, the strain tends to focus in the regions near the pentagons and heptagons, and this results in more prominent localized cusps and saddle points. Rather elegant toroidal structures with D_{nh} and D_{nd} symmetry are produced, depending on whether the various paired heptagon/pentagon sets which lie at opposite ends of the tube are aligned or are offset. In general, they probably lie is fairly randomly disposed positions. Chiral structures can be produced by off-setting the pentagons and heptagons. In the D_{5d} structure shown in Fig. 11 which was developed for the basic study, the walls are fluted between the heptagons at opposite ends of the inner tube and the pentagons of the outer wall rim[17]. It is interesting to note that in the computer images the localized cusping leads to variations in the smoothness of the image generated by the rim, though it still appears to be quite elliptical when viewed at an angle[17]. The observed image appears to exhibit variations that are consistent with the localized cusps as the model predicts.

In this study, we note that epitaxial graphitization is achieved by heat treatment of the apparently mainly amorphous material which surrounds a single-walled nanotube[17]. As well as bulk graphitization, localized hemi-toroidal structures that connect adjacent walls have been identified and appear to be fairly common in this type of material. This type of infrastructure may be important as it suggests that double walls may form fairly readily. Indeed, the observations suggest that pure carbon rim-sealed structures may be readily produced by heat treatment, suggesting that the future fabrication of stabilized double-walled nanoscale graphite tubes in which dangling bonds have been eliminated is a feasible objective. It will be interesting to prove the relative reactivities of these structures for their possible future applications in nanoscale devices (e.g., as quantum wire supports). Although the curvatures of the rims appear to be quite tight, it is clear from the abundance of loop images observed, that the occurrence of such turnovers between concentric cylinders with a gap spacing close to the standard graphite interlayer spacing is relatively common. Interestingly, the edges of the toroidal structures appear to be readily visible and this has allowed us to confirm the relationship between opposing loops. Bulges in the loops of the kind observed are simulated theoretically[17].

Once one layer has formed (the primary nanotube core), further secondary layers appear to deposit with various degrees of epitaxial coherence. When inhomogeneous deposition occurs in PCNTs, the thickening has a characteristic spindle shape, which may be a consequence of non-carbon impurities which impede graphitization (see below)—this is not the case for ACNTs were growth takes place in an essentially all-carbon atmosphere, except, of course, for the rare gas. These spindles probably include the appropriate number of pentagons as required by variants of Euler's Law. Hypothetical structural models for these spindles are depicted in Fig. 12. It is possible that similar two-stage growth processes occur in the case of ACNTs but, in general, the secondary growth appears to be intrinsically highly epitaxial. This may be because in the ACNT growth case only carbon atoms are involved and there are fewer (non-graphitizing) alternative accretion pathways available. It is likely that epitaxial growth control factors will be rather weak when secondary deposition is very fast, and so thin layers may result in poorly ordered graphitic structure in the thicker sections. It appears that graphitization of this secondary deposit that occurs upon heat treatment may be partly responsible for the fine structure such as compartmentalization, as well as basic tip morphology[17].

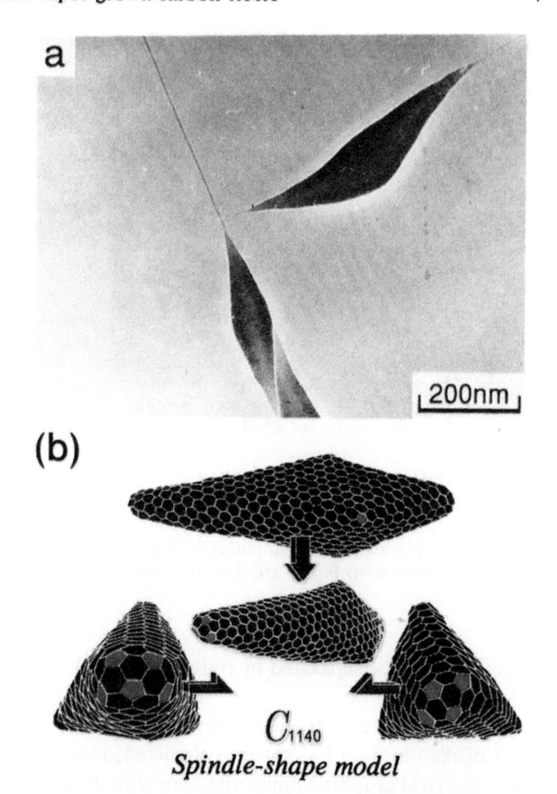

Fig. 12. As-grown PCNTs with partially thickened spindle shape (a) and the proposed structural model for spindle particles including 12 pentagons in hexagon cage (b).

7. VGCFs DERIVED FROM NANOTUBES

In Fig. 13 is shown the 002 lattice images of an "as-formed" very thin VGCF. The innermost core diameter (ca. 20 nm as indicated by arrows) has two layers; it is rather straight and appears to be the primary nanotube. The outer carbon layers, with diameters ca. 3–4 nm, are quite uniformly stacked parallel to the central core with 0.35 nm spacing. From the difference in structure as well as the special features in the mechanical strength (as in Fig. 7) it might appear possible that the two intrinsically different types of material

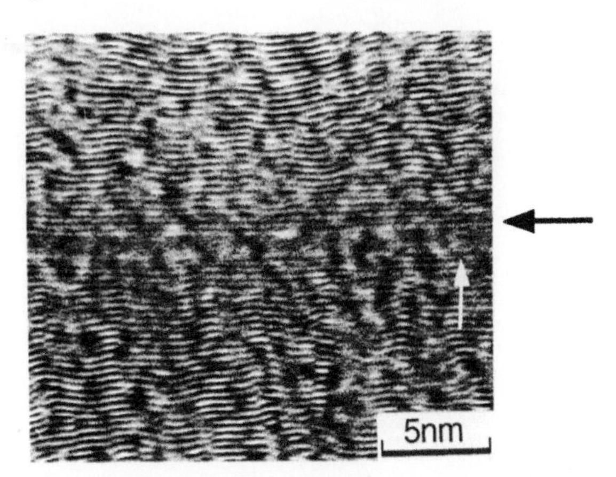

Fig. 13. HRTEM image of an as-grown thick PCNT. 002 lattice image demonstrates the innermost hollow core (core diam. 2.13 nm) presumably corresponding to the "as-formed" nanotube. The straight and continuous innermost two fringes similar to Fig. 5 are seen (arrow).

involved might be separated by pulverizing the VGCF material.

In Fig. 14a, a ca. 10 μm diameter VGCF that has been broken in liquid nitrogen is depicted, revealing the cylindrical graphitic nanotube core with diameter

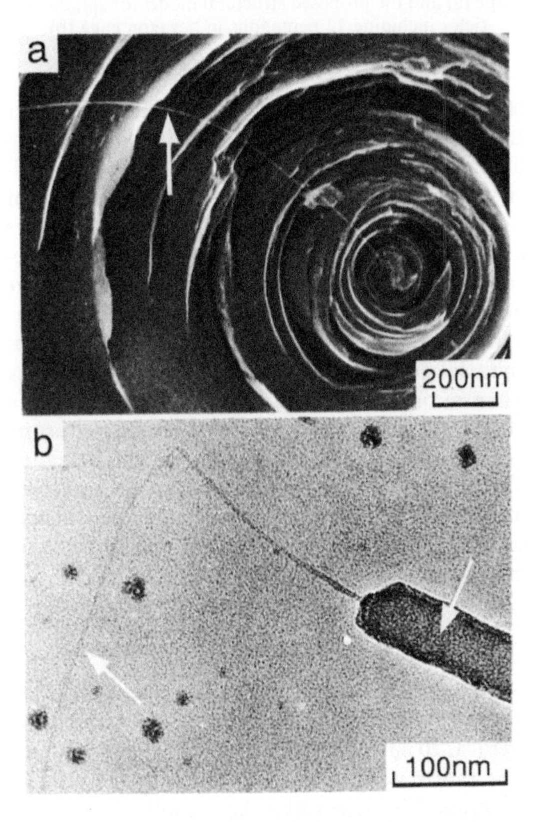

Fig. 14. PCNTs (white arrow) appeared after breakage of VGCF, (a) FE-SEM image of broken VGCF, cut in liquid nitrogen and (b) HRTEM image showing the broken part observed in very thin VGCF. The nanotube is clearly observed and this indicates that thin VGCF grow from nanometer core by thickening.

of ca. 10 nm (white arrow), observed by field emission scanning electron microscopy (FE-SEM)[25]. It is, thus, suggested that at least some of the VGCFs start as nanotube cores, which act as a substrate for subsequent thickening by deposition of secondary pyrolytic carbon material, as in the catalytically primarily grown hollow fiber. In Fig. 14b is also shown the TEM image corresponding to the extruded nanotube from a very thin fiber. It is clearly observed that the exposed nanotube is continuing into the fiber as a central hollow core, as indicated by the white arrow in the figure. It is interesting that, as indicated before (in Fig. 14a), the core is more flexible than the pyrolytic part, which is more fragile.

8. CONCLUSION

Pyrolytic carbon nanotubes (PCNTs), which grow during hydrocarbon pyrolysis, appear to have structures similar to those obtained by arc/discharge techniques using graphite electrodes (ACNTs). The PCNTs tend to exhibit a characteristic thickening feature due to secondary pyrolytic carbon deposition. Various tip morphologies are observed, but the one most frequently seen has a 20° opening angle, suggesting that, in general, the graphene conical tips possess a cluster of five pentagons that may be actively involved in tube growth. PCNTs with spindle-like shapes and that have conical caps at both ends are also observed, for which a structural model is proposed. The spindle-like structures observed for the secondary growth thickening that occurs in PCNTs may be a consequence of the lower carbon content present in the growth atmosphere than occurs in the case of ACNT growth. Possible structural models for these spindles have been discussed. The longitudinal growth of nanotubes appears to occur at the hemi-spherical active tips and this process has been discussed on the basis of a closed cap mechanism[9,11]. The PCNTs are interesting, not only from the viewpoint of the fundamental perspective that they are very interesting giant fullerene structures, but also because they promise to be applications in novel strategically important materials in the near future. PCNT production appears, at this time, more readily susceptible to process control than is ACNT production and, thus, their possible value as fillers in advanced composites is under investigation.

Acknowledgements—Japanese authors are indebted to M. S. Dresselhaus and G. Dresselhaus of MIT and to A. Oberlin of Laboratoire Marcel Mathieu (CNRS) for their useful discussions and suggestions. HWK thanks D. R. M. Walton for help and the Royal Society and the SERC (UK) for support. Part of the work by ME is supported by a grant-in-aid for scientific research in priority area "carbon cluster" from the Ministry of Education, Science and Culture, Japan.

REFERENCES

1. S. Iijima, *Nature* **354**, 56 (1991).
2. H. W. Kroto, J. R. Heath, S. C .O'Brien, R. F. Curl, and R. E. Samlley, *Nature* **318**, 162 (1985).

3. T. W. Ebbesen and P. M. Ajayan, *Nature* **358**, 220 (1992).
4. M. Endo, H. Fijiwara, and E. Fukunaga, *18th Meeting Japanese Carbon Society*, (1991) p. 34.
5. M. Endo, H. Fujiwara, and E. Fukunaga, *2nd C60 Symposium in Japan*, (1992) p. 101.
6. M. Endo, K. Takeuchi, S. Igarashi, and K. Kobori, *19th Meeting Japanese Carbon Society*, (1992) p. 192.
7. M. Endo, K. Takeuchi, S. Igarashi, K. Kobori, and M. Shiraishi, *Mat. Res. Soc. Spring Meet* (1993) p.S2.2.
8. M. Endo, K. Takeuchi, S. Igarashi, K. Kobori, M. Shiraishi, and H. W. Kroto, *Mat. Res. Soc. Fall Meet.* **G2.1** (1994).
9. M. Endo, K. Takeuchi, S. Igarashi, K. Kobori, M. Shiraishi, and H. W. Kroto, *J. Phys. Chem. Solids* **54**, 1841 (1993).
10. M. Endo, *Chemtech* **18**, 568 (1988).
11. M. Endo and H. W. Kroto, *J. Phys. Chem.* **96**, 6941 (1992).
12. H. W. Kroto, K. Prassides, R. Taylor, D. R. M. Walton, and M. Endo, *International Conference Solid State Devices and Materials of The Japan Society of Applied Physics* (1993), p. 104.
13. S. Iijima, *Mat. Sci. Eng.* **B19**, 172 (1993).
14. P. M. Ajayan, T. W. Ebbesen, T. Ichihashi, S. Iijima, K. Tanigaki, and H. Hiura, *Nature* **362**, 522 (1993).
15. M. S. Dresselhaus, G. Dresselhaus, K. Sugihara, I. L. Spain, H. A. Goldberg, In *Graphite Fibers and Filaments*, (edited by M. Cardona) pp. 244–286. Berlin, Springer.
16. J. S. Speck, M. Endo, and M. S. Dresselhaus, *J. Crystal Growth* **94**, 834 (1989).
17. A. Sarkar, H. W. Kroto, and M. Endo (in preparation).
18. H. Hiura, T. W. Ebbesen, J. Fujita, K. Tanigaki, and T. Takada, *Nature* **367**, 148 (1994).
19. M. Ge and K. Sattler, *Mat. Res. Soc. Spring Meet.* **S1.3**, 360 (1993).
20. G. Ulmer, E. E. B. Cambel, R. Kuhnle, H. G. Busmann, and I. V. Hertel, *Chem. Phys. Letts.* **182**, 114 (1991).
21. S. W. McElvaney, M. N. Ross, N. S. Goroff, and F. Diederich, *Science* **259**, 1594 (1993).
22. R. Saito, G. Dresselhaus, M. Fujita, and M. S. Dresselhaus, *4th NEC Symp. Phys. Chem. Nanometer Scale Mats.* (1992).
23. S. Iijima, *Gordon Conference on the Chemistry of Hydrocarbon Resources*, Hawaii (1994).
24. A. Sarker, H. W. Kroto, and M. Endo (to be published).
25. M. Endo, K. Takeuchi, K. Kobori, K. Takahashi, and H. W. Kroto (in preparation).

ELECTRIC EFFECTS IN NANOTUBE GROWTH

DANIEL T. COLBERT and RICHARD E. SMALLEY
Rice Quantum Institute and Departments of Chemistry and Physics, MS 100,
Rice University, Houston, TX 77251-1892, U.S.A.

(*Received* 3 *April* 1995; *accepted* 7 *April* 1995)

Abstract—We present experimental evidence that strongly supports the hypothesis that the electric field of the arc plasma is essential for nanotube growth in the arc by stabilizing the open tip structure against closure. By controlling the temperature and bias voltage applied to a single nanotube mounted on a macroscopic electrode, we find that the nanotube tip closes when heated to a temperature similar to that in the arc *unless* an electric field is applied. We have also developed a more refined awareness of "open" tips in which adatoms bridge between edge atoms of adjacent layers, thereby lowering the exothermicity in going from the open to the perfect dome-closed tip. Whereas realistic fields appear to be insufficient by themselves to stabilize an open tip with its edges completely exposed, the field-induced energy lowering of a tip having adatom spot-welds can, and indeed in the arc does, make the open tip stable relative to the closed one.

Key Words—Nanotubes, electric field, arc plasma.

1. INTRODUCTION

As recounted throughout this special issue, significant advances in illuminating various aspects of nanotube growth have been made[1,2] since Iijima's eventful discovery in 1991;[3] these advances are crucial to gaining control over nanotube synthesis, yield, and properties such as length, number of layers, and helicity. The carbon arc method Iijima used remains the principle method of producing bulk amounts of quality nanotubes, and provides key clues for their growth there and elsewhere. The bounty of nanotubes deposited on the cathode (Ebbesen and Ajayan have found that up to 50% of the deposited carbon is tubular[4]) is particularly puzzling when one confronts the evidence of Ugarte[5] that tubular objects are energetically less stable than spheroidal onions.

It is largely accepted that nanotube growth occurs at an appreciable rate only at open tips. With this constraint, the mystery over tube growth in the arc redoubles when one realizes that the cathode temperature (~3000°C) is well above that required to anneal carbon vapor to spheroidal closed shells (fullerenes and onions) with great efficiency. The impetus to close is, just as for spheroidal fullerenes, elimination of the dangling bonds that unavoidably exist in any open structure by incorporation of pentagons into the hexagonal lattice. Thus, a central question in the growth of nanotubes in the arc is: How do they stay open?

One of us (RES) suggested over two years ago[6] that the resolution to this question lies in the electric field inherent to the arc plasma. As argued then, neither thermal nor concentration gradients are close to the magnitudes required to influence tip annealing, and trace impurities such as hydrogen, which might keep the tip open, should have almost no chemisorption residence time at 3000°C. The fact that well-formed nanotubes are found only in the cathode deposit, where the electric field concentrates, and never in the soots condensed from the carbon vapor exiting the arcing region, suggest a vital role for the electric field. Furthermore, the field strength at the nanotube tips is very large, due both to the way the plasma concentrates most of the potential drop in a very short distance above the cathode, and to the concentrating effects of the field at the tips of objects as small as nanotubes. The field may be on the order of the strength required to break carbon-carbon bonds, and could thus dramatically effect the tip structure.

In the remaining sections of this paper, we describe the experimental results leading to confirmation of the stabilizing role of the electric field in arc nanotube growth. These include: relating the plasma structure to the morphology of the cathode deposit, which revealed that the integral role of nanotubes in sustaining the arc plasma is their field emission of electrons into the plasma; studying the field emission characteristics of isolated, individual arc-grown nanotubes; and the discovery of a novel production of nanotubes that significantly alters the image of the "open" tip that the arc electric field keeps from closing.

2. NANOTUBES AS FIELD EMITTERS

Defects in arc-grown nanotubes place limitations on their utility. Since defects appear to arise predominantly due to sintering of adjacent nanotubes in the high temperature of the arc, it seemed sensible to try to reduce the extent of sintering by cooling the cathode better[2]. The most vivid assay for the extent of sintering is the oxidative heat purification treatment of Ebbesen and coworkers[7], in which amorphous carbon and shorter nanoparticles are etched away before nanotubes are substantially shortened. Since, as we proposed, most of the nanoparticle impurities originated as broken fragments of sintered nanotubes, the amount of remaining material reflects the degree of sintering.

Our examinations of oxygen-purified deposits led to construction of a model of nanotube growth in the arc in which the nanotubes play an active role in sustaining the arc plasma, rather than simply being a passive product[2]. Imaging unpurified nanotube-rich arc deposit from the top by scanning electron microscopy (SEM) revealed a roughly hexagonal lattice of 50-micron diameter circles spaced ~50 microns apart. After oxidative treatment the circular regions were seen to have etched away, leaving a hole. More strikingly, when the deposit was etched after being cleaved vertically to expose the inside of the deposit, SEM imaging showed that columns the diameter of the circles had been etched all the way from the top to the bottom of the deposit, leaving only the intervening material. Prior SEM images of the column material (zone 1) showed that the nanotubes there were highly aligned in the direction of the electric field (also the direction of deposit growth), whereas nanotubes in the surrounding region (zone 2) lay in tangles, unaligned with the field[2]. Since zone 1 nanotubes tend to be in much greater contact with one another, they are far more susceptible to sintering than those in zone 2, resulting in the observed preferential oxidative etch of zone 1.

These observations consummated in a growth model that confers on the millions of aligned zone 1 nanotubes the role of field emitters, a role they play so effectively that they are the dominant source of electron injection into the plasma. In response, the plasma structure, in which current flow becomes concentrated above zone 1, enhances and sustains the growth of the field emission source—that is, zone 1 nanotubes. A convection cell is set up in order to allow the inert helium gas, which is swept down by collisions with carbon ions toward zone 1, to return to the plasma. The helium flow carries unreacted carbon feedstock out of zone 1, where it can add to the growing zone 2 nanotubes. In the model, it is the size and spacing of these convection cells in the plasma that determine the spacing of the zone 1 columns in a hexagonal lattice.

3. FIELD EMISSION FROM AN ATOMIC WIRE

Realization of the critical importance played by emission in our arc growth model added impetus to investigations already underway to characterize nanotube field emission behavior in a more controlled manner. We had begun working with individual nanotubes in the hope of using them as seed crystals for controlled, continuous growth (this remains an active goal). This required developing techniques for harvesting nanotubes from arc deposits, and attaching them with good mechanical and electrical connection to macroscopic manipulators[2,8,9]. The resulting nanoelectrode was then placed in a vacuum chamber in which the nanotube tip could be heated by application of Ar^+-laser light (514.5 nm) while the potential bias was controlled relative to an opposing electrode, and if desired, reactive gases could be introduced.

Two classes of emission behavior were found. An inactivated state, in which the emission current increased upon laser heating at a fixed potential bias, was consistent with well understood thermionic field emission models. Figure 1a displays the emission current as the laser beam is blocked and unblocked, revealing a 300-fold thermal enhancement upon heating. Etching the nanotube tip with oxygen while the tube was laser heated to 1500°C and held at −75 V bias produced an activated state with exactly the opposite behavior, shown in Fig. 2b; the emission current *increased* by nearly two orders of magnitude when the laser beam was blocked! Once we eliminated the possibility that species chemisorbed on the tip might be responsible for this behavior, the explanation had to invoke a structure built only of carbon whose sharpness would concentrate the field, thus enhancing the emission current. As a result of these studies[9], a dramatic and unexpected picture has emerged of the nanotube as field emitter, in which the emitting source is an atomic wire composed of a single chain of carbon atoms that has been unraveled from the tip by the force of the applied electric field (see Fig. 2). These carbon wires can be pulled out from the end of the nanotube only once the ragged edges of the nanotube layers have been exposed. Laser irradiation causes the chains to be clipped from the open tube ends, resulting in low emission when the laser beam is unblocked, but fresh ones are pulled out once the laser is blocked. This unraveling behavior is reversible and reproducible.

4. THE STRUCTURE OF AN OPEN NANOTUBE TIP

A portion of our ongoing work focusing on spheroidal fullerenes, particularly metallofullerenes, utilized the same method of production as was originally used in the discovery of fullerenes, the laser-vaporization method, except for the modification of placing the flow tube in an oven to create better annealing conditions for fullerene formation. Since we knew that at the typical 1200°C oven temperature, carbon clusters readily condensed and annealed to spheroidal fullerenes (in yields close to 40%), we were astonished to find, upon transmission electron micrographic examination of the collected soots, multiwalled nanotubes with few or no defects up to 300 nm long[10]! How, we asked ourselves, was it possible for a nanotube precursor to remain open under conditions known to favor its closing, especially considering the absence of extrinsic agents such as a strong electric field, metal particles, or impurities to hold the tip open for growth and elongation?

The only conclusion we find tenable is that an *intrinsic* factor of the nanotube was stabilizing it against closure, specifically, the bonding of carbon atoms to edge atoms of adjacent layers, as illustrated in Fig. 2. Tight-binding calculations[11] indicate that such sites are energetically preferred over direct addition to the hexagonal lattice of a single layer by as much as 1.5 eV

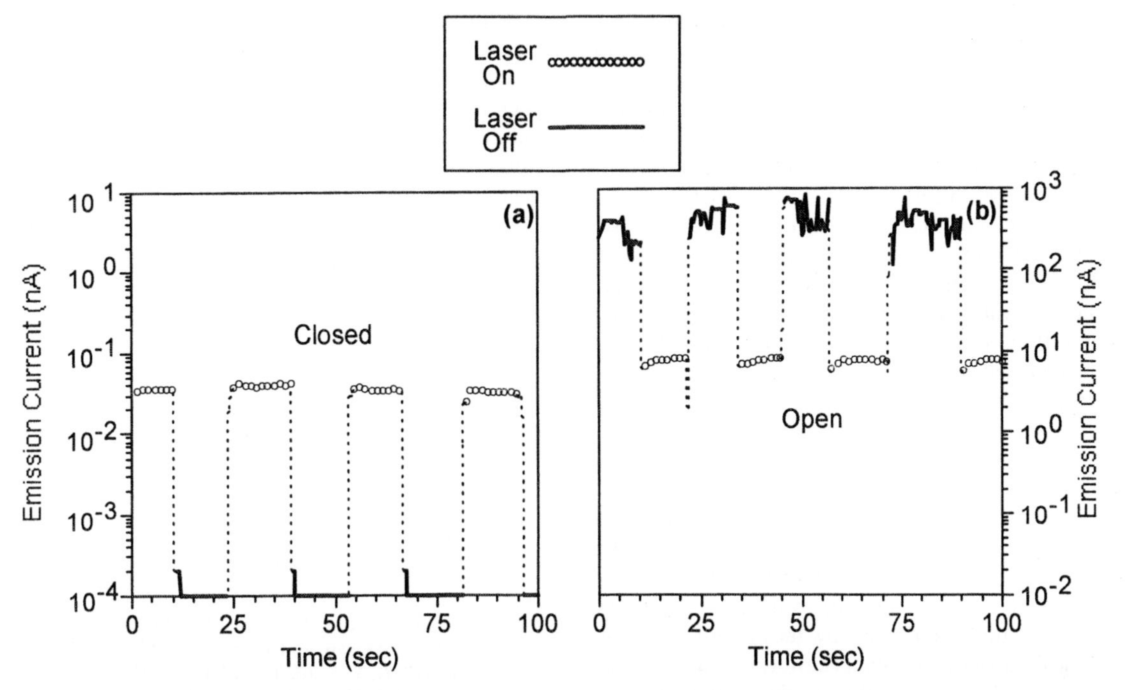

Fig. 1. Field emission data from a mounted nanotube. An activated nanotube emits a higher current when heated by the laser than when the laser beam is blocked (a). When activated by exposing the nanotube to oxygen while heating the tip, this behavior is reversed, and the emission current increases dramatically when the laser is blocked. The activated state can also be achieved by laser heating while maintaining a bias voltage of −75 V. Note that the scale of the two plots is different; the activated current is always higher than the inactivated current. As discussed in the text, these data led to the conclusion that the emitting feature is a chain of carbon atoms pulled from a single layer of the nanotube—an atomic wire.

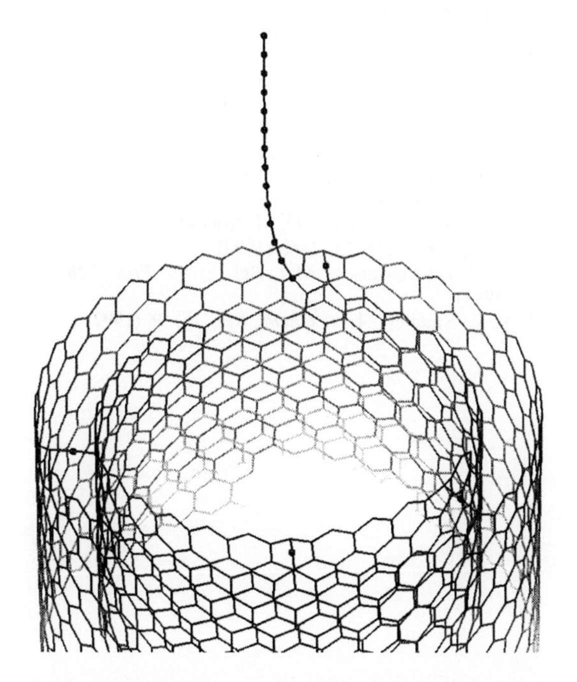

Fig. 2. A graphic of a nanotube showing a pulled-out atomic wire and several stabilizing spot-welds. Only two layers have been shown for clarity, although typical multiwalled nanotubes have 10–15 layers. The spot-weld adatoms shown between layers stabilize the open tip conformation against closure. The atomic wire shown was previously part of the hexagonal lattice of the inner layer. It is prevented from pulling out further by the spot-weld at its base.

per adatom. We also knew at this time that the electric field of the arc was not by itself sufficient to stabilize an open tip having no spot-welds against closure[12], so we now regard these adatom "spot-welds" as a necessary ingredient to explain growth of nanotubes in the oven laser-vaporization method as well as in the arc, and probably other existing methods of nanotube production.

5. ELECTRIC FIELD STABILIZATION OF AN OPEN NANOTUBE TIP

The proposal that the essential feature of arc growth was the high electric field that concentrates at the growing nanotube tip prompted ab initio structure calculations[12,13] to assess this hypothesis quantitatively. These calculations, which were performed for *single-walled* nanotubes in high applied electric fields, showed that field-induced lowering of the open tip energy is not sufficient to make the open conformation more stable than the closed tip at any field less than 10 V/Å. Whereas single-walled objects certainly anneal and close readily at 1200°C to form spheroidal fullerenes[14,15], open *multi*walled species have other alternatives, and thus may be quite different in this respect. In particular, for multiwalled species, adatom spot-welds may be sufficiently stabilizing to allow growth and elongation before succumbing to the closed conformation.

In the absence of an electric field, the dome-closed conformation must be the most stable tip structure, even when spot-welds are considered, since only the perfectly dome-closed tip has no dangling bonds (i.e., it is a true hemifullerene). At the 3000°C temperature of the arc, the rate of tip annealing should be so fast that it is sure to find its most stable structure (i.e., to close as a dome). Clear evidence of this facile closure is the fact that virtually all nanotubes found in the arc deposit are dome-closed. (Even stronger evidence is the observation of only dome-closed nanotubes made at 1200°C by the oven laser vaporization method.) Such considerations constituted the original motivation for the electric field hypothesis.

Armed with these results, a direct test of the hypothesis using a single mounted nanotube in our vacuum apparatus was sensible. A dome-closed nanotube harvested from the arc deposit gave inactivated state behavior at −75 V bias. Maintaining the bias voltage at −75 V, the nanotube was irradiated for about 30 seconds with sufficient intensity to sublime some carbon from the tip (~3000°C). Now the nanotube exhibited typical activated emission behavior, indicating an open tip from which long carbon chains were pulled, constituting the emitting structures described in section 3 above. When the nanotube was reheated at 0 V bias, the tip was re-closed. Subsequent heating to 3000°C at −75 V bias re-opened the tip. These results can only be explained by the electric field's providing the necessary stabilization to keep the tip open.

The structure calculations on single-walled tubes show that the stabilizing effect of the field is at most about 10% of that required to lower the energy of the open below that of the closed tip, before reaching unrealistically strong field strengths. However, with our enhanced understanding of the structure of nanotube tips, much of the energy lowering is achieved by the adatom spot-welds (not included in the calculations), leaving less of an energy gap for the electric field effect to bridge. We emphasize that spot-welds alone cannot be sufficient to render the open tip stable relative to the dome-closed one, since the latter structure is the only known way to eliminate all the energetically costly dangling bonds. An electric field is necessary.

With expanding knowledge about the ways nanotubes form and behave, and as their special properties are increasingly probed, the time is fast approaching when nanotubes can be put to novel uses. Their size and electrical properties suggest their use as nanoprobes, for instance, as nanoelectrodes for probing the chemistry of living cells on the nanometer scale. The atomic wire may be an unrivaled cold field emission source of coherent electrons. Such potential uses offer the prospect of opening up new worlds of investigation into previously unapproachable domains.

Acknowledgements—This work was supported by the Office of Naval Research, the National Science Foundation, the Robert A. Welch Foundation, and used equipment designed for study of fullerene-encapsulated catalysts supported by the Department of Energy, Division of Chemical Sciences.

REFERENCES

1. T. W. Ebbesen, *Ann. Rev. Mat. Sci.* **24**, 235 (1994); S. Iijima, P. M. Ajayan, and T. Ichihashi, *Phys. Rev. Lett.* **69**, 3100 (1992); Y. Saito, T. Yoshikawa, M. Inagaki, M. Tomita, and T. Hayashi, *Chem. Phys. Lett.* **204**, 277 (1993).
2. D. T. Colbert *et al.*, *Science* **266**, 1218 (1994).
3. S. Iijima, *Nature* **354**, 56 (1991).
4. T. W. Ebbesen and P. M. Ajayan, *Nature* **358**, 220 (1992).
5. D. Ugarte, *Chem. Phys. Lett.* **198**, 596 (1992); D. Ugarte, *Nature* **359**, 707 (1992).
6. R. E. Smalley, *Mat. Sci. Eng.* **B19**, 1 (1993).
7. T. W. Ebbesen, P. M. Ajayan, H. Hiura, and K. Tanigaki, *Nature* **367**, 519 (1994).
8. A. G. Rinzler, J. H. Hafner, P. Nikolaev, D. T. Colbert, and R. E. Smalley, *MRS Proceedings* **359** (1995).
9. A. G. Rinzler *et al.*, in preparation.
10. T. Guo *et al.*, submitted for publication.
11. J. Jund, S. G. Kim, and D. Tomanek, in preparation; C. Xu and G. E. Scuseria, in preparation.
12. L. Lou, P. Nordlander, and R. E. Smalley, *Phys. Rev. Lett.* (in press).
13. A. Maiti, C. J. Brabec, C. M. Roland, and J. Bernholc, *Phys. Rev. Lett.* **73**, 2468 (1994).
14. R. E. Haufler *et al.*, *Mat. Res. Symp. Proc.* **206**, 627 (1991).
15. R. E. Smalley, *Acct. Chem. Res.* **25**, 98 (1992).

CATALYTIC PRODUCTION AND PURIFICATION OF NANOTUBULES HAVING FULLERENE-SCALE DIAMETERS

V. Ivanov,[a,*] A. Fonseca,[a] J. B.Nagy,[a†] A. Lucas,[a] P. Lambin,[a] D. Bernaerts[b] and X. B. Zhang[b]

[a]Institute for Studies of Interface Science, Facultés Universitaires Notre Dame de la Paix, 61 rue de Bruxelles, B-5000 Namur, Belgium

[b]EMAT, University of Antwerp (RUCA), Groenenborgerlaan 171, B-2020 Antwerp, Belgium

(*Received* 25 *July* 1994; *accepted in revised form* 13 *March* 1995)

Abstract—Carbon nanotubules were produced in a large amount by catalytic decomposition of acetylene in the presence of various supported transition metal catalysts. The influence of different parameters such as the nature of the support, the size of active metal particles and the reaction conditions on the formation of nanotubules was studied. The process was optimized towards the production of nanotubules having the same diameters as the fullerene tubules obtained from the arc-discharge method. The separation of tubules from the substrate, their purification and opening were also investigated.

Key Words—Nanotubules, fullerenes, catalysis.

1. INTRODUCTION

The catalytic growth of graphitic carbon nanofibers during the decomposition of hydrocarbons in the presence of either supported or unsupported metals, has been widely studied over the last years[1–6]. The main goal of these studies was to avoid the formation of "filamentous" carbon, which strongly poisons the catalyst. More recently, carbon tubules of nanodiameter were found to be a byproduct of arc-discharge production of fullerenes[7]. Their calculated unique properties such as high mechanical strength[8], their capillary properties[9] and their remarkable electronic structure[10–12] suggest a wide range of potential uses in the future. The catalytically produced filaments can be assumed to be analogous to the nanotubules obtained from arc-discharge and hence to possess similar properties[5], they can also be used as models of fullerene nanotubes. Moreover, advantages over arc-discharge fibers include a much larger length (up to 50 μm) and a relatively low price because of simpler preparation. Unfortunately, carbon filaments usually obtained in catalytic processes are rather thick, the thickness being related to the size of the active metal particles. The graphite layers of as-made fibres contain many defects. These filaments are strongly covered with amorphous carbon, which is a product of the thermal decomposition of hydrocarbons[13]. The catalytic formation of thin nanotubes was previously reported[14]. In this paper we present the detailed description of the catalytic deposition of carbon on various well-dispersed metal catalysts. The process has been optimized towards the large scale nanotubes production. The synthesis of the nanotubules of various diameters, length and structure as dependent on the parameters of the method is studied in detail. The elimination of amorphous carbon is also investigated.

2. EXPERIMENTAL

The catalytic decomposition of acetylene was carried out in a flow reactor at atmospheric pressure. A ceramic boat containing 20–100 mg of the catalyst was placed in a quartz tube (inner diameter 4–10 mm, length 60–100 cm). The reaction mixture of 2.5–10% C_2H_2 (Alphagaz, 99.6%) in N_2 (Alphagaz, 99.99%) was passed over the catalyst bed at a rate of 0.15–0.59 mol C_2H_2 $g^{-1}h^{-1}$ for several hours at temperatures in the range 773–1073 K.

The catalysts were prepared by the following methods. Graphite supported samples containing 0.5–10 wt% of metal were prepared by impregnation of natural graphite flakes (Johnson–Matthey, 99.5%) with the solutions of the metal salts in the appropriate concentrations: Fe or Co oxalate (Johnson–Matthey), Ni or Cu acetate (Merck). Catalysts deposited on SiO_2 were obtained by porous impregnation of silica gel (with pores of 9 nm, S_{sp} 600 m^2g^1, Janssen Chimica) with aqueous solutions of Fe(III) or Co(II) nitrates in the appropriate amounts to obtain 2.5 wt% of metal or by ion-exchange-precipitation of the same silica gel with 0.015 M solution of Co(II) nitrate (Merck) following a procedure described in Ref. [15]. The catalyst prepared by the latter method had 2.1 wt% of Co. All samples were dried overnight at 403 K and then calcined for 2 hours at 773 K in flowing nitrogen and reduced in a flow of 10% H_2 in N_2 at 773 K for 8 hours.

Zeolite-supported Co catalyst was synthesized by solid-state ion exchange using the procedure described by Kucherov and Slinkin[16, 17]. CoO

*To whom all correspondence should be addressed.

†Permanent address: Laboratory of Organic Catalysis, Chemistry Department, Moscow State University, 119899, Moscow, Russia.

was mixed in an agath morter with HY zeolite. The product was pressed, crushed, dried overnight at 403 K and calcined in air for 1 hour at 793 K, then for 1 hour at 1073 K and after cooling for 30 minutes in flowing nitrogen, the catalyst was reduced in a flow of 10% H_2 in N_2 for 3 hours at 673 K. The concentration of CoO was calculated in order to obtain 8 wt% of Co in the zeolite.

The list of studied catalysts and some characteristics are given in Table 1.

The samples were examined before and after catalysis by SEM (Philips XL 20) and HREM by both a JEOL 200 CX operating at 200 kV and a JEOL 4000 EX operating at 400 kV. The specimens for TEM were either directly glued on copper grids or dispersed in acetone by ultrasound, then dropped on the holey carbon grids.

^{1}H-NMR studies were performed on a Bruker MSL-400 spectrometer operating in the Fourier transform mode, using a static multinuclei probehead operating at 400.13 MHz. A pulse length of 1 μs is used for the ^{1}H 90° flip angle and the repetition time used (1 second) is longer than five times T_{1Z} (^{1}H) of the analyzed samples.

3. RESULTS AND DISCUSSION

3.1 *Catalyst support*

The influence of the support on the mechanism of filament formation was previously described[1–4]. The growth process was shown to be strongly dependent on the catalyst–support interaction. In the first stage of our studies we performed the acetylene decomposition reaction over graphite supported metals. This procedure was reported in Ref. [13] as promising to obtain a large amount of long nanotubes. The reaction was carried out in the presence of either Cu, Ni, Fe or Co supported particles. All of these metals showed a remarkable activity in filament formation (Fig. 1). The structure of the filaments was different on the various metals. We have observed the formation of hollow structures on the surface of Co and Fe catalysts. On Cu and Ni, carbon was deposited in the form of irregular fibres. The detailed observation showed fragments of turbostratic graphite sheets on the latter catalysts. The tubular filaments on Fe- and Co-graphite sometimes possessed well-crystalline graphite layers. In the same growth batch we also observed a large amount of non-hollow filaments with a structure similar to that observed on Cu and Ni catalysts.

In general, encapsulated metal particles were observed on all graphite-supported catalysts. According to Ref. [4] it can be the result of a rather weak metal–graphite interaction. We mention the existence of two types of encapsulated metal particles: those enclosed in filaments (Fig. 1) and those encapsulated by graphite. It is interesting to note that graphite layers were parallel to the surface of the encapsulated particles.

As was found in Ref. [13], the method of catalytic decomposition of acetylene on graphite-supported catalysts provides the formation of very long (50 μm) tubes. We also observed the formation of filaments up to 60 μm length on Fe- and Co-graphite. In all cases these long tubules were rather thick. The thickness varied from 40 to 100 nm. Note that the dispersion of metal particles varied in the same range. Some metal aggregates of around 500 nm in diameter were also found after the procedure of catalyst pretreatment (Fig. 2). Only a very small amount of thin (20–40 nm diameter) tubules was observed.

The as-produced filaments were very strongly covered by amorphous carbon produced by thermal pyrolysis of acetylene. The amount of amorphous carbon varied with the reaction conditions. It increased with increasing reaction temperature and with the percentage of acetylene in the reaction mixture. Even in optimal conditions not less than 50% of the carbon was deposited in the form of amorphous carbon in accordance with[13].

As it was established by Geus *et al.*[18, 19] the decrease of the rate of carbon deposition is a positive factor for the growth of fibres on metal catalysts. SiO_2 is an inhibitor of carbon condensation as was shown in Ref. [20]. This support also provides possibilities for the stabilization of metal dispersion. Co and Fe, i.e. the metals that give the best results for the tubular condensation of carbon on graphite support, were introduced on the surface of silica gel

Table 1. Method of preparation and metal content of the catalysts

Sample	Pore Ø (Å)	Method of preparation	Metal (wt%)	Metal particle diameter[a] (nm)
Co–graphite	—	Impregnation	0.5–10	10–100
Fe–graphite	—	Impregnation	2.5	⩾100
Ni–graphite	—	Impregnation	2.5	⩾100
Cu–graphite	—	Impregnation	2.5	⩾100
Co–SiO_2	90	Ion exchange precipitation	2.1	2–20[b]
Co–HY	7.5	Solid state ion exchange	8	1–50
Co–SiO_2–1	40	Pore impregnation	2.5	10–100
Co–SiO_2–2	90	Pore impregnation	2.5	10–100
Fe–SiO_2	90	Pore impregnation	2.5	10–100

[a]Measured by SEM and TEM.
[b]The distribution of the particles was also measured (Fig. 6).

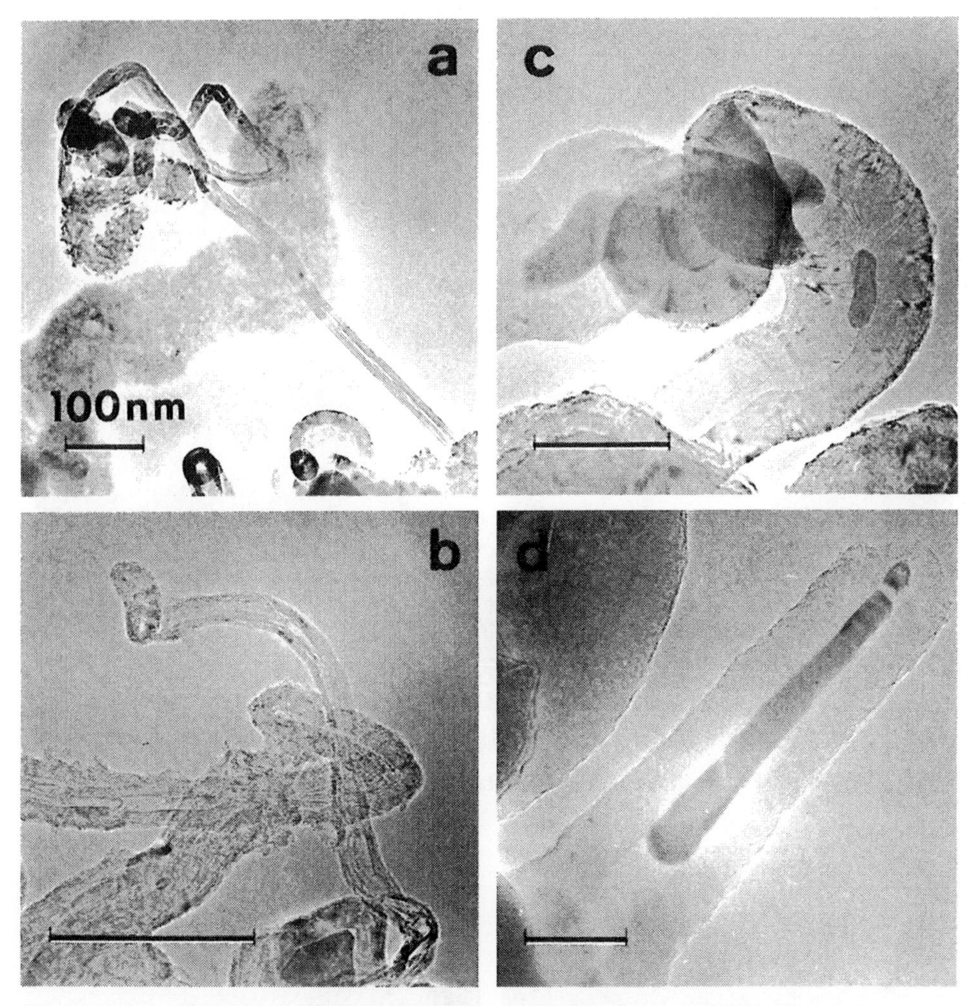

Fig. 1. Carbon filaments grown after acetylene decomposition at 973 K for 5 hours on (a) Co(2.5%)–graphite; (b) Fe–graphite; (c) Ni–graphite; (d) Cu–graphite.

by different methods. Both metals showed very similar catalytic behaviour. Carbon was deposited on these catalysts mostly in the form of filaments. TEM images of the tubules obtained on these catalysts are given in Fig. 3. Most of the filaments produced on silica-supported catalysts were tubular, with well-resolved graphite layers. Nevertheless non-tubular filaments also grow in these conditions. We observed that the relative quantity of well-graphitized tubules was higher on Co-silica than on Fe-silica catalyst.

As in the case of graphite-supported catalysts, some metal particles were also encapsulated by the deposited carbon (Fig. 4). However, the amount of encapsulated metal was much less. Differences in the nature of encapsulation were observed. Almost all encapsulated metal particles on silica-supported catalysts were found inside the tubules (Fig. 4(a)). The probable mechanism of this encapsulation was precisely described elsewhere[21]. We supposed that they were catalytic particles that became inactive after introduction into the tubules during the growth process. On the other hand, the formation of graphite layers around the metal in the case of graphite-supported catalysts can be explained on the basis of models proposed earlier[4,18,22]. The metal outside of the support is saturated by the carbon produced by hydrocarbon decomposition, possibly in the form of "active" carbides. The latter then decomposes on the surface of the metal, producing graphite layers. Such a situation is typical for catalysts with a weak metal–support interaction, as in the case of graphite.

The zeolite support was used to create very finely dispersed metal clusters. Metals can be localized in the solid-state exchanged zeolites in the small cages, supercages or intercrystalline spaces. In fact, in accordance with previously observed data[23], hydrogenation of as-made catalysts led to the migration of metal to the outer surface of the zeolite HY. The sizes of metal crystallites varied in our catalyst from 1 to 50 nm. We suppose that because of steric limitations only the metal particles at the outer surface and in supercages could be available for filament growth. The hydrocarbon decomposition over Co–HY provides the formation of different graphite-related structures (it should be noted that only a small amount of amorphous carbon was observed). Similar to the previous catalysts, nanotubules of various radius and metal particles encapsulated by graphite were found

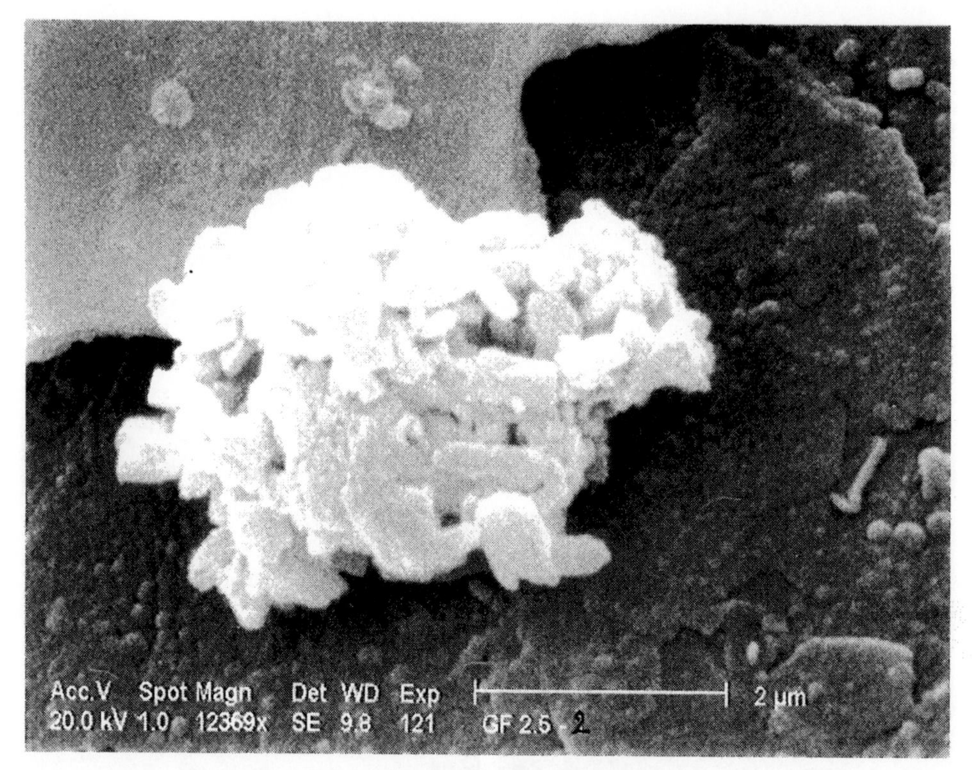

Fig. 2. SEM image of the surface of Fe–graphite after pretreatment before catalysis.

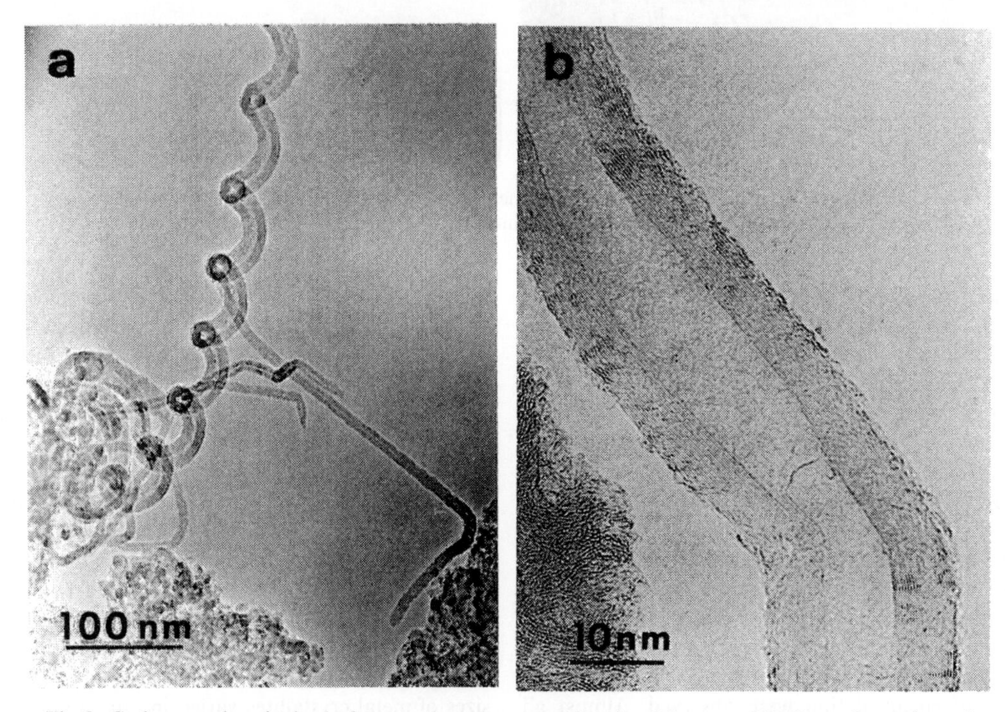

Fig. 3. Carbon species obtained after acetylene decomposition for 5 hours at 973 K on the surface of silica-supported catalysts made by pore impregnation: (a) Co–SiO_2–1; (b) Co–SiO_2–2.

on Co–zeolite (Fig. 5(a)). The thickness of tubules was also dependent on the metal dispersion. Only on this catalyst we could observe extremely thin (approximately 4 nm diameter) graphite tubules with the walls composed of 2–3 layers. Their amount, however, was very small, owing to the small quantity of highly dispersed Co that remains in zeolite structure after the reduction by hydrogen.

3.2 *The influence of reaction parameters*

3.2.1 *The rate of hydrocarbon flow.* The study of the reaction conditions was performed on Co–

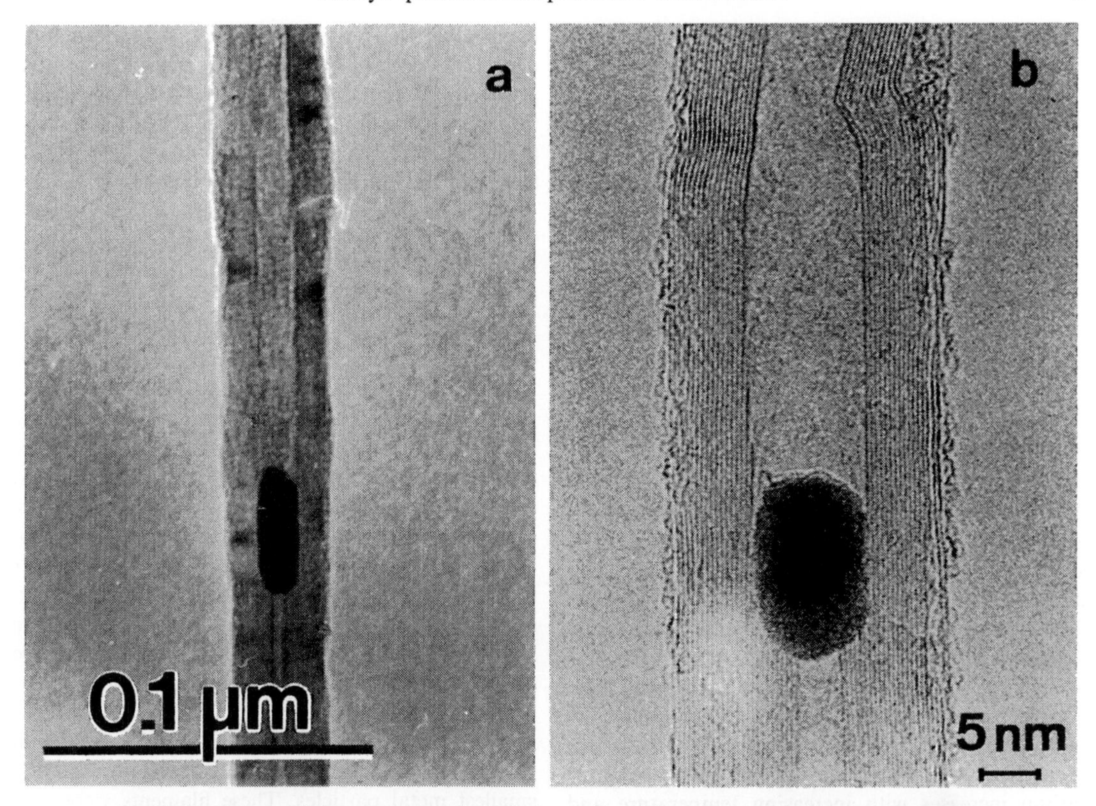

Fig. 4. Catalytic particles encapsulated in tubules on Co–SiO_2: (a) low magnification; (b) HREM.

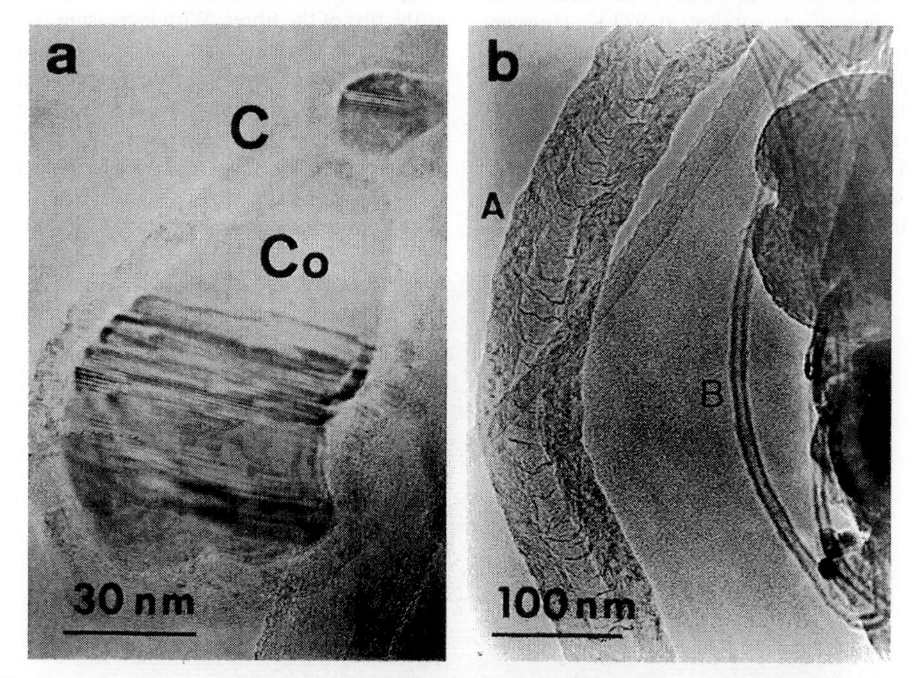

Fig. 5. Acetylene decomposition on Co-HY (973 K, 30 minutes): (a) encapsulated metal particle; (b) carbon filaments (A) and tubules of small diameters (B) on the surface of the catalyst.

silica catalyst synthesized by the ion-exchange-precipitation method. This method leads to a better dispersion and a sharper size distribution than porous impregnation (Fig. 6). The rate of acetylene was varied from 0.15 to 0.59 mol C_2H_2 g^{-1} h^{-1}. The amount of condensed carbon at 973 K changed as a function of hydrocarbon rate from 20 to 50 wt% relatively. The optimal C_2H_2 rate was found to be 0.34 mol g^{-1} h^{-1}. At this rate, 40–50 wt% of carbon precipitates on the catalyst surface after 30 minutes of reaction. The increase of acetylene rate leads to an increase of the amount of amorphous carbon (equal to 2 wt% on the external walls of the tubules, for optimal conditions).

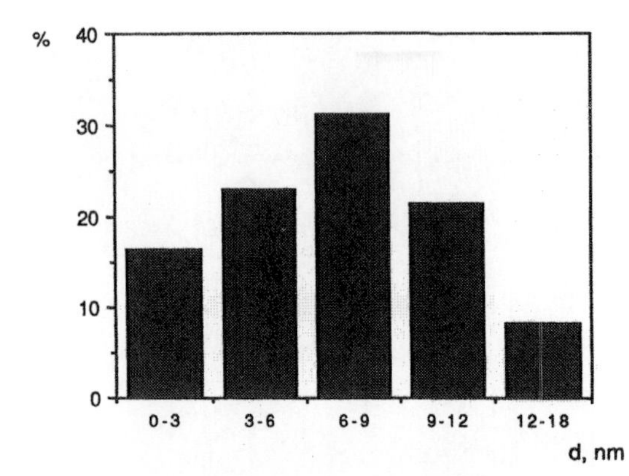

Fig. 6. Size-distribution of metal crystallites on the surface of Co–silica made by precipitation-ion-exchange method.

3.2.2 *Reaction temperature.* The reaction temperature was varied in the range 773–1073 K. The formation of filament structures was observed at all studied temperatures. As has already been mentioned, the graphitization of carbon into the tubular structures on metal-supported catalysts is generally accompanied by the formation of amorphous carbon. Both processes are temperature dependent. The filaments grown at low temperature (773 K) are relatively free of amorphous carbon. The amount of amorphous carbon increases with increasing temperature and represents about 10% of all carbon condensed on the external surface of the catalyst at 973 K. However, crystallinity of the graphite layers in tubules also strongly depends on the reaction temperature being the lowest at low temperature.

The average length of the tubules is not strongly influenced by temperature. However, the amorphous carbon on the outer layers of filaments produced under optimal conditions is often deposited in fragments (Fig. 7). Thus, we suppose that the formation of graphite tubules in these conditions is a very rapid process and the thermal pyrolysis leading to the formation of amorphous carbon does not have a great influence. Hence, carbon nanotubules, *quasi*-free from amorphous carbon, are formed.

3.2.3 *Reaction time.* Two series of experiments were performed in order to study the influence of the reaction time on the characteristics of surface carbon structures. In the first series, the hydrocarbon deposition was periodically stopped, the catalyst was cooled down under flowing nitrogen and it was removed from the furnace. After taking a small part of the reaction mixture for TEM analysis, the remaining amount of the catalyst was put back into the furnace and the hydrocarbon deposition was further carried out under the same conditions. In the second series, different portions of catalyst were treated by hydrocarbon for different times. The results were similar for both series of catalysts. Typical images of carbon surface structures grown during different times are shown in Fig. 8. In accordance with Ref. [4] we observed the dependence of the rate of filament formation on the size of the catalytic particles. In the first (1 minute) reaction period, mostly very thin carbon filaments were observed as grown on the smallest metal particles. These filaments were very irregular and the metal particles were generally found at the tips of the fibres. With increasing reaction time the amount of well-graphitized tubules progressively increased. At the same time the average length of the nanotubules increased. We need, however, to note that a relation exists between the lengths of the tubules and their diameters. The longest tubules are also the thickest. For instance, the tubules of 30–60 μm length have diameters of 35–40 nm corre-

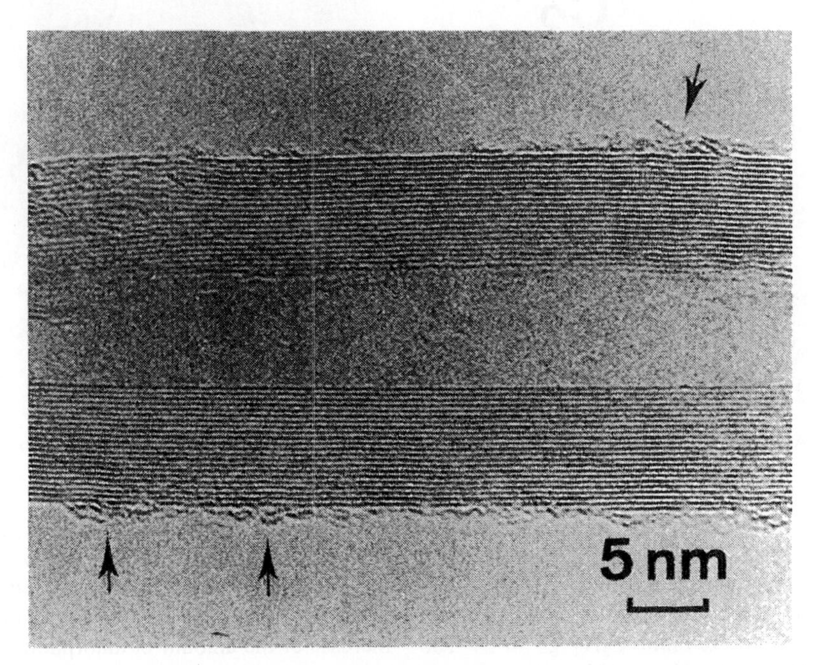

Fig. 7. Graphite nanotubule on $Co–SiO_2$ with the fragments of amorphous carbon (arrowed) at the external surface.

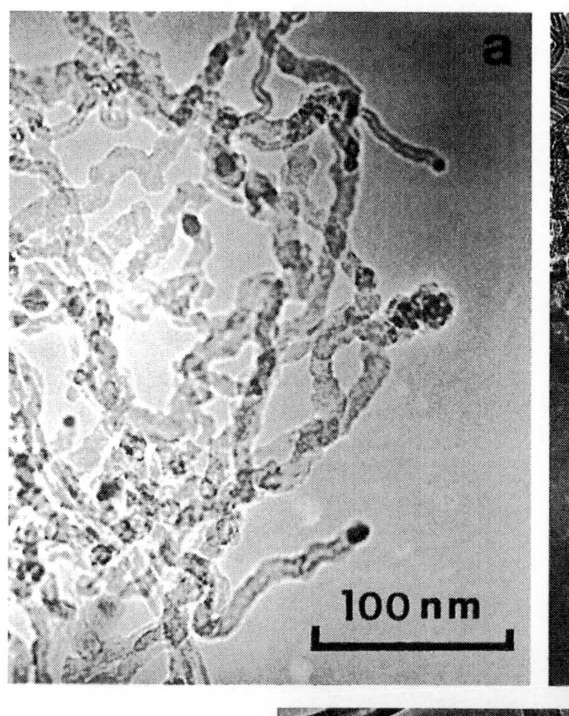

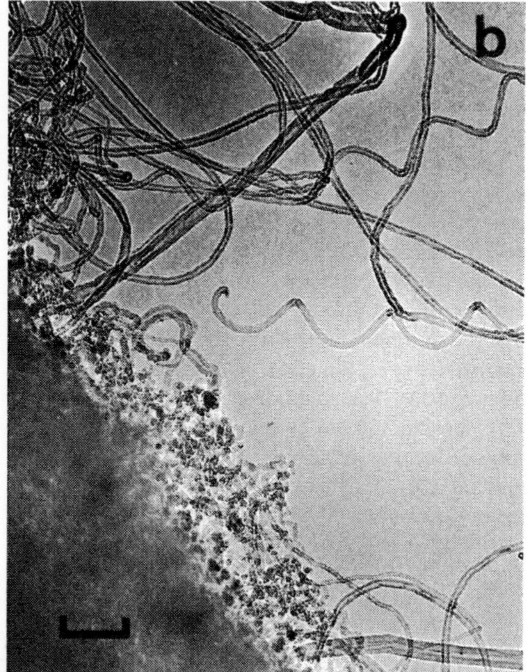

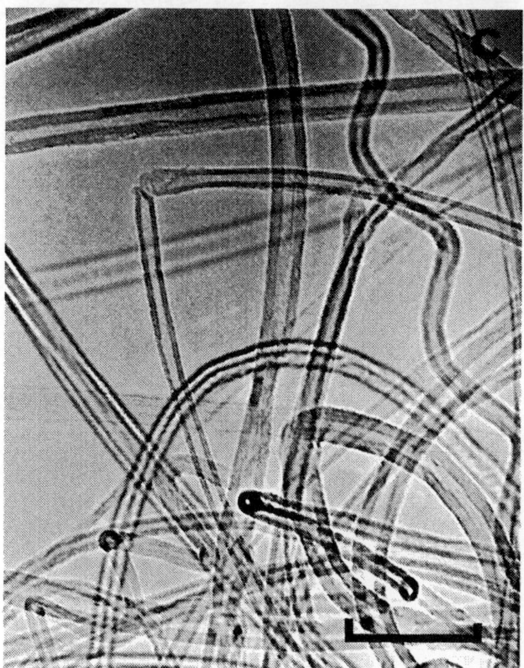

Fig. 8. Surface carbon species produced in acetylene decomposition on Co–SiO_2 at 973 K after different reaction times: (a) 3 minutes; (b) 5 minutes; (c) 20 minutes.

sponding to 40–50 layers of graphite. The formation of amorphous carbon is negligible after small reaction times (30 minutes). Then its relative amount increases. The nanotube formation was found to be the same on the catalyst after 5 and 48 hours of reaction. However, the amount of amorphous carbon on the latter sample was much higher (20 and 50%, respectively).

We propose the following scheme for carbon deposition on Co–silica. Under optimal conditions, during the first period (30 minutes) of the reaction the rate of filament growth is high because of the contribution of the small metal particles. Practically all precipitated carbon is consumed in filament formation, which is the preferred process. The length of the narrow tubules is increasing faster than the length of the thick tubules, the rate of material consumption being roughly constant in the system. After some reaction time needed for the production of the fibres on the largest metal particles, amorphous pyrolytic carbon is also produced.

Interestingly, the growth of filaments continued

after cooling and reheating the carbon deposited sample. Metal particles even found near the tips of the tubules were always covered by graphite layers. It supports the model of an "extrusion" of the carbon tubules from the surface of active particles[24].

3.2.4 *Influence of hydrogen.* The influence of the presence of hydrogen in the reaction mixture on the formation of nanofibres has been shown in various papers[1–3, 18, 25, 26]. It was postulated that the presence of H_2 decreases the rate of hydrocarbon decomposition and as a result favours the process of carbon polycondensation over the production of filaments. However, the addition of hydrogen into the mixture of acetylene and nitrogen did not give major effects on the tubule formation in our case. We suppose that the activity of metal nanoparticles on our Co–SiO_2 catalyst was high enough to provide filament formation without hydrogen addition. It differs from the previous investigations, which were performed on metal-supported catalysts with larger and, thus, less active metal aggregates.

It is also important to point out that pure cobalt oxide, alone or finely dispersed in SiO_2 (i.e. Co–SiO_2, Co–SiO_2–1 and Co–SiO_2–2 in Table 1), zeolite HY, fullerene (i.e. C_{60}/C_{70} : 80/20) is at least as effective as the reduced oxides for the production of nanotubules in our experimental conditions. In fact, the catalysts studied in this work are also active if the hydrogenation step is not performed. This important point, is presently being investigated in our laboratory in order to elucidate the nature of the active catalyst (probably a metal carbide) for the production of nanotubules.

3.3 *The amount of hydrogen in filaments*

As it can be observed from the high resolution images of tubules (Fig. 9(a)) their graphitic structure is generally defective. The defects can be of different

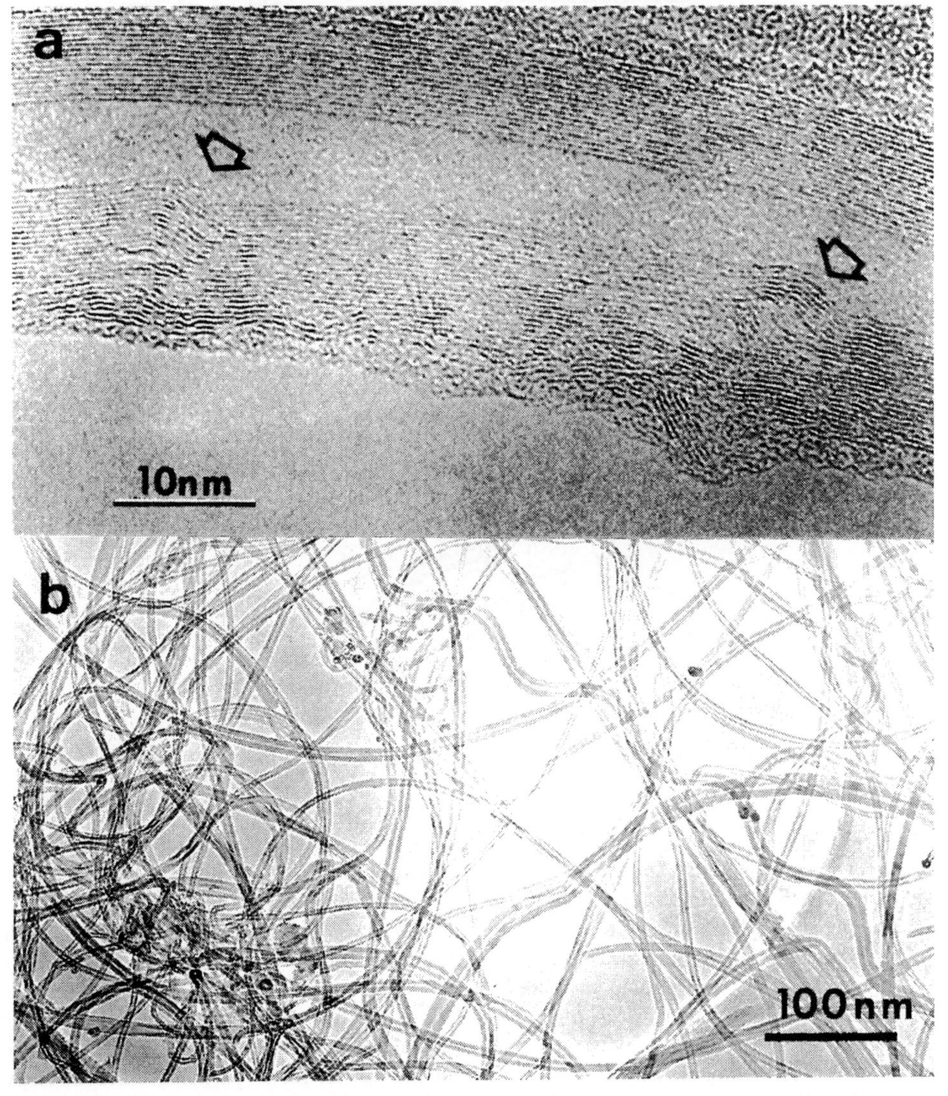

Fig. 9. Carbon nanotubules on Co–SiO_2: (a) HREM image showing defects in tubules; (b) helical tubules of various pitches between the straight tubules.

Table 2. Hydrogen content measured by quantative ^{1}H-NMR

Sample	T_{1Z} (^{1}H)[a] (ms)	Hydrogen (wt%)
Coronene (reference)	520	4.03
Catalyst Co–SiO_2	20	1.33
Evacuated catalyst Co–SiO_2	960	0.26
Carbonated catalyst $CoSiO_2$	50	1.80
Evacuated carbonated catalyst Co–SiO_2	160	1.26[b]

[a] ^{1}H longitudinal relaxation time, measured by the inversion-recovery technique, at 293 K.

[b] As 50 wt% of hydrocarbons are deposited on the catalyst, the hydrogen content of the hydrocarbons is: 2(1.26−0.13)=2.26 wt%.

natures. The most interesting ones are regular defects leading to the formation of helices. The helical tubules are 6–10% of the total amount of filaments as estimated from the microscopy observations (Fig. 9(b)). The mechanism of helices formation was discussed elsewhere[27] and a model based on the regular introduction of pentagon–heptagon pairs was proposed. The presence of stress causes the formation of "kinks" in the graphite layers. This kind of defect was also well described[21]. There are also defects in the graphite layers, which are typical for turbostratic graphite. We already mentioned that the formation of these kinds of defect strongly decreases with the increase of reaction temperature. The free valencies of carbon in such defects can be compensated by the formation of C—H bonds. The carbon layers produced on the surface of silica-supported catalysts after hydrocarbon decomposition always have chemically bonded hydrogen (up to 2 wt% in some cases).

We performed a quantitative ^{1}H-NMR study using coronene ($C_{24}H_{12}$) as external standard. The ^{1}H-NMR spectra of the catalyst samples before and after reaction are given in Fig. 10. The static proton spectrum gives a broad band at 6.9 ppm similar to that obtained on the reference sample (7.2 ppm). The sharp proton band at 7.9 ppm before the catalysis can be related to the small amount of hydroxyl groups still remaining on the surface of silica after temperature treatment before reaction. This amount was taken into account in the calculations of the hydrogen content in carbon species on the surface (Table 2). The total quantity of hydrogen is approximately 2 wt%. This amount can be sufficient to saturate all free carbon vacancies in sp^3 defects of graphitic structure.

3.4 *Gasification of nanotubes*

The carbon deposited catalysts were treated both by oxidation and hydrogenation at temperatures in the range of 873–1173 K for various exposure times. Some results of oxidation treatment are presented in

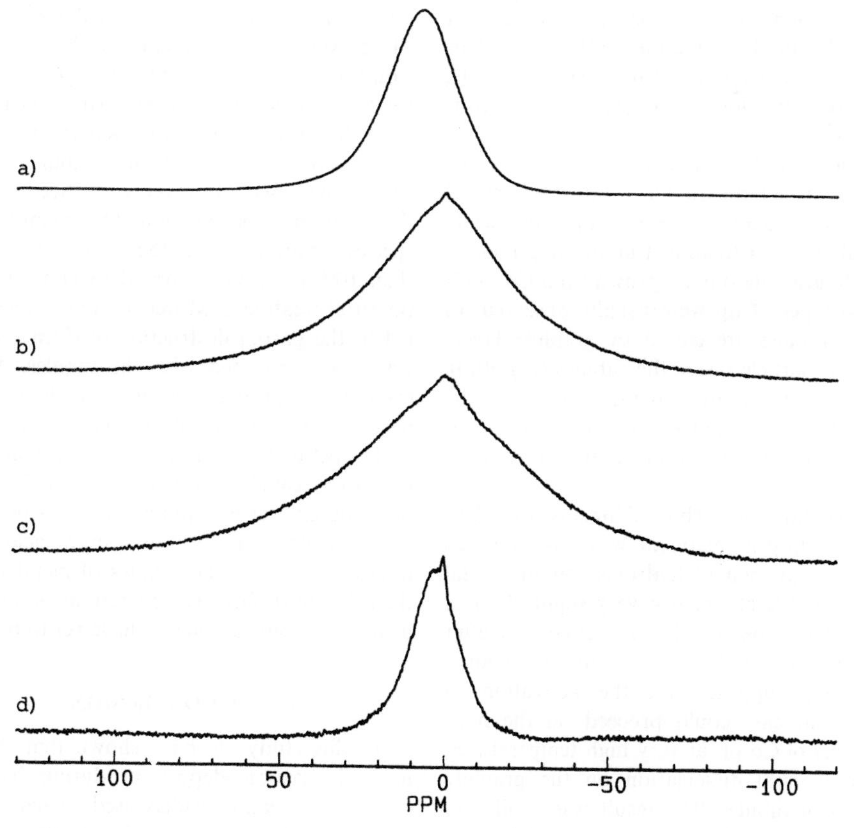

Fig. 10. ^{1}H NMR spectra: (a) coronene; (b) Co–SiO_2 covered by carbon nanotubules; (c) Co–SiO_2 covered by carbon nanotubules and evacuated to 10^{-5} torr for the NMR measurement; (d) Co–SiO_2 evacuated to 10^{-5} torr for the NMR measurement.

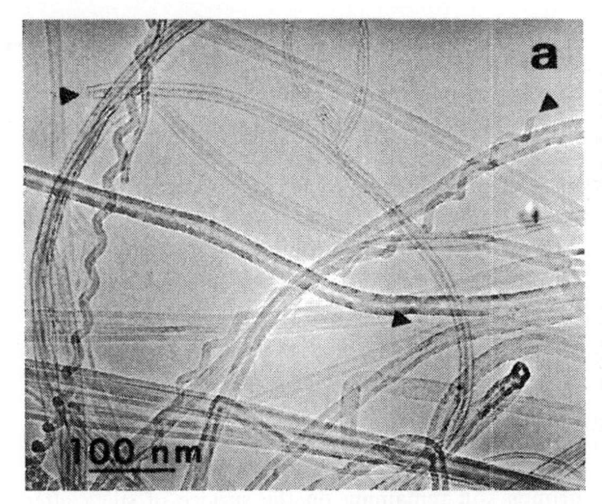

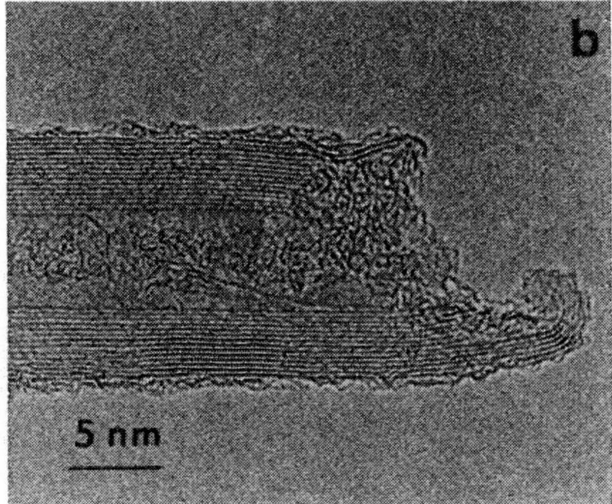

Fig. 11. Tips of carbon nanotubules grown on Co–SiO_2 (acetylene reaction at 973 K, 30 minutes after oxidation in air for 30 minutes at 873 K: (a) low magnification; (b) HREM.

Fig. 11. The loss of carbon rapidly increases with the increase of temperature. Heating of the catalysts in open air for 30 minutes at 973 K leads to the total elimination of carbon from the surface. The gasification of amorphous carbon proceeds more rapidly than that of filaments. The tubules obtained after oxidation of carbon-deposited catalysts during 30 minutes at 873 K are almost free from amorphous carbon. The process of gasification of nanotubules on the surface of the catalyst is easier in comparison with the oxidation of nanotubes containing soot obtained by the arc-discharge method[28, 29]. This can be easily explained, in agreement with Ref. [30], by the surface activation of oxygen of the gaseous phase on Co–SiO_2 catalyst.

The gasification of graphite layers proceeds more easily at the tips of the tubules and at structural defects. Typical images of the tips of catalytically produced tubules after treatment in air are presented in Fig. 11. On graphite tubules grown from Co–SiO_2 catalyst, two types of tip were usually observed. In the first, the tubules are closed by graphite layers with the metal particle inside the tubules (Fig. 4(a)). In the second type, more generally observed, the tubules are closed with amorphous carbon. The opening of tubules during oxidation could proceed on both types of tip.

The gasification of carbon filaments by high-temperature hydrogen treatment was postulated as involving the activation of hydrogen on the metal surface[31–33]. We observed a very slight effect of catalyst hydrogenation, which was visible only after the treatment of carbon-deposited catalyst for 5 hours at 1173 K. We suppose that the activation of hydrogen in our case could proceed on the non-covered centers of Co or, at very high temperatures, it could be thermal dissociation on the graphite surface layers of tubules. The result was similar to that of oxidation but the process proceeded much slower. We called it "gentle" gasification and we believe that this method of thinning of the nanotubules could be preferable in comparison with oxidation, because of the easier control in the former case.

3.5 *Product purification*

For the physico-chemical measurements and practical utilisation in some cases the purification of nanotubules is necessary. In our particular case, purification means the separation of filaments from the substrate–silica support and Co particles.

The carbon-containing catalyst was treated by ultra-sound (US) in acetone at different conditions. The power of US treatment, and the time and regime (constant or pulsed), were varied. Even the weakest treatments made it possible to extract the nanotubules from the catalyst. With the increase of the time and the power of treatment the amount of extracted carbon increased. However, we noticed limitations of this method of purification. The quantity of carbon species separated from the substrate was no more than 10% from all deposited carbon after the most powerful treatment. Moreover, the increase of power led to the partial destruction of silica grains, which were then extracted with the tubules. As a result, even in the optimal conditions the final product was never completely free of silica (Fig. 12).

For better purification, the tubule-containing catalyst was treated by HF (40%) over 72 hours. The resulting extract was purer than that obtained after US treatment. The addition of nitric acid also makes it possible to free the tubules of metal particles on the external surface. The conditions of the acid treatment and tubule extraction have yet to be optimized.

4. CONCLUSIONS

In this study we have shown that the catalytic method—carbon deposition during hydrocarbons conversion—can be widely used for nanotubule production methods. By variation of the catalysts and reaction conditions it is possible to optimize the process towards the preferred formation of hollow

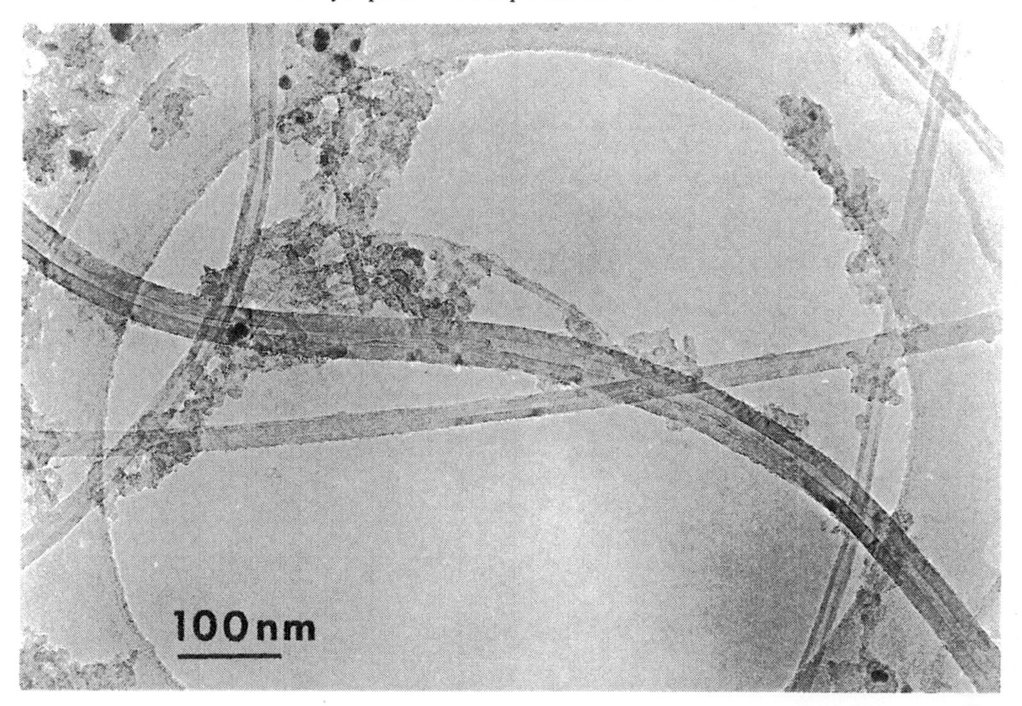

Fig. 12. Carbon nanotubules after separation from the substrate by ultra-sound treatment. Note the SiO_2 grains attached to the tubules.

carbon filaments. The deposited carbon in these filaments has a graphitic structure, i.e. the as-made tubules can be assumed to be analogous to the tubules obtained by the arc-discharge fullerene production. By the use of catalysts with very finely dispersed metal or metal oxide particles on the surface, it was possible to produce nanotubules having diameters of the same range as nanotubules obtained by the arc-discharge method. We can thus suppose that the former tubules will possess all the properties predicted for the fullerene tubules.

The catalytic method, as was shown in this study, has some advantages over arc-discharge tubule production. First, the yield of nanotubules in the catalytic production is higher than in the arc-discharge. It is possible to optimize the method for the deposition of almost all of the carbon in the form of tubular filaments. In the arc-discharge production the amount of tubules in the soot is usually no more than 25%. Isolation of nanotubules is also easier in the case of catalytic production. They can be separated from the substrate by the combination of various methods (ultra-sound treatment, chemical treatment). The high percentage of tubules in the product (only tubules are seen by TEM on the catalyst surface) makes possible their effective purification by gasification, either by oxidation or hydrogenation. The former treatment can also be used for the opening of nanotubules.

The characteristics of nanotubules obtained by catalytic reaction are better controlled than in the arc-discharge method. By varying the active particles on the surface of the catalyst the nanotubule diameters can be adjusted. The length of the tubules is dependent on the reaction time; very long tubules, even up to 60 μm, can be produced.

The catalytic method provides the basis for synthesis of carbon tubules of a large variety of forms. Straight tubules, as well as bent and helically wound tubules, were observed. The latter regular helices of fullerene diameter can be of special interest from both theoretical and practical points of view.

Acknowledgements—This text presents research results of the Belgian Programme on Inter University Poles of Attraction initiated by the Belgian State Prime Minister's Office of Science Policy Programming. The scientific responsibility is assumed by the authors. A. Fonseca and D. Bernaerts acknowledge, respectively, the Région Wallonne and the National Found for Scientific Research, for financial support. Thanks are due to Prof. G. Van Tendeloo and Prof. J. Van Landuyt for useful discussions and for their continued interest in this research.

REFERENCES

1. A. Sacco, Jr, F. W. A. H. Geurts, G. A. Jablonski, S. Lee and R. A. Gately, *J. Catal* **119**, 322 (1989).
2. G. A. Jablonski, F. W. Geurts and A. Sacco, Jr, *Carbon* **30**, 87 (1992).
3. P. E. Nolan and D. C. Lynch, *Carbon* **32**, 477 (1994).
4. R. T. K. Baker, *Carbon* **27**, 315 (1989).
5. N. M. Rodriguez, *J. Mater. Res.* **8**, 3233 (1993).
6. A. Oberlin, M. Endo and T. Koyama, *J. Cryst. Growth* **32**, 335 (1976).
7. S. Iijima, *Nature* **354**, 56 (1991).
8. R. E. Smalley, Proc. R. A. Welch Found. Conf. Chem. Res. XXXVI, Houston, Texas, 26–27 October 1992, p. 161.
9. M. R. Pederson and J. Q. Broughton, *Phys. Rev. Lett.* **69**, 2689 (1992).
10. J. W. Mintmire, B. I. Dunlap and C. T. White, *Phys. Rev. Lett.* **68**, 631 (1992).

11. N. Hamada, S. Sawada and A. Oshiyama, *Phys. Rev. Lett.* **68**, 1579 (1992).
12. K. Harigaya and M. Fujita, *Phys. Rev.* **B47**, 16563 (1993).
13. M. José-Yacamàn, M. Miki-Yoshida, L. Rendon and J. G. Santiesteban, *Appl. Phys. Lett.* **62**, 657 (1993).
14. V. Ivanov, J. B. Nagy, P. Lambin, A. Lucas, X. B. Zhang, X. F. Zhang, D. Bernaerts, G. Van Tendeloo, S. Amelinckx and J. Van Landuyt, *Chem. Phys. Lett.* **223**, 329 (1994).
15. J. C. Lee, D. L. Trimm, M. A. Kohler, M. S. Wainwright and N. W. Cant, *Catal. Today* **2**, 643 (1988).
16. A. V. Kucherov and A. A. Slinkin, *Zeolites* **6**, 175 (1986).
17. A. V. Kucherov and A. A. Slinkin, *Zeolites* **7**, 38 (1987).
18. A. J. H. M. Kock, P. K. de Bokx, E. Boellaard, W. Klop and J. W. Geus, *J. Catal.* **96**, 468 (1985).
19. J. W. Geus, private communication.
20. R. T. K. Baker and J. J. Chludzinski, Jr, *J. Catal.* **64**, 464 (1980).
21. D. Bernaerts, X. B. Zhang, X. F. Zhang, G. Van Tendeloo, S. Amelinckx, J. Van Landuyt, V. Ivanov and J. B. Nagy, *Phil. Mag.* **71**, 605 (1995).
22. L. S. Lobo and M. D. Franco, *Catal. Today* **7**, 247 (1990).
23. A. A. Slinkin, M. I. Loktev, I. V. Michin, V. A. Plachotnik, A. L. Klyachko and A. M. Rubinstein, *Kinet. Katal.* **20**, 181 (1979).
24. S. Amelinckx, X. B. Zhang, D. Bernaerts, X. F. Zhang, V. Ivanov and J. B.Nagy, *Science*, **265**, 635 (1995).
25. M. S. Kim, N. M. Rodriguez and R. T. K. Baker, *J. Catal.* **131**, 60 (1991).
26. A. Sacco, P. Thacker, T. N. Chang and A. T. S. Chiang, *J. Catal.* **85**, 224 (1984).
27. X. B. Zhang, X. F. Zhang, D. Bernaerts, G. Van Tendeloo, S. Amelinckx, J. Van Landuyt, V. Ivanov, J. B. Nagy, Ph. Lambin and A. A. Lucas, *Europhys. Lett.* **27**, 141 (1994).
28. P. M. Ajayan, T. W. Ebbesen, T. Ichihashi, S. Iijima, K. Tanigaki and H. Hiura, *Nature* **362**, 522 (1993).
29. T. W. Ebbesen, P. M. Ajayan, H. Hiura and K. Tanigaki, *Nature* **367**, 519 (1994).
30. R. T. K. Baker and R. D. Sherwood, *J. Catal.* **70**, 198 (1981).
31. A. Tomita and Y. Tamai, *J. Catal.* **27**, 293 (1972).
32. A. Tomita and Y. Tamai, *J. Phys. Chem.* **78**, 2254 (1974).
33. J. L. Figueiredo, C. A. Bernardo, J. J. Chludzinski and R. T. K. Baker, *J. Catal.* **110**, 127 (1988).

PHYSICS OF CARBON NANOTUBES

M. S. DRESSELHAUS,[1] G. DRESSELHAUS,[2] and R. SAITO[3]
[1]Department of Electrical Engineering and Computer Science and Department of Physics, Massachusetts Institute of Technology, Cambridge, Massachusetts 02139, U.S.A.
[2]Francis Bitter National Magnet Laboratory, Massachusetts Institute of Technology, Cambridge, Massachusetts 02139, U.S.A.
[3]Department of Electronics-Engineering, University of Electro-Communications, Tokyo 182, Japan

(*Received* 26 *October* 1994; *accepted* 10 *February* 1995)

Abstract—The fundamental relations governing the geometry of carbon nanotubes are reviewed, and explicit examples are presented. A framework is given for the symmetry properties of carbon nanotubes for both symmorphic and non-symmorphic tubules which have screw-axis symmetry. The implications of symmetry on the vibrational and electronic structure of 1D carbon nanotube systems are considered. The corresponding properties of double-wall nanotubes and arrays of nanotubes are also discussed.

Key Words—Single-wall, multi-wall, vibrational modes, chiral nanotubes, electronic bands, tubule arrays.

1. INTRODUCTION

Carbon nanotube research was greatly stimulated by the initial report of observation of carbon tubules of nanometer dimensions[1] and the subsequent report on the observation of conditions for the synthesis of large quantities of nanotubes[2,3]. Since these early reports, much work has been done, and the results show basically that carbon nanotubes behave like rolled-up cylinders of graphene sheets of sp^2 bonded carbon atoms, except that the tubule diameters in some cases are small enough to exhibit the effects of one-dimensional (1D) periodicity. In this article, we review simple aspects of the symmetry of carbon nanotubules (both monolayer and multilayer) and comment on the significance of symmetry for the unique properties predicted for carbon nanotubes because of their 1D periodicity.

Of particular importance to carbon nanotube physics are the many possible symmetries or geometries that can be realized on a cylindrical surface in carbon nanotubes without the introduction of strain. For 1D systems on a cylindrical surface, translational symmetry with a screw axis could affect the electronic structure and related properties. The exotic electronic properties of 1D carbon nanotubes are seen to arise predominately from intralayer interactions, rather than from interlayer interactions between multilayers within a single carbon nanotube or between two different nanotubes. Since the symmetry of a single nanotube is essential for understanding the basic physics of carbon nanotubes, most of this article focuses on the symmetry properties of single layer nanotubes, with a brief discussion also provided for two-layer nanotubes and an ordered array of similar nanotubes.

2. FUNDAMENTAL PARAMETERS AND RELATIONS FOR CARBON NANOTUBES

In this section, we summarize the fundamental parameters for carbon nanotubes, give the basic relations governing these parameters, and list typical numerical values for these parameters.

In the theoretical carbon nanotube literature, the focus is on single-wall tubules, cylindrical in shape with caps at each end, such that the two caps can be joined together to form a fullerene. The cylindrical portions of the tubules consist of a single graphene sheet that is shaped to form the cylinder. With the recent discovery of methods to prepare single-walled nanotubes[4,5], it is now possible to test the predictions of the theoretical calculations.

It is convenient to specify a general carbon nanotubule in terms of the tubule diameter d_t and the chiral angle θ, which are shown in Fig. 1. The chiral vector $\mathbf{C}_h$ is defined in Table 1 in terms of the integers (n,m) and the basis vectors $\mathbf{a}_1$ and $\mathbf{a}_2$ of the honeycomb lattice, which are also given in the table in terms of rectangular coordinates. The integers (n,m) uniquely determine d_t and θ. The length L of the chiral vector $\mathbf{C}_h$ (see Table 1) is directly related to the tubule diameter d_t. The chiral angle θ between the $\mathbf{C}_h$ direction and the zigzag direction of the honeycomb lattice $(n,0)$ (see Fig. 1) is related in Table 1 to the integers (n,m).

We can specify a single-wall C_{60}-derived carbon nanotube by bisecting a C_{60} molecule at the equator and joining the two resulting hemispheres with a cylindrical tube having the same diameter as the C_{60} molecule, and consisting of the honeycomb structure of a single layer of graphite (a graphene layer). If the C_{60} molecule is bisected normal to a five-fold axis, the "armchair" tubule shown in Fig. 2 (a) is formed, and if the C_{60} molecule is bisected normal to a 3-fold axis, the "zigzag" tubule in Fig. 2(b) is formed[6]. Armchair and zigzag carbon nanotubules of larger diameter, and having correspondingly larger caps, can likewise be defined, and these nanotubules have the general appearance shown in Figs. 2(a) and (b). In addition, a large number of chiral carbon nanotubes can be formed for $0 < |\theta| < 30°$, with a screw axis along

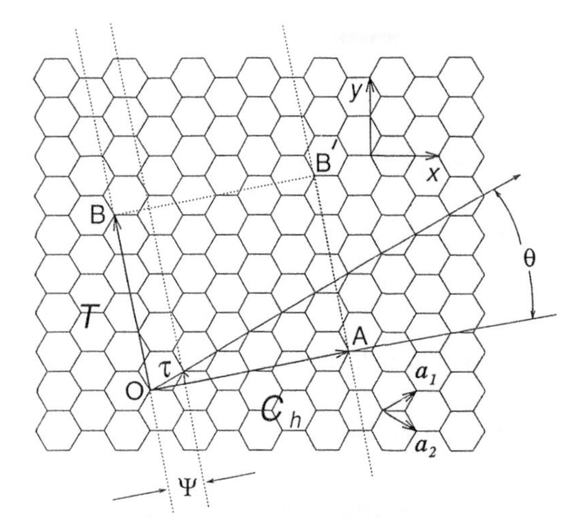

Fig. 1. The 2D graphene sheet is shown along with the vector which specifies the chiral nanotube. The chiral vector **OA** or $\mathbf{C}_h = n\boldsymbol{a}_1 + m\boldsymbol{a}_2$ is defined on the honeycomb lattice by unit vectors $\boldsymbol{a}_1$ and $\boldsymbol{a}_2$ and the chiral angle θ is defined with respect to the zigzag axis. Along the zigzag axis $\theta = 0°$. Also shown are the lattice vector $\mathbf{OB} = \mathbf{T}$ of the 1D tubule unit cell, and the rotation angle ψ and the translation τ which constitute the basic symmetry operation $R = (\psi|\tau)$. The diagram is constructed for $(n,m) = (4,2)$.

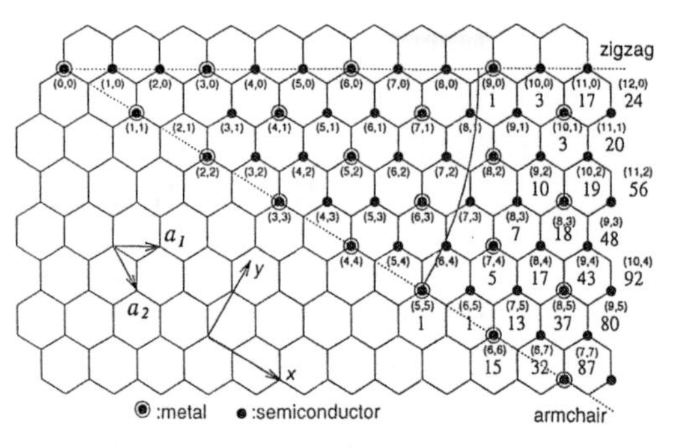

Fig. 3. The 2D graphene sheet is shown along with the vector which specifies the chiral nanotube. The pairs of integers (n,m) in the figure specify chiral vectors $\mathbf{C}_h$ (see Table 1) for carbon nanotubes, including zigzag, armchair, and chiral tubules. Below each pair of integers (n,m) is listed the number of distinct caps that can be joined continuously to the cylindrical carbon tubule denoted by (n,m)[6]. The circled dots denote metallic tubules and the small dots are for semiconducting tubules.

the axis of the tubule, and with a variety of hemispherical caps. A representative chiral nanotube is shown in Fig. 2(c).

The unit cell of the carbon nanotube is shown in Fig. 1 as the rectangle bounded by the vectors $\mathbf{C}_h$ and **T**, where **T** is the 1D translation vector of the nanotube. The vector **T** is normal to $\mathbf{C}_h$ and extends from the origin to the first lattice point **B** in the honeycomb lattice. It is convenient to express **T** in terms of the integers (t_1, t_2) given in Table 1, where it is seen that the length of **T** is $\sqrt{3}L/d_R$ and d_R is either equal to the highest common divisor of (n,m), denoted by d, or to $3d$, depending on whether $n - m = 3dr$, r being an integer, or not (see Table 1). The number of carbon atoms per unit cell n_c of the 1D tubule is $2N$, as given in Table 1, each hexagon (or unit cell) of the honeycomb lattice containing two carbon atoms.

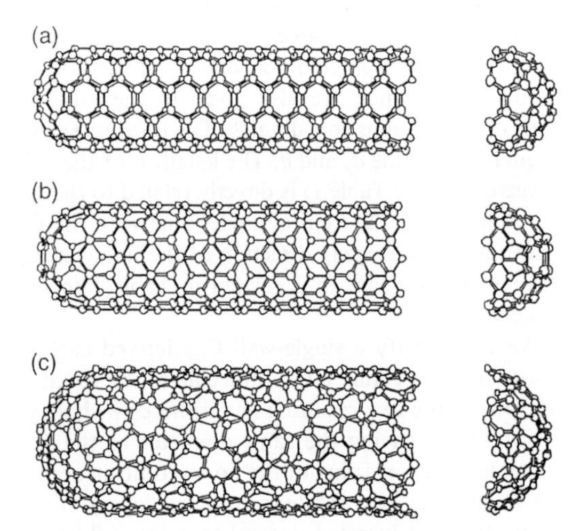

Fig. 2. By rolling up a graphene sheet (a single layer of carbon atoms from a 3D graphite crystal) as a cylinder and capping each end of the cylinder with half of a fullerene molecule, a "fullerene-derived tubule," one layer in thickness, is formed. Shown here is a schematic theoretical model for a single-wall carbon tubule with the tubule axis OB (see Fig. 1) normal to: (a) the $\theta = 30°$ direction (an "armchair" tubule), (b) the $\theta = 0°$ direction (a "zigzag" tubule), and (c) a general direction B with $0 < |\theta| < 30°$ (a "chiral" tubule). The actual tubules shown in the figure correspond to (n,m) values of: (a) (5,5), (b) (9,0), and (c) (10,5).

Figure 3 shows the number of distinct caps that can be formed theoretically from pentagons and hexagons, such that each cap fits continuously on to the cylinders of the tubule, specified by a given (n,m) pair. Figure 3 shows that the hemispheres of C_{60} are the smallest caps satisfying these requirements, so that the diameter of the smallest carbon nanotube is expected to be 7 Å, in good agreement with experiment[4,5]. Figure 3 also shows that the number of possible caps increases rapidly with increasing tubule diameter.

Corresponding to selected and representative (n,m) pairs, we list in Table 2 values for various parameters enumerated in Table 1, including the tubule diameter d_t, the highest common divisors d and d_R, *the length* L of the chiral vector $\mathbf{C}_h$ in units of a (where a is the length of the 2D lattice vector), the length of the 1D translation vector **T** of the tubule in units of a, and the number of carbon hexagons per 1D tubule unit cell N. Also given in Table 2 are various symmetry parameters discussed in section 3.

3. SYMMETRY OF CARBON NANOTUBES

In discussing the symmetry of the carbon nanotubes, it is assumed that the tubule length is much larger than its diameter, so that the tubule caps can be neglected when discussing the physical properties of the nanotubes.

The symmetry groups for carbon nanotubes can be either symmorphic [such as armchair (n,n) and zigzag

Table 1. Parameters of carbon nanotubes

Symbol	Name	Formula	Value
$a_{\text{C-C}}$	carbon-carbon distance		1.421 Å (graphite)
a	length of unit vector	$\sqrt{3}a_{\text{C-C}}$	2.46 Å
$\mathbf{a}_1, \mathbf{a}_2$	unit vectors	$\left(\frac{\sqrt{3}}{2}, \frac{1}{2}\right)a, \left(\frac{\sqrt{3}}{2}, -\frac{1}{2}\right)a$	in (x,y) coordinates
$\mathbf{b}_1, \mathbf{b}_2$	reciprocal lattice vectors	$\left(\frac{1}{\sqrt{3}}, 1\right)\frac{2\pi}{a}, \left(\frac{1}{\sqrt{3}}, -1\right)\frac{2\pi}{a}$	in (x,y) coordinates
$\mathbf{C}_h$	chiral vector	$\mathbf{C}_h = n\mathbf{a}_1 + m\mathbf{a}_2 \equiv (n,m)$	n,m: integers
L	circumference of nanotube	$L = \lvert\mathbf{C}_h\rvert = a\sqrt{n^2 + m^2 + nm}$	$0 \le \lvert m\rvert \le n$
d_t	diameter of nanotube	$d_t = \frac{L}{\pi} = \frac{\sqrt{n^2 + m^2 + nm}}{\pi}a$	
θ	chiral angle	$\sin\theta = \frac{\sqrt{3}m}{2\sqrt{n^2 + m^2 + nm}}$	$0 \le \lvert\theta\rvert \le 30°$
		$\cos\theta = \frac{2n + m}{2\sqrt{n^2 + m^2 + nm}}$	
		$\tan\theta = \frac{\sqrt{3}m}{2n + m}$	
d	the highest common divisor of (n,m)		
d_R	the highest common divisor of $(2n + m, 2m + n)$	$d_R = \begin{cases} d & \text{if } n - m \text{ not a multiple of } 3d \\ 3d & \text{if } n - m \text{ a multiple of } 3d. \end{cases}$	
$\mathbf{T}$	translational vector of 1D unit cell	$\mathbf{T} = t_1\mathbf{a}_1 + t_2\mathbf{a}_2 \equiv (t_1, t_2)$	t_1, t_2: integers
		$t_1 = \frac{2m + n}{d_R}$	
		$t_2 = -\frac{2n + m}{d_R}$	
T	length of $\mathbf{T}$	$T = \frac{\sqrt{3}L}{d_R}$	
N	number of hexagons per 1D unit cell	$N = \frac{2(n^2 + m^2 + nm)}{d_R}$	$2N \equiv n_C$/unit cell
$\mathbf{R}$	symmetry vector‡	$\mathbf{R} = p\mathbf{a}_1 + q\mathbf{a}_2 \equiv (p,q)$	p,q: integers†
		$d = mp - nq,\ 0 \le p \le n/d,\ 0 \le q \le m/d$	
M	number of 2π revolutions	$M = [(2n + m)p + (2m + n)q]/d_R$	M: integer
		$N\mathbf{R} = M\mathbf{C}_h + d\mathbf{T}$	
R	basic symmetry operation‡	$R = (\psi \vert \tau)$	
ψ	rotation operation	$\psi = 2\pi\frac{M}{N}, \left(\chi = \frac{\psi L}{2\pi}\right)$	ψ: radians
τ	translation operation	$\tau = \frac{d\mathbf{T}}{N}$	τ, χ: length

† (p,q) are uniquely determined by $d = mp - nq$, subject to conditions stated in table, except for zigzag tubes for which $\mathbf{C}_h = (n,0)$, and we define $p = 1$, $q = -1$, which gives $M = 1$.
‡$\mathbf{R}$ and R refer to the same symmetry operation.

$(n,0)$ tubules], where the translational and rotational symmetry operations can each be executed independently, or the symmetry group can be non-symmorphic (for a general nanotube), where the basic symmetry operations require both a rotation ψ and translation τ and is written as $R = (\psi|\tau)$[7]. We consider the symmorphic case in some detail in this article, and refer the reader to the paper by Eklund *et al.*[8] in this volume for further details regarding the non-symmorphic space groups for chiral nanotubes.

The symmetry operations of the infinitely long armchair tubule ($n = m$), or zigzag tubule ($m = 0$), are described by the symmetry groups D_{nh} or D_{nd} for even or odd n, respectively, since inversion is an element of D_{nd} only for odd n, and is an element of D_{nh} only for even n[9]. Character tables for the D_n groups

Table 2. Values for characterization parameters for selected carbon nanotubes labeled by (n,m)[7]

(n,m)	d	d_R	d_t (Å)	L/a	T/a	N	$\psi/2\pi$	τ/a	M
(5,5)	5	15	6.78	$\sqrt{75}$	1	10	1/10	1/2	1
(9,0)	9	9	7.05	9	$\sqrt{3}$	18	1/18	$\sqrt{3}/2$	1
(6,5)	1	1	7.47	$\sqrt{91}$	$\sqrt{273}$	182	149/182	$\sqrt{3/364}$	149
(7,4)	1	3	7.55	$\sqrt{93}$	$\sqrt{31}$	62	17/62	$1/\sqrt{124}$	17
(8,3)	1	1	7.72	$\sqrt{97}$	$\sqrt{291}$	194	71/194	$\sqrt{3/388}$	71
(10,0)	10	10	7.83	10	$\sqrt{3}$	20	1/20	$\sqrt{3}/2$	1
(6,6)	6	18	8.14	$\sqrt{108}$	1	12	1/12	1/2	1
10,5)	5	5	10.36	$\sqrt{175}$	$\sqrt{21}$	70	1/14	$\sqrt{3/28}$	5
(20,5)	5	15	17.95	$\sqrt{525}$	$\sqrt{7}$	70	3/70	$1/(\sqrt{28})$	3
(30,15)	15	15	31.09	$\sqrt{1575}$	$\sqrt{21}$	210	1/42	$\sqrt{3/28}$	5
⋮	⋮	⋮	⋮	⋮	⋮	⋮	⋮	⋮	⋮
(n,n)	n	$3n$	$\sqrt{3}na/\pi$	$\sqrt{3}n$	1	$2n$	$1/2n$	1/2	1
$(n,0)$	n	n	na/π	n	$\sqrt{3}$	$2n$	$1/2n$	$\sqrt{3}/2$	1

are given in Table 3 (for odd $n = 2j + 1$) and in Table 4 (for even $n = 2j$), where j is an integer. Useful basis functions are listed in Table 5 for both the symmorphic groups (D_{2j} and D_{2j+1}) and non-symmorphic groups $C_{N/\Omega}$ discussed by Eklund *et al.*[8].

Upon taking the direct product of group D_n with the inversion group which contains two elements (E,i), we can construct the character tables for $D_{nd} = D_n \otimes i$ from Table 3 to yield D_{5d}, D_{7d}, ...for symmorphic tubules with odd numbers of unit cells around the circumference [(5,5), (7,7), ... armchair tubules and (9,0), (11,0), ... zigzag tubules]. Likewise, the character table for $D_{nh} = D_n \otimes \sigma_h$ can be obtained from Table 4 to yield D_{6h}, D_{8h}, ... for even n. Table 4 shows two additional classes for group D_{2j} relative to group $D_{(2j+1)}$, because rotation by π about the main symmetry axis is in a class by itself for groups D_{2j}. Also the n two-fold axes nC_2' form a class and represent two-fold rotations in a plane normal to the main symmetry axis $C_{\phi j}$, while the nC_2'' dihedral axes, which are bisectors of the nC_2' axes, also form a class for group D_n when n is an even integer. Correspondingly, there are two additional one-dimensional representations B_1 and B_2 in D_{2j} corresponding to the two additional classes cited above.

The symmetry groups for the chiral tubules are Abelian groups. The corresponding space groups are non-symmorphic and the basic symmetry operations

Table 3. Character table for group $D_{(2j+1)}$

$\mathcal{R}$	E	$2C_{\phi_j}^1$	$2C_{\phi_j}^2$	...	$2C_{\phi_j}^j$	$(2j+1)C_2'$
A_1	1	1	1	...	1	1
A_2	1	1	1	...	1	−1
E_1	2	$2\cos\phi_j$	$2\cos 2\phi_j$	...	$2\cos j\phi_j$	0
E_2	2	$2\cos 2\phi_j$	$2\cos 4\phi_j$	...	$2\cos 2j\phi_j$	0
⋮	⋮	⋮	⋮	⋮	⋮	⋮
E_j	2	$2\cos j\phi_j$	$2\cos 2j\phi_j$	...	$2\cos j^2\phi_j$	0

where $\phi_j = 2\pi/(2j+1)$ and j is an integer.

$R = (\psi|\tau)$ require translations $\boldsymbol{\tau}$ in addition to rotations ψ. The irreducible representations for all Abelian groups have a phase factor ϵ, consistent with the requirement that all h symmetry elements of the symmetry group commute. These symmetry elements of the Abelian group are obtained by multiplication of the symmetry element $R = (\psi|\tau)$ by itself an appropriate number of times, since $R^h = E$, where E is the identity element, and h is the number of elements in the Abelian group. We note that N, the number of hexagons in the 1D unit cell of the nanotube, is not always equal h, particularly when $d \neq 1$ and $d_R \neq d$.

To find the symmetry operations for the Abelian group for a carbon nanotube specified by the (n,m) integer pair, we introduce the basic symmetry vector $\mathbf{R} = p\mathbf{a}_1 + q\mathbf{a}_2$, shown in Fig. 4, which has a very important physical meaning. The projection of $\mathbf{R}$ on the $\mathbf{C}_h$ axis specifies the angle of rotation ψ in the basic symmetry operation $R = (\psi|\tau)$, while the projection of $\mathbf{R}$ on the $\mathbf{T}$ axis specifies the translation $\boldsymbol{\tau}$. In Fig. 4 the rotation angle ψ is shown as $\chi = \psi L/2\pi$. If we translate $\mathbf{R}$ by (N/d) times, we reach a lattice point B'' (see Fig. 4). This leads to the relation $N\mathbf{R} = M\mathbf{C}_h + d\mathbf{T}$ where the integer M is interpreted as the integral number of 2π cycles of rotation which occur after N rotations of ψ. Explicit relations for $\mathbf{R}$, ψ, and τ are contained in Table 1. If d the largest common divisor of (n,m) is an integer greater than 1, than (N/d) translations of $\mathbf{R}$ will translate the origin O to a lattice point B'', and the projection $(N/d)\mathbf{R}\cdot\mathbf{T} = T^2$. The total rotation angle ψ then becomes $2\pi(M/d)$ when $(N/d)\mathbf{R}$ reaches a lattice point B''. Listed in Table 2 are values for several representative carbon nanotubes for the rotation angle ψ in units of 2π, and the translation length τ in units of lattice constant a for the graphene layer, as well as values for M.

From the symmetry operations $R = (\psi|\tau)$ for tubules (n,m), the non-symmorphic symmetry group of the chiral tubule can be determined. Thus, from a symmetry standpoint, a carbon tubule is a one-dimensional crystal with a translation vector $\mathbf{T}$ along the cylinder axis, and a small number N of carbon

Table 4. Character table for group $D_{(2j)}$

$\mathcal{R}$	E	C_2	$2C^1_{\phi_j}$	$2C^2_{\phi_j}$	...	$2C^{j-1}_{\phi_j}$	$(2j)C'_2$	$(2j)C''_2$
A_1	1	1	1	1	...	1	1	1
A_2	1	1	1	1	...	1	−1	−1
B_1	1	−1	1	1	...	1	1	−1
B_2	1	−1	1	1	...	1	−1	1
E_1	2	−2	$2\cos\phi_j$	$2\cos 2\phi_j$	...	$2\cos(j-1)\phi_j$	0	0
E_2	2	2	$2\cos 2\phi_j$	$2\cos 4\phi_j$	...	$2\cos 2(j-1)\phi_j$	0	0
⋮	⋮	⋮	⋮	⋮	⋮	⋮	⋮	⋮
E_{j-1}	2	$(-1)^{j-1}2$	$2\cos(j-1)\phi_j$	$2\cos 2(j-1)\phi_j$	...	$2\cos(j-1)^2\phi_j$	0	0

where $\phi_j = 2\pi/(2j)$ and j is an integer.

hexagons associated with the 1D unit cell. The phase factor ϵ for the nanotube Abelian group becomes $\epsilon = \exp(2\pi iM/N)$ for the case where (n,m) have no common divisors (i.e., $d=1$). If $M=1$, as for the case of zigzag tubules as in Fig. 2(b) $N\mathbf{R}$ reach a lattice point after a 2π rotation.

As seen in Table 2, many of the chiral tubules with $d=1$ have large values for M; for example, for the (6,5) tubule, $M=149$, while for the (7,4) tubule, $M=17$. Thus, many 2π rotations around the tubule axis are needed in some cases to reach a lattice point of the 1D lattice. A more detailed discussion of the symmetry properties of the non-symmorphic chiral groups is given elsewhere in this volume[8].

Because the 1D unit cells for the symmorphic groups are relatively small in area, the number of phonon branches or the number of electronic energy bands associated with the 1D dispersion relations is relatively small. Of course, for the chiral tubules the 1D unit cells are very large, so that the number of phonon branches and electronic energy bands is also large. Using the transformation properties of the atoms within the unit cell ($\chi^{\text{atom sites}}$) and the transformation properties of the 1D unit cells that form an Abelian group, the symmetries for the dispersion relations for phonon are obtained[9,10]. In the case of π energy bands, the number and symmetries of the distinct energy bands can be obtained by the decomposition of the equivalence transformation ($\chi^{\text{atom sites}}$) for the atoms for the 1D unit cell using the irreducible representations of the symmetry group.

Table 5. Basis functions for groups $D_{(2j)}$ and $D_{(2j+1)}$

Basis function		$D_{(2j)}$	$D_{(2j+1)}$	$C_{N/\Omega}$
(x^2+y^2, z^2)		A_1	A_1	A
	z	A_2	A_1	A
	R_z	A_2	A_2	A
(xz, yz)	$\left.\begin{matrix}(x,y)\\(R_x,R_y)\end{matrix}\right\}$	E_1	E_1	E_1
(x^2-y^2, xy)		E_2	E_2	E_2
		⋮	⋮	⋮

We illustrate some typical results below for electrons and phonons. Closely related results are given elsewhere in this volume[8,11].

The phonon dispersion relations for (n,0) zigzag tubules have $4\times 3n = 12n$ degrees of freedom with 60 phonon branches, having the symmetry types (for n odd, and D_{nd} symmetry):

$$\begin{aligned}\Gamma_n^{\text{vib}} &= 3A_{1g} + 3A_{1u} + 3A_{2g} + 3A_{2u}\\ &\quad + 6\,E_{1g} + 6E_{1u} + 6E_{2g} + 6E_{2u} \\ &\quad + \ldots + 6E_{[(n-1)/2]g} + 6E_{[(n-1)/2]u}.\end{aligned} \tag{1}$$

Of these many modes there are only 7 nonvanishing modes which are infrared-active ($2A_{2u} + 5E_{1u}$) and 15 modes that are Raman-active. Thus, by increasing the diameter of the zigzag tubules, modes with different symmetries are added, though the number and symmetry of the optically active modes remain the

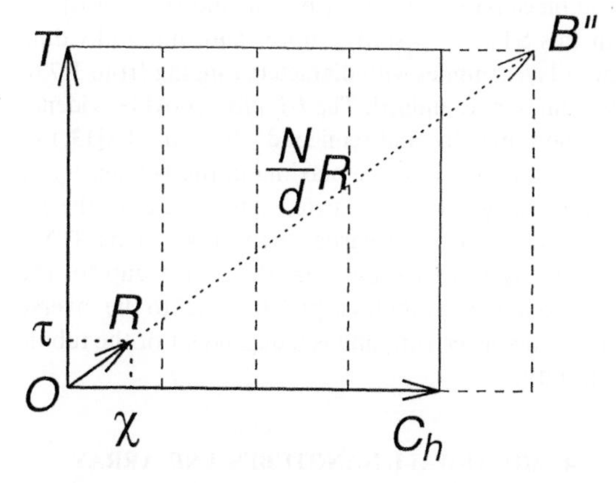

Fig. 4. The relation between the fundamental symmetry vector $\mathbf{R} = p\mathbf{a}_1 + q\mathbf{a}_2$ and the two vectors of the tubule unit cell for a carbon nanotube specified by (n,m) which, in turn, determine the chiral vector $\mathbf{C}_h$ and the translation vector $\mathbf{T}$. The projection of $\mathbf{R}$ on the $\mathbf{C}_h$ and $\mathbf{T}$ axes, respectively, yield ψ (or χ) and $\boldsymbol{\tau}$ (see text). After (N/d) translations, $\mathbf{R}$ reaches a lattice point B''. The dashed vertical lines denote normals to the vector $\mathbf{C}_h$ at distances of L/d, $2L/d$, $3L/d, \ldots, L$ from the origin.

same. This is a symmetry-imposed result that is generally valid for all carbon nanotubes.

Regarding the electronic structure, the number of energy bands for $(n,0)$ zigzag carbon nanotubes is $2n$, the number of carbon atoms per unit cell, with symmetries

$$\chi_n^{\text{electronic}} = A_{1g} + A_{1u} + A_{2g} + A_{2u} + 2E_{1g} + 2E_{1u} + 2E_{2g} + 2E_{2u} + \ldots + 2E_{[(n-1/2]g} + 2E_{[(n-1)/2]u}. \quad (2)$$

A symmetry-imposed band degeneracy occurs for the $E_{[(n-3)/2]g}$ and $E_{[(n-3/2]u}$ bands at the Fermi level, when $n = 3r$, r being an integer, thereby giving rise to zero gap tubules with metallic conduction. On the other hand, when $n \neq 3r$, a bandgap and semiconducting behavior results. Independent of whether the tubules are conducting or semiconducting, each of the $[4 + 2(n-1)]$ energy bands is expected to show a $(E - E_0)^{-1/2}$ type singularity in the density of states at its band extremum energy E_0[10].

The most promising present technique for carrying out sensitive measurements of the electronic properties of individual tubules is scanning tunneling spectroscopy (STS) because of the ability of the tunneling tip to probe most sensitively the electronic density of states of either a single-wall nanotube[12], or the outermost cylinder of a multi-wall tubule or, more generally, a bundle of tubules. With this technique, it is further possible to carry out both STS and scanning tunneling microscopy (STM) measurements at the same location on the same tubule and, therefore, to measure the tubule diameter concurrently with the STS spectrum.

Although still preliminary, the study that provides the most detailed test of the theory for the electronic properties of the 1D carbon nanotubes, thus far, is the combined STM/STS study by Olk and Heremans[13]. In this STM/STS study, more than nine individual multilayer tubules with diameters ranging from 1.7 to 9.5 nm were examined. The *I-V* plots provide evidence for both metallic and semiconducting tubules[13,14]. Plots of dI/dV indicate maxima in the 1D density of states, suggestive of predicted singularities in the 1D density of states for carbon nanotubes. This STM/STS study further shows that the energy gap for the semiconducting tubules is proportional to the inverse tubule diameter $1/d_t$, and is independent of the tubule chirality.

4. MULTI-WALL NANOTUBES AND ARRAYS

Much of the experimental observations on carbon nanotubes thus far have been made on multi-wall tubules[15–19]. This has inspired a number of theoretical calculations to extend the theoretical results initially obtained for single-wall nanotubes to observations in multilayer tubules. These calculations for multi-wall tubules have been informative for the interpretation of experiments, and influential for suggesting new research directions. The multi-wall calculations have been predominantly done for double-wall tubules, although some calculations have been done for a four-walled tubule[16–18] and also for nanotube arrays [16,17].

The first calculation for a double-wall carbon nanotube[15] was done using the tight binding technique, which sensitively includes all symmetry constraints in a simplified Hamiltonian. The specific geometrical arrangement that was considered is the most commensurate case possible for a double-layer nanotube, for which the ratio of the chiral vectors for the two layers is 1:2, and in the direction of translational vectors, the ratio of the lengths is 1:1. Because the C_{60}-derived tubule has a radius of 3.4 Å, which is close to the interlayer distance for turbostratic graphite, this geometry corresponds to the minimum diameter for a double-layer tubule. This geometry has many similarities to the AB stacking of graphite. In the double-layer tubule with the diameter ratio 1:2, the interlayer interaction γ_1 involves only half the number of carbon atoms as in graphite, because of the smaller number of atoms on the inner tubule. Even though the geometry was chosen to give rise to the most commensurate interlayer stacking, the energy dispersion relations are only weakly perturbed by the interlayer interaction.

More specifically, the calculated energy band structure showed that two coaxial zigzag nanotubes that would each be metallic as single-wall nanotubes yield a metallic double-wall nanotube when a weak interlayer coupling between the concentric nanotubes is introduced. Similarly, two coaxial semiconducting tubules remain semiconducting when the weak interlayer coupling is introduced[15]. More interesting is the case of coaxial metal-semiconductor and semiconductor-metal nanotubes, which also retain their individual metallic and semiconducting identities when the weak interlayer interaction is turned on. On the basis of this result, we conclude that it might be possible to prepare metal-insulator device structures in the coaxial geometry without introducing any doping impurities[20], as has already been suggested in the literature[10,20,21].

A second calculation was done for a two-layer tubule using density functional theory in the local density approximation to establish the optimum interlayer distance between an inner (5,5) armchair tubule and an outer armchair (10,10) tubule. The result of this calculation yielded a 3.39 Å interlayer separation [16,17], with an energy stabilization of 48 meV/carbon atom. The fact that the interlayer separation is about halfway between the graphite value of 3.35 Å and the 3.44 Å separation expected for turbostratic graphite may be explained by interlayer correlation between the carbon atom sites both along the tubule axis direction and circumferentially. A similar calculation for double-layered hyper-fullerenes has also been carried out, yielding an interlayer spacing of 3.524 Å for $C_{60}@C_{240}$ with an energy stabilization of 14 meV/C atom for this case[22]. In the case of the double-layered hyper-fullerene, there is a greatly reduced possibility for interlayer correlations, even if C_{60} and C_{240} take the same I_h axes. Further, in the case of C_{240}, the molecule deviates from a spherical shape to an icosahedron shape. Because of the curvature, it is expected that the spherically averaged interlayer spacing between the double-layered hyper-fullerenes is greater than that for turbostratic graphite.

In addition, for two coaxial armchair tubules, estimates for the translational and rotational energy barriers (of 0.23 meV/atom and 0.52 meV/atom, respectively) were obtained, suggesting significant translational and rotational interlayer mobility of ideal tubules at room temperature[16,17]. Of course, constraints associated with the cap structure and with defects on the tubules would be expected to restrict these motions. The detailed band calculations for various interplanar geometries for the two coaxial armchair tubules basically confirm the tight binding results mentioned above[16,17].

Further calculations are needed to determine whether or not a Peierls distortion might remove the coaxial nesting of carbon nanotubes. Generally 1D metallic bands are unstable against weak perturbations which open an energy gap at E_F and consequently lower the total energy, which is known as the Peierls instability[23]. In the case of carbon nanotubes, both in-plane and out-of-plane lattice distortions may couple with the electrons at the Fermi energy. Mintmire and White have discussed the case of in-plane distortion and have concluded that carbon nanotubes are stable against a Peierls distortion in-plane at room temperature[24], though the in-plane distortion, like a Kekulé pattern, will be at least 3 times as large a unit cell as that of graphite. The corresponding chiral vectors satisfy the condition for metallic conduction ($n - m = 3r, r$:integer). However, if we consider the direction of the translational vector **T**, a symmetry-lowering distortion is not always possible, consistent with the boundary conditions for the general tubules[25]. On the other hand, out-of-plane vibrations do not change the size of the unit cell, but result in a different site energy for carbon atoms on A and B sites for carbon nanotube structures[26]. This situation is applicable, too, if the dimerization is of the "quinone" or chain-like type, where out-of-plane distortions lead to a perturbation approaching the limit of 2D graphite. Further, Harigaya and Fujita[27,28] showed that an in-plane alternating double-single bond pattern for the carbon atoms within the 1D unit cell is possible only for several choices of chiral vectors.

Solving the self-consistent calculation for these types of distortion, an energy gap is always opened by the Peierls instability. However, the energy gap is very small compared with that of normal 1D cosine energy bands. The reason why the energy gap for 1D tubules is so small is that the energy gain comes from only one of the many 1D energy bands, while the energy loss due to the distortion affects all the 1D energy bands. Thus, the Peierls energy gap decreases exponentially with increasing number of energy bands N[24,26–28]. Because the energy change due to the Peierls distortion is zero in the limit of 2D graphite, this result is consistent with the limiting case of $N = \infty$. This very small Peierls gap is, thus, negligible at finite temperatures and in the presence of fluctuations arising from 1D conductors. Very recently, Viet *et al.* showed[29] that the in-plane and out-of-plane distortions do not occur simultaneously, but their conclusions regarding the Peierls gap for carbon nanotubes are essentially as discussed above.

The band structure of four concentric armchair tubules with 10, 20, 30, and 40 carbon atoms around their circumferences (external diameter 27.12 Å) was calculated, where the tubules were positioned to minimize the energy for all bilayered pairs[17]. In this case, the four-layered tubule remains metallic, similar to the behavior of two double-layered armchair nanotubes, except that tiny band splittings form.

Inspired by experimental observations on bundles of carbon nanotubes, calculations of the electronic structure have also been carried out on arrays of (6,6) armchair nanotubes to determine the crystalline structure of the arrays, the relative orientation of adjacent nanotubes, and the optimal spacing between them. Figure 5 shows one tetragonal and two hexagonal arrays that were considered, with space group symmetries $P4_2/mmc$ (D_{4h}^9), $P6/mmm$ (D_{6h}^1), and $P6/mcc$ (D_{6h}^2), respectively[16,17,30]. The calculation shows

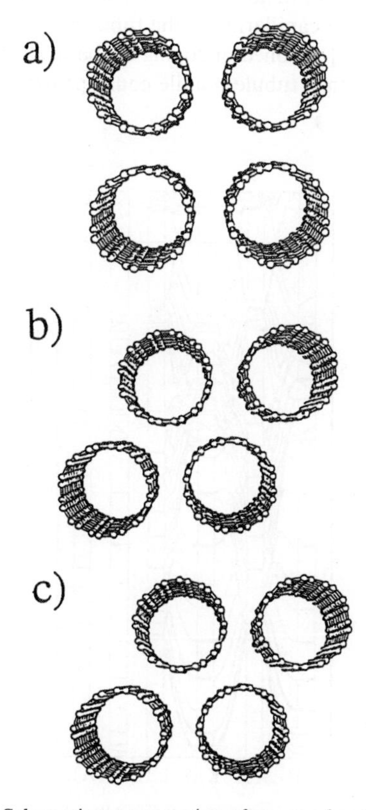

Fig. 5. Schematic representation of arrays of carbon nanotubes with a common tubule axial direction in the (a) tetragonal, (b) hexagonal I, and (c) hexagonal II arrangements. The reference nanotube is generated using a planar ring of twelve carbon atoms arranged in six pairs with the D_{6h} symmetry [16,17,30].

that the hexagonal $P6/mcc$ (D_{6h}^2) space group has the lowest energy, leading to a gain in cohesive energy of 2.4 meV/C atom. The orientational alignment between tubules leads to an even greater gain in cohesive energy (3.4 eV/C atom). The optimal alignment between tubules relates closely to the ABAB stacking of graphite, with an inter-tubule separation of 3.14 Å at closest approach, showing that the curvature of the tubules lowers the minimum interplanar distance (as is also found for fullerenes where the corresponding distance is 2.8 Å). The importance of the inter-tubule interaction can be seen in the reduction in the inter-tubule closest approach distance to 3.14 Å for the $P6/mcc$ (D_{6h}^2) structure, from 3.36 Å and 3.35 Å, respectively, for the tetragonal $P4_2/mmc$ (D_{4h}^9) and $P6/mmm$ (D_{6h}^1) space groups. A plot of the electron dispersion relations for the most stable case is given in Fig. 6[16,17,30], showing the metallic nature of this tubule array by the degeneracy point between the H and K points in the Brillouin zone between the valence and conduction bands. It is expected that further calculations will consider the interactions between nested nanotubes having different symmetries, which on physical grounds should interact more weakly, because of a lack of correlation between near neighbors.

Modifications of the conduction properties of semiconducting carbon nanotubes by B (p-type) and N (n-type) substitutional doping has also been discussed[31] and, in addition, electronic modifications by filling the capillaries of the tubes have also been proposed[32]. Exohedral doping of the space between nanotubes in a tubule bundle could provide yet another mechanism for doping the tubules. Doping of the nanotubes by insertion of an intercalate species between the layers of the tubules seems unfavorable because the interlayer spacing is too small to accommodate an intercalate layer without fracturing the shells within the nanotube.

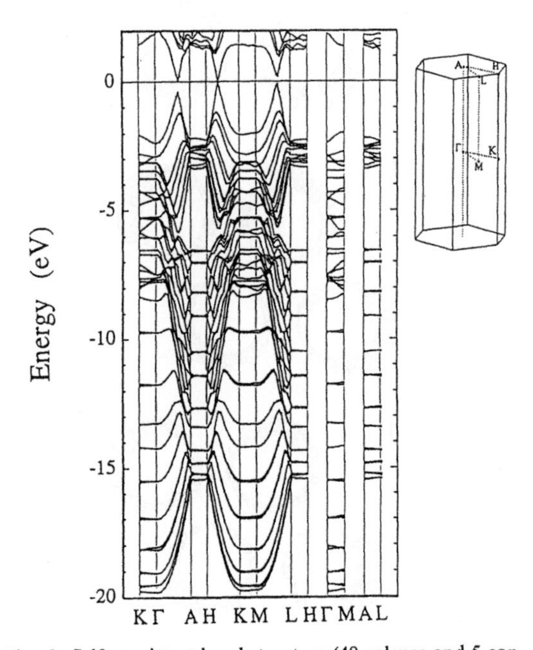

Fig. 6. Self-consistent band structure (48 valence and 5 conduction bands) for the hexagonal II arrangement of nanotubes, calculated along different high-symmetry directions in the Brillouin zone. The Fermi level is positioned at the degeneracy point appearing between K-H, indicating metallic behavior for this tubule array[17].

No superconductivity has yet been found in carbon nanotubes or nanotube arrays. Despite the prediction that 1D electronic systems cannot support superconductivity[33,34], it is not clear that such theories are applicable to carbon nanotubes, which are tubular with a hollow core and have several unit cells around the circumference. Doping of nanotube bundles by the insertion of alkali metal dopants between the tubules could lead to superconductivity. The doping of individual tubules may provide another possible approach to superconductivity for carbon nanotube systems.

5. DISCUSSION

This journal issue features the many unusual properties of carbon nanotubes. Most of these unusual properties are a direct consequence of their 1D quantum behavior and symmetry properties, including their unique conduction properties[11] and their unique vibrational spectra[8].

Regarding electrical conduction, carbon nanotubes show the unique property that the conductivity can be either metallic or semiconducting, depending on the tubule diameter d_t and chiral angle θ. For carbon nanotubes, metallic conduction can be achieved without the introduction of doping or defects. Among the tubules that are semiconducting, their band gaps appear to be proportional to $1/d_t$, independent of the tubule chirality. Regarding lattice vibrations, the number of vibrational normal modes increases with increasing diameter, as expected. Nevertheless, following from the 1D symmetry properties of the nanotubes, the number of infrared-active and Raman-active modes remains independent of tubule diameter, though the vibrational frequencies for these optically active modes are sensitive to tubule diameter and chirality[8]. Because of the restrictions on momentum transfer between electrons and phonons in the electron-phonon interaction for carbon nanotubes, it has been predicted that the interaction between electrons and longitudinal phonons gives rise only to intraband scattering and not interband scattering. Correspondingly, the interaction between electrons and transverse phonons gives rise only to interband electron scattering and not to intraband scattering[35].

These properties are illustrative of the unique behavior of 1D systems on a rolled surface and result from the group symmetry outlined in this paper. Observation of 1D quantum effects in carbon nanotubes requires study of tubules of sufficiently small diameter to exhibit measurable quantum effects and, ideally, the measurements should be made on single nanotubes, characterized for their diameter and chirality. Interesting effects can be observed in carbon nanotubes for diameters in the range 1–20 nm, depending on the property under investigation. To see 1D effects, faceting should be avoided, insofar as facets lead to 2D behavior, as in graphite. To emphasize the possibility of semiconducting properties in non-defective carbon nanotubes, and to distinguish between conductors and semiconductors of similar diameter, experiments should be done on nanotubes of the smallest possible diameter. To demonstrate experimentally the high density of electronic states expected for 1D systems, experiments should ideally be carried out on single-walled tubules of small diameter. However, to demonstrate magnetic properties in carbon nanotubes with a magnetic field normal to the tubule axis, the tubule diameter should be large compared with the Landau radius and, in this case, a tubule size of $\sim$10 nm would be more desirable, because the magnetic localization within the tubule diameter would otherwise lead to high field graphitic behavior.

The ability of experimentalists to study 1D quantum behavior in carbon nanotubes would be greatly enhanced if the purification of carbon tubules in the synthesis process could successfully separate tubules of a given diameter and chirality. A new method for producing mass quantities of carbon nanotubes under controlled conditions would be highly desirable, as is now the case for producing commercial quantities of carbon fibers. It is expected that nano-techniques for manipulating very small quantities of material of nm size[14,36] will be improved through research of carbon nanotubes, including research capabilities involving the STM and AFM techniques. Also of interest will be the bonding of carbon nanotubes to the other surfaces, and the preparation of composite or multilayer systems that involve carbon nanotubes. The unbelievable progress in the last 30 years of semiconducting physics and devices inspires our imagination about future progress in 1D systems, where carbon nanotubes may become a benchmark material for study of 1D systems about a cylindrical surface.

Acknowledgements—We gratefully acknowledge stimulating discussions with T. W. Ebbesen, M. Endo, and R. A. Jishi. We are also in debt to many colleagues for assistance. The research at MIT is funded by NSF grant DMR-92-01878. One of the authors (RS) acknowledges the Japan Society for the Promotion of Science for supporting part of his joint research with MIT. Part of the work by RS is supported by a Grant-in-Aid for Scientific Research in Priority Area "Carbon Cluster" (Area No. 234/05233214) from the Ministry of Education, Science and Culture, Japan.

REFERENCES

1. S. Iijima, *Nature* (London) **354**, 56 (1991).
2. T. W. Ebbesen and P. M. Ajayan, *Nature* (London) **358**, 220 (1992).
3. T. W. Ebbesen, H. Hiura, J. Fujita, Y. Ochiai, S. Matsui, and K. Tanigaki, *Chem. Phys. Lett.* **209**, 83 (1993).
4. D. S. Bethune, C. H. Kiang, M. S. de Vries, G. Gorman, R. Savoy, J. Vazquez, and R. Beyers, *Nature* (London) **363**, 605 (1993).
5. S. Iijima and T. Ichihashi, *Nature* (London) **363**, 603 (1993).
6. M. S. Dresselhaus, G. Dresselhaus, and R. Saito, *Phys. Rev. B* **45**, 6234 (1992).
7. R. A. Jishi, M. S. Dresselhaus, and G. Dresselhaus, *Phys. Rev. B* **47**, 16671 (1993).
8. P. C. Eklund, J. M. Holden, and R. A. Jishi, *Carbon* **33**, 959 (1995).
9. R. A. Jishi, L. Venkataraman, M. S. Dresselhaus, and G. Dresselhaus, *Chem. Phys. Lett.* **209**, 77 (1993).
10. Riichiro Saito, Mitsutaka Fujita, G. Dresselhaus, and M. S. Dresselhaus, *Mater. Sci. Engin.* **B19**, 185 (1993).
11. J. W. Mintmire and C. T. White, *Carbon* **33**, 893 (1995).
12. S. Wang and D. Zhou, *Chem. Phys. Lett.* **225**, 165 (1994).
13. C. H. Olk and J. P. Heremans, *J. Mater. Res.* **9**, 259 (1994).
14. J. P. Issi *et al.*, *Carbon* **33**, 941 (1995).
15. R. Saito, G. Dresselhaus, and M. S. Dresselhaus, *J. Appl. Phys.* **73**, 494 (1993).
16. J. C. Charlier and J. P. Michenaud, *Phys. Rev. Lett.* **70**, 1858 (1993).
17. J. C. Charlier, *Carbon Nanotubes and Fullerenes*. PhD thesis, Catholic University of Louvain, Department of Physics, May 1994.
18. Ph. Lambin, L. Philippe, J. C. Charlier, and J. P. Michenaud, *Comput. Mater. Sci.* **2**, 350 (1994).
19. Ph. Lambin, L. Philippe, J. C. Charlier, and J. P. Michenaud, In *Proceedings of the Winter School on Fullerenes* (Edited by H. Kuzmany, J. Fink, M. Mehring, and S. Roth), Kirchberg Winter School, Singapore, World Scientific Publishing Co., Ltd. (1994).
20. R. Saito, M. Fujita, G. Dresselhaus, and M. S. Dresselhaus, *Appl. Phys. Lett.* **60**, 2204 (1992).
21. M. S. Dresselhaus, G. Dresselhaus, and Riichiro Saito, *Mater. Sci. Engin.* **B19**, 122 (1993).
22. Y. Yosida, *Fullerene Sci. Tech.* **1**, 55 (1993).
23. R. E. Peierls, In *Quantum Theory of Solids*. London, Oxford University Press (1955).
24. J. W. Mintmire, *Phys. Rev B* **43**, 14281 (June 1991).
25. Kikuo Harigaya, *Chem. Phys. Lett.* **189**, 79 (1992).
26. R. Saito, M. Fujita, G. Dresselhaus, and M. S. Dresselhaus, In *Electrical, Optical and Magnetic Properties of Organic Solid State Materials, MRS Symposia Proceedings, Boston*. Edited by L. Y. Chiang, A. F. Garito, and D. J. Sandman, vol. 247, p. 333, Pittsburgh, PA, Materials Research Society Press (1992).
27. K. Harigaya and M. Fujita, *Phys. Rev. B* **47**, 16563 (1993).
28. K. Harigaya and M. Fujita, *Synth. Metals* **55**, 3196 (1993).
29. N. A. Viet, H. Ajiki, and T. Ando, *ISSP Technical Report* **2828** (1994).
30. J. C. Charlier, X. Gonze, and J. P. Michenaud, *Europhys. Lett.* **29**, 43 (1994).
31. J. Y. Yi and J. Bernholc, *Phys. Rev. B* **47**, 1708 (1993).
32. T. W. Ebbesen, *Annu. Rev. Mater. Sci.* **24**, 235 (1994).
33. H. Fröhlich, *Phys. Rev.* **79**, 845 (1950).
34. H. Fröhlich, *Proc. Roy. Soc. London* **A215**, 291 (1952).
35. R. A. Jishi, M. S. Dresselhaus, and G. Dresselhaus, *Phys. Rev. B* **48**, 11385 (1993).
36. L. Langer, L. Stockman, J. P. Heremans, V. Bayot, C. H. Olk, C. Van Haesendonck, Y. Bruynseraede, and J. P. Issi, *J. Mat. Res.* **9**, 927 (1994).

ELECTRONIC AND STRUCTURAL PROPERTIES OF CARBON NANOTUBES

J. W. MINTMIRE and C. T. WHITE
Chemistry Division, Naval Research Laboratory, Washington, DC 20375-5342, U.S.A.

(*Received* 12 *October* 1994; *accepted in revised form* 15 *February* 1995)

Abstract—Recent developments using synthetic methods typical of fullerene production have been used to generate graphitic nanotubes with diameters on the order of fullerene diameters: "carbon nanotubes." The individual hollow concentric graphitic nanotubes that comprise these fibers can be visualized as constructed from rolled-up single sheets of graphite. We discuss the use of helical symmetry for the electronic structure of these nanotubes, and the resulting trends we observe in both band gap and strain energy versus nanotube radius, using both empirical and first-principles techniques. With potential electronic and structural applications, these materials appear to be appropriate synthetic targets for the current decade.

Key Words—Carbon nanotube, electronic properties, structural properties, strain energy, band gap, band structure, electronic structure.

1. INTRODUCTION

Less than four years ago Iijima[1] reported the novel synthesis based on the techniques used for fullerene synthesis[2,3] of substantial quantities of multiple-shell graphitic nanotubes with diameters of nanometer dimensions. These nanotube diameters were more than an order of magnitude smaller than those typically obtained using routine synthetic methods for graphite fibers[4,5]. This work has been widely confirmed in the literature, with subsequent work by Ebbesen and Ajayan[6] demonstrating the synthesis of bulk quantities of these materials. More recent work has further demonstrated the synthesis of abundant amounts of single-shell graphitic nanotubes with diameters on the order of one nanometer[7–9]. Concurrent with these experimental studies, there have been many theoretical studies of the mechanical and electronic properties of these novel fibers[10–30]. Already, theoretical studies of the individual hollow concentric graphitic nanotubes, which comprise these fibers, predict that these nanometer-scale diameter nanotubes will exhibit conducting properties ranging from metals to moderate bandgap semiconductors, depending on their radii and helical structure[10–22]. Other theoretical studies have focused on structural properties and have suggested that these nanotubes could have high strengths and rigidity resulting from their graphitic and tubular structure[23–30]. The metallic nanotubes—termed serpentine[23]—have also been predicted to be stable against a Peierls distortion to temperatures far below room temperature[10]. The fullerene nanotubes show the promise of an array of all-carbon structures that exhibits a broad range of electronic and structural properties, making these materials an important synthetic target for the current decade.

Herein, we summarize some of the basic electronic and structural properties expected of these nanotubes from theoretical grounds. First we will discuss the basic structures of the nanotubes, define the nomenclature used in the rest of the manuscript, and present an analysis of the rotational and helical symmetries of the nanotube. Then, we will discuss the electronic structure of the nanotubes in terms of applying Born-von Karman boundary conditions to the two-dimensional graphene sheet. We will then discuss changes introduced by treating the nanotube realistically as a three-dimensional system with helicity, including results both from all-valence empirical tight-binding results and first-principles local-density functional (LDF) results.

2. NANOTUBE STRUCTURE AND SYMMETRY

Each single-walled nanotube can be viewed as a conformal mapping of the two-dimensional lattice of a single sheet of graphite (graphene), depicted as the honeycomb lattice of a single layer of graphite in Fig. 1, onto the surface of a cylinder. As pointed out by Iijima[1], the proper boundary conditions around the cylinder can only be satisfied if one of the Bravais lattice vectors of the graphite sheet maps to a circumference around the cylinder. Thus, each real lattice vector of the two-dimensional hexagonal lattice (the Bravais lattice for the honeycomb) defines a different way of rolling up the sheet into a nanotube. Each such lattice vector, $\mathbf{B}$, can be defined in terms of the two primitive lattice vectors $\mathbf{R}_1$ and $\mathbf{R}_2$ and a pair of integer indices $[n_1, n_2]$, such that $\mathbf{B} = n_1\mathbf{R}_1 + n_2\mathbf{R}_2$, with Fig. 2 depicting an example for a [4,3] nanotube. The point group symmetry of the honeycomb lattice will make many of these equivalent, however, so truly unique nanotubes are only generated using a one-twelfth irreducible wedge of the Bravais lattice. Within this wedge, only a finite number of nanotubes can be constructed with a circumference below any given value.

The construction of the nanotube from a conformal mapping of the graphite sheet shows that each nanotube can have up to three inequivalent (by point

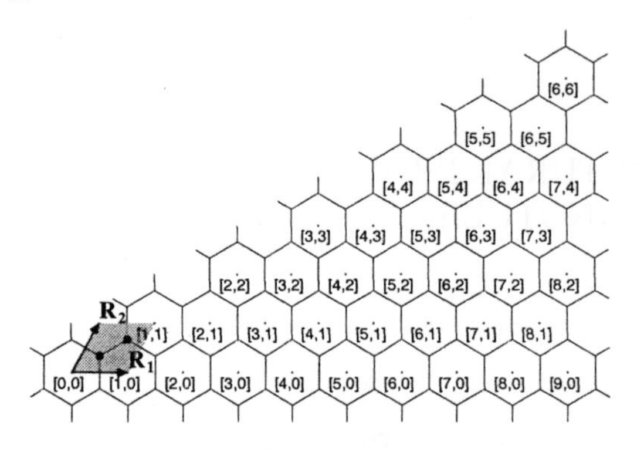

Fig. 1. Two-dimensional honeycomb lattice of graphene; primitive lattice vectors $\mathbf{R}_1$ and $\mathbf{R}_2$ are depicted outlining primitive unit cell.

group symmetry) helical operations derived from the primitive lattice vectors of the graphite sheet. Thus, while *all* nanotubes have a helical structure, nanotubes constructed by mapping directions equivalent to lattice translation indices of the form $[n,0]$ and $[n,n]$, to the circumference of the nanotube will possess a reflection plane. These high-symmetry nanotubes will, therefore, be achiral[12–14,23]. For convenience, we have labeled these special structures based on the shapes made by the most direct continuous path of bonds around the circumference of the nanotube[23]. Specifically, the $[n,0]$-type structures were labeled as sawtooth and the $[n,n]$-type structures as serpentine. For all other conformations inequivalent to these two

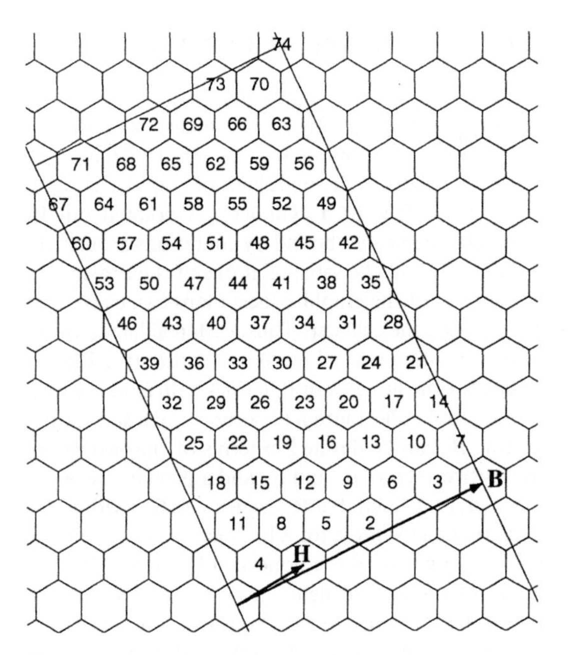

Fig. 2. Depiction of conformal mapping of graphene lattice to [4,3] nanotube. **B** denotes [4,3] lattice vector that transforms to circumference of nanotube, and **H** transforms into the helical operator yielding the minimum unit cell size under helical symmetry. The numerals indicate the ordering of the helical steps necessary to obtain one-dimensional translation periodicity.

sets, the nanotubes are chiral and have three inequivalent helical operations.

Because real lattice vectors can be found that are normal to the primitive real lattice vectors for the graphite sheet, each nanotube thus generated can be shown to be translationally periodic down the nanotube axis[12–14,23]. However, even for relatively small diameter nanotubes, the minimum number of atoms in a translational unit cell can be quite large. For example, for the [4,3] nanotube ($n_1 = 4$ and $n_2 = 3$) then the radius of the nanotube is less than 0.3 nm, but the translational unit cell contains 148 carbon atoms as depicted in Fig. 2. The rapid growth in the number of atoms that can occur in the minimum translational unit cell makes recourse to the helical and any higher point group symmetry of these nanotubes practically mandatory in any comprehensive study of their properties as a function of radius and helicity. These symmetries can be used to reduce to two the number of atoms necessary to generate any nanotube[13,14]; for example, reducing the matrices that have to be diagonalized in a calculation of the nanotube's electronic structure to a size no larger than that encountered in a corresponding electronic structure calculation of two-dimensional graphene.

Before we can analyze the electronic structure of a nanotube in terms of its helical symmetry, we need to find an appropriate helical operator $\mathcal{S}(h,\varphi)$, representing a screw operation with a translation h units along the cylinder axis in conjunction with a rotation φ radians about this axis. We also wish to find the operator $\mathcal{S}$ that requires the minimum unit cell size (i.e., the smallest set of carbon atoms needed to generate the entire nanotube using $\mathcal{S}$) to minimize the computational complexity of calculating the electronic structure. We can find this helical operator $\mathcal{S}(h,\varphi)$ by first finding the real lattice vector $\mathbf{H} = m_1\mathbf{R}_1 + m_2\mathbf{R}_2$ in the honeycomb lattice that will transform to $\mathcal{S}(h,\varphi)$ under conformal mapping, such that $h = |\mathbf{H}\times\mathbf{B}|/|\mathbf{B}|$ and $\varphi = 2\pi(\mathbf{H}\cdot\mathbf{B})/|\mathbf{B}|^2$. An example of $\mathbf{H}$ is depicted in Fig. 2 for the [4,3] nanotube. If we construct the cross product $\mathbf{H}\times\mathbf{B}$ of $\mathbf{H}$ and $\mathbf{B}$, the magnitude $|\mathbf{H}\times\mathbf{B}|$ will correspond to the area "swept out" by the helical operator $\mathcal{S}$. Expanding, we find

$$|\mathbf{H}\times\mathbf{B}| = (m_1 n_2 - m_2 n_1)|\mathbf{R}_1\times\mathbf{R}_2| \qquad (1)$$

where $|\mathbf{R}_1\times\mathbf{R}_2|$ is just the area per two-carbon unit cell of the primitive graphene lattice. Thus $(m_1 n_2 - m_2 n_1)$ gives the number of carbon pairs required for the unit cell under helical symmetry with a helical operator generated using $\mathbf{H}$ and the conformal mapping. Given a choice for n_1 and n_2, then $(m_1 n_2 - m_2 n_1)$ must equal some integer value

$$(m_1 n_2 - m_2 n_1) = \pm N. \qquad (2)$$

Integer arithmetic then shows that the smallest magnitude nonzero value of N will be given by the greatest common divisor of n_1 and n_2. Thus, we can easily determine the helical operator $\mathbf{H}\rightarrow\mathcal{S}(h,\varphi)$ that yields the minimum unit cell (primitive helical motif) size of $2N$ atoms, where N is the greatest common divisor of n_1 and n_2.

Next, consider the possible rotational symmetry around the helical axis. This operation is of special interest because all the rotation operations around the chain axis will commute with the helical operator $\mathcal{S}(h,\varphi)$, allowing solutions of any one-electron Hamiltonian model to be block-diagonalized into the irreducible representations of both the helical operation and the rotation operator simultaneously. We observe that the highest order rotational symmetry (and smallest possible rotation angle for such an operator) will be given by a C_N rotation operator, where N remains the greatest common divisor of n_1 and n_2[13]. This can be seen straightforwardly by considering the real lattice vector $\mathbf{B}_N$,

$$\mathbf{B}_N = (n_1/N)\mathbf{R}_1 + (n_2/N)\mathbf{R}_2. \quad (3)$$

The conformal mapping will transform this lattice operation $\mathbf{B}_N$ to a rotation of $2\pi/N$ radians around the nanotube axis, thus generating a C_N subgroup.

The rotational and helical symmetries of a nanotube defined by $\mathbf{B}$ can then be seen by using the corresponding helical and rotational symmetry operators $\mathcal{S}(h,\varphi)$ and C_N to generate the nanotube[13,14]. This is done by first introducing a cylinder of radius $|\mathbf{B}|/2\pi$. The two carbon atoms located at $d \equiv (\mathbf{R}_1 + \mathbf{R}_2)/3$ and 2d in the [0,0] unit cell of Fig. 1 are then mapped to the surface of this cylinder. The first atom is mapped to an arbitrary point on the cylinder surface, which requires that the position of the second be found by rotating this point by $2\pi(d\cdot\mathbf{B})/|\mathbf{B}|^2$ radians about the cylinder axis in conjunction with a translation $|\mathbf{d}\times\mathbf{B}|/|\mathbf{B}|$ units along this axis. The positions of these first two atoms can then be used to generate $2(N-1)$ additional atoms on the cylinder surface by $(N-1)$ successive $2\pi/N$ rotations about the cylinder axis. Altogether, these $2N$ atoms complete the specification of the helical motif which corresponds to an area on the cylinder surface given by $N|\mathbf{R}_1 \times \mathbf{R}_2|$. This helical motif can then be used to tile the remainder of the nanotube by repeated operation of the helical operation defined by $\mathcal{S}(h,\varphi)$ generated by the $\mathbf{H}$ defined using eqn (2).

3. ELECTRONIC STRUCTURE OF CARBON NANOTUBES

We will now discuss the electronic structure of single-shell carbon nanotubes in a progression of more sophisticated models. We shall begin with perhaps the simplest model for the electronic structure of the nanotubes: a Hückel model for a single graphite sheet with periodic boundary conditions analogous to those imposed by the rolling up of the nanotube. We then continue by analyzing a structurally correct nanotube within a Slater-Koster sp^3 tight-binding model using the helical symmetry of the nanotube. We then finish by discussing results for both geometry optimization and band structure using first-principles local-density functional methods.

3.1 *Graphene model*

One of the simplest models for the electronic structure of the states near the Fermi level in the nanotubes is that of a single sheet of graphite (graphene) with periodic boundary conditions[10–16]. Let us consider a Slater-Koster tight-binding model[31] and assume that the nearest-neighbor matrix elements are the same as those in the planar graphene (i.e., we neglect curvature effects). The electronic structure of the nanotube will then be basically that of graphene, with the additional imposition of Born-von Karman boundary conditions on the electronic states, so that they are periodic under translations of the circumference vector $\mathbf{B}$. The electronic structure of the π-bands near the Fermi level then reduces to a Hückel model with one parameter, the $V_{pp\pi}$ hopping matrix element. This model can be easily solved, and the one-electron eigenvalues can be given as a function of the two-dimensional wavevector $\mathbf{k}$ in the standard hexagonal Brillouin zone of graphene in terms of the primitive lattice vectors $\mathbf{R}_1$ and $\mathbf{R}_2$:

$$\varepsilon(\mathbf{k}) = V_{pp\pi}\sqrt{3 + 2\cos\mathbf{k}\cdot\mathbf{R}_1 + 2\cos\mathbf{k}\cdot\mathbf{R}_2 + 2\cos\mathbf{k}\cdot(\mathbf{R}_1 - \mathbf{R}_2)} \quad (4)$$

The Born-von Karman boundary conditions then restrict the allowed electronic states to those in the graphene Brillouin zone that satisfy

$$\mathbf{k}\cdot\mathbf{B} = 2\pi m, \quad (5)$$

that m an integer[10,14,15]. In terms of the two-dimensional Brillouin zone of graphene, the allowed states will lie along parallel lines separated by a spacing of $2\pi/|\mathbf{B}|$. Fig. 3a depicts a set of these lines for the [4,3] nanotube. The reduced dimensionality of these one-electron states is analogous to those found in quantum confinement systems; we might expect that the multiple crossings of the Brillouin zone edge would lead to the creation of multiple gaps in the electronic density-of-states (DOS) as we conceptually transform graphene into a nanotube. However, as depicted in Fig. 3b, these one-electron states are actually continuous in an extended Brillouin zone picture. A direct consequence of our conclusion in the previous section—that the helical unit cell for the [4,3] nanotube would have only two carbons, the same as in graphene itself—is that the π-band in graphene will transform into a single band, rather than several.

Initially, we showed that the set of serpentine nanotubes [n,n] are metallic. Soon thereafter, Hamada *et al.*[15] and Saito *et al.*[16,17] pointed out that the periodicity condition in eqn (5) further groups the remaining nanotubes (those that cannot be constructed from the condition $n_1 = n_2$) into one set that has moderate band gaps and a second set with small band gaps. The graphene model predicts that the second set of nanotubes would have zero band gaps, but the symmetry breaking introduced by curvature effects results in small, but nonzero, band gaps. To demonstrate this point, consider the standard reciprocal lattice vectors $\mathbf{K}_1$ and $\mathbf{K}_2$ for the graphene lattice given by $\mathbf{K}_i \cdot \mathbf{R}_j = 2\pi\delta_{ij}$. The band structure given by eqn (4) will have a band crossing (i.e., the occupied band and the unoccupied band will touch at zero energy) at the corners $\bar{K}$ of the hexagonal Brillouin zone, as depicted in Fig. 3. These six corners $\bar{K}$ of the central Brillouin zone are given by the vectors $\mathbf{k}_{\bar{K}} \equiv \pm(\mathbf{K}_1 - \mathbf{K}_2)/3$, $\pm(2\mathbf{K}_1 + \mathbf{K}_2)/3$, and $\pm(\mathbf{K}_1 + 2\mathbf{K}_2)/3$. Given our earlier definition of $\mathbf{B}$, $\mathbf{B} = n_1\mathbf{R}_1 + n_2\mathbf{R}_2$, a nanotube will have zero band gap (within the graphene model) if and only if $\mathbf{k}_{\bar{K}}$ satisfies eqn (5), leading to the condition $n_1 - n_2 = 3m$, where m is an integer. As we shall see later, when we include curvature effects in the electronic Hamiltonian, these "*metallic*" nanotubes will actually fall into two categories: the serpentine nanotubes that are truly metallic by symmetry[10,14], and quasimetallic with small band gaps[15–18].

In addition to the zero band gap condition, we have examined the behavior of the electron states in the vicinity of the gap to estimate the band gap for the moderate band gap nanotubes[13,14]. Consider a wave vector $\mathbf{k}$ in the vicinity of the band crossing point $\bar{K}$ and define $\Delta\mathbf{k} = \mathbf{k} - \mathbf{k}_{\bar{K}}$, with $\Delta k \equiv |\Delta\mathbf{k}|$. The function $\varepsilon(\mathbf{k})$ defined in eqn (4) has a cusp in the vicinity of $\bar{K}$, but $\varepsilon^2(\mathbf{k})$ is well-behaved and can be expanded in a Taylor expansion in Δk. Expanding $\varepsilon^2(\mathbf{k})$, we find

$$\varepsilon^2(\Delta k) = \tfrac{3}{4} V_{pp\pi}^2 a^2 \Delta k^2, \tag{6}$$

where $a = |\mathbf{R}_1| = |\mathbf{R}_2|$ is the lattice spacing of the honeycomb lattice. The allowed nanotube states satisfying eqn (5) lie along parallel lines as depicted in Fig. 2 with a spacing of $2\pi/B$, where $B \equiv |\mathbf{B}|$. For the nonmetallic case, the smallest band gap for the nanotube will occur at the nearest allowed point to $\bar{K}$, which will lie one third of the line spacing from $\bar{K}$. Thus, using $\Delta k = 2\pi/3B$, we find that the band gap equals[13,14]

$$E_g = 2\varepsilon(\Delta k) = \frac{2\pi}{\sqrt{3}} \frac{V_{pp\pi} a}{B} = \frac{V_{pp\pi} r_{CC}}{R_T}, \tag{7}$$

where r_{CC} is the carbon-carbon bond distance ($r_{CC} = a/\sqrt{3} \sim 1.4$ Å) and R_T is the nanotube radius ($R_T = B/2\pi$). Similar results were also obtained by Ajiki and Ando[18].

3.2 *Using helical symmetry*

The previous analysis of the electronic structure of the carbon nanotubes assumed that we could neglect curvature effects, treating the nanotube as a single graphite sheet with periodic boundary conditions imposed analogous to those in a quantum confinement problem. This level approximation allowed us to analyze the electronic structure in terms of the two-dimensional band structure of graphene. Going beyond this level of approximation will require explicit treatment of the helical symmetry of the nanotube for practical computational treatment of the electronic structure problem. We have already discussed the ability to determine the optimum choice of helical operator $\mathcal{S}$ in terms of a graphene lattice translation $\mathbf{H}$. We now examine some of the consequences of using helical symmetry and its parallels with standard band structure theory. Because the symmetry group generated by the screw operation $\mathcal{S}$ is isomorphic with the one-dimensional translation group, Bloch's theorem can be generalized so that the one-electron wavefunctions will transform under $\mathcal{S}$ according to

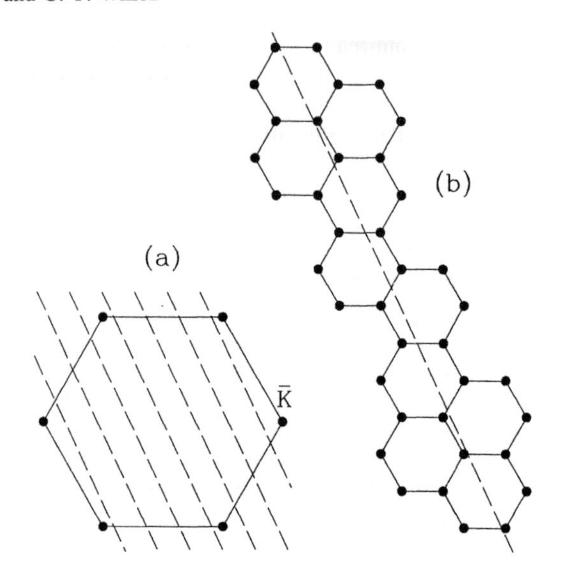

Fig. 3. (a) Depiction of central Brillouin zone and allowed graphene sheet states for a [4,3] nanotube conformation. Note Fermi level for graphene occurs at $\bar{K}$ points at vertices of hexagonal Brillouin zone. (b) Extended Brillouin zone picture of [4,3] nanotube. Note that top left hexagon is equivalent to bottom right hexagon.

$$\mathcal{S}^m \psi_i(\mathbf{r};\kappa) = e^{i\kappa m}\psi_i(\mathbf{r};\kappa). \tag{8}$$

The quantity κ is a dimensionless quantity which is conventionally restricted to a range of $-\pi < \kappa \leq \pi$, a central Brillouin zone. For the case $\varphi = 0$ (i.e., $\mathcal{S}$ a pure translation), κ corresponds to a normalized quasi-momentum for a system with one-dimensional translational periodicity (i.e., $\kappa \equiv kh$, where k is the traditional wavevector from Bloch's theorem in solid-state band-structure theory). In the previous analysis of helical symmetry, with $\mathbf{H}$ the lattice vector in the graphene sheet defining the helical symmetry generator, κ in the graphene model corresponds similarly to the product $\kappa = \mathbf{k}\cdot\mathbf{H}$ where $\mathbf{k}$ is the two-dimensional quasimomentum vector of graphene.

Before we continue our description of electronic structure methods for the carbon nanotubes using helical symmetry, let us reconsider the metallic and quasi-metallic cases discussed in the previous section in more detail. The graphene model suggests that a metallic state will occur where two bands cross, and that the Fermi level will be pinned to the band crossing. In terms of band structure theory however, if these two bands belong to the same irreducible representation of a point group of the nuclear lattice that also leaves the point in the Brillouin zone invariant, then rather than touching (and being degenerate in energy) these one-electron eigenfunctions will mix and lead to an avoided crossing. Only if the two eigenfunctions belong to different irreducible representations of the point group can they be degenerate. For graphene, the high symmetry of the honeycomb lattice allows the degeneracy of the highest-occupied and lowest-unoccupied states at the corners $\bar{K}$ of the hexagonal Brillouin zone in graphene. Rolling up graphene into a nanotube breaks this symmetry, and we must ask what point group symmetries are left that can allow a degeneracy at the band crossing rather than an avoided crossing. For the nanotubes, the appropriate symmetry operations that leave an entire band in the Brillouin zone invariant are the C_n rotation operations around the helical axis and reflection planes that contain the helical axis. We see from the graphene model that a reflection plane will generally be necessary to allow a degeneracy at the Fermi level, because the highest-occupied and lowest-unoccupied states will share the same irreducible representation of the rotation group. To demonstrate this, consider the irreducible representations of the rotation group. The different irreducible representations transform under the generating rotation (of $2\pi/N$ radians) with a phase factor an integer multiple $2\pi m/N$, where $m = 0, \ldots, N-1$. Within the graphene model, each allowed state at quasimomentum $\mathbf{k}$ will transform under the rotation by the phase factor given by $\mathbf{k}\cdot\mathbf{B}/N$, and by eqn (5) we see that the phase factor at $\bar{K}$ is just $2\ \pi m/N$. The eigenfunctions predicted using the graphene model are therefore already members of the irreducible representations of the rotation point group. Furthermore, the eigenfunctions at a given Brillouin zone point $\mathbf{k}$ in the graphene model must be members of the same irreducible representation of the rotation point group.

For the nanotubes, then, the appropriate symmetries for an allowed band crossing are only present for the serpentine ($[n,n]$) and the sawtooth ($[n,0]$) conformations, which will both have C_{nv} point group symmetries that will allow band crossings, and with rotation groups generated by the operations equivalent by conformal mapping to the lattice translations $\mathbf{R}_1 + \mathbf{R}_2$ and $\mathbf{R}_1$, respectively. However, examination of the graphene model shows that only the serpentine nanotubes will have states of the correct symmetry (i.e., different parities under the reflection operation) at the $\bar{K}$ point where the bands can cross. Consider the $\bar{K}$ point at $(\mathbf{K}_1 - \mathbf{K}_2)/3$. The serpentine case always satisfies eqn (5), and at the $\bar{K}$ points the one-electron wave functions transform under the generator of the rotation group C_n with a phase factor given by $\mathbf{k}_{\bar{K}}\cdot(\mathbf{R}_1 + \mathbf{R}_2) = 0$. This irreducible representation of the C_n group is split under reflection into the two irreducible representations a_1 and a_2 of the C_{nv} group that are symmetric and antisymmetric, respectively, under the reflection plane; the states at $\bar{K}$ will belong to these two separate irreducible representations. Thus, the serpentine nanotubes are always metallic because of symmetry if the Hamiltonian allows sufficient bandwidth for a crossing, as is normally the case[10]. The sawtooth nanotubes, however, present a different picture. The one-electron wave functions at $\bar{K}$ transform under the generator of the rotation group for this nanotube with a phase factor given by $\mathbf{k}_{\bar{K}}\cdot\mathbf{R}_1 = 2\pi/3$. This phase factor will belong to one of the e representations of the C_{nv} group, and the states at $\bar{K}$ in the graphene Brillouin zone will therefore belong to the same symmetry group. This will lead to an avoided crossing. Therefore, the band gaps of the non-serpentine nanotubes that satisfy eqn (5) are not truly metallic but only small band gap systems, with band gaps we estimate from empirical and first-principles calculation to be of the order of 0.1 eV or less.

Now, let us return to our discussion of carrying out an electronic structure calculation for a nanotube using helical symmetry. The one-electron wavefunctions ψ_i can be constructed from a linear combination of Bloch functions φ_j, which are in turn constructed from a linear combination of nuclear-centered functions $\chi_j(\mathbf{r})$,

$$\psi_i(\mathbf{r};\kappa) = \sum_j c_{ji}(\kappa)\varphi_j(\mathbf{r};\kappa) \tag{9}$$

$$\varphi_j(\mathbf{r};\kappa) = \sum_m e^{-i\kappa m}\mathcal{S}^m\chi_j(\mathbf{r}). \tag{10}$$

As the next step in including curvature effects beyond the graphene model, we have used a Slater-Koster parameterization[31] of the carbon valence states—which we have parameterized[32,33] to earlier LDF band structure calculations[34] on polyacetylene—in the empirical tight-binding calculations. Within the notation in ref. [31] our tight-binding parameters are given by $V_{ss\sigma} = -4.76$ eV, $V_{sp\sigma} = 4.33$ eV, $V_{pp\sigma} = 4.37$ eV, and $V_{pp\pi} = -2.77$ eV.[33] We choose the diagonal term for the carbon p orbital, $\varepsilon_p = 0$ which results in the s diagonal term of $\varepsilon_s = -6.0$ eV. This tight-binding model reproduces first-principles band structures qualitatively quite well. As an example, Fig. 4 depicts both Slater-Koster tight-binding results and first-principles LDF results[10,12] for the band structure of the [5,5] serpentine nanotube within helical symmetry. All bands have been labeled for the LDF results according to the four irreducible representations of the C_{5v} point group: the rotationally invariant a_1 and a_2 representations, and the doubly-degenerate e_1 and e_2 representation. As noted in our discussion for the ser-

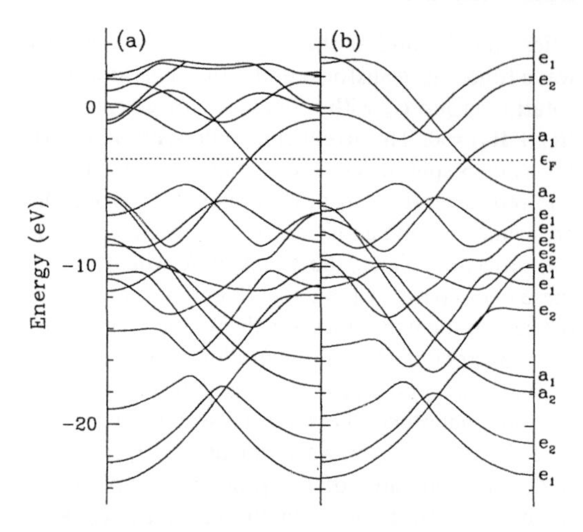

Fig. 4. (a) Slater-Koster valence tight-binding and (b) first-principles LDF band structures for [5,5] nanotube. Band structure runs from left at helical phase factor $\kappa = 0$ to right at $\kappa = \pi$. Fermi level ε_F for Slater-Koster results has been shifted to align with LDF results.

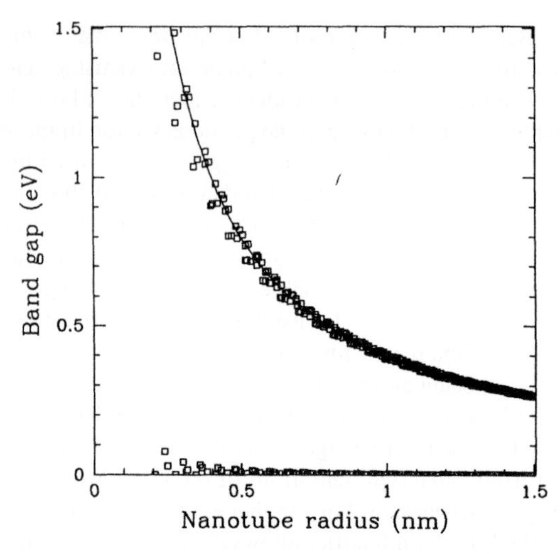

Fig. 5. Band gap as a function of nanotube radius calculated using empirical tight-binding Hamiltonian. Solid line gives estimate using Taylor expansion of graphene sheet results in eqn. (7).

pentine nanotubes, this band structure is metallic with a band crossing of the a_1 and a_2 bands.

Within the Slater-Koster approximation, we can easily test the validity of the approximations made in eqn (7) based on the graphene model. In Fig. 5 we depict the band gaps using the empirical tight-binding method for nanotube radii less than 1.5 nm. The nonmetallic nanotubes ($n_1 - n_2 \neq 3\ m$) are shown in the upper curve where we have also depicted a solid line showing the estimated band gap for the nonmetallic nanotubes using $|V_{pp\pi}| r_{CC}/R_T$, with $V_{pp\pi}$ as given above and $r_{CC} = 1.44$ Å. We see excellent agreement between the estimate based on the graphene sheet and the exact calculations for the tight-binding model with curvature introduced, with the disagreement increasing with increasing curvature at the smaller radii.

3.3 *First-principles methods*

We have extended the linear combination of Gaussian-type orbitals local-density functional approach to calculate the total energies and electronic structures of helical chain polymers[35]. This method was originally developed for molecular systems[36–40], and extended to two-dimensionally periodic systems[41,42] and chain polymers[34]. The one-electron wavefunctions here ψ_i are constructed from a linear combination of Bloch functions φ_j, which are in turn constructed from a linear combination of nuclear-centered Gaussian-type orbitals $\chi_j(\mathbf{r})$ (in this case, products of Gaussians and the real solid spherical harmonics). The one-electron density matrix is given by

$$\rho(\mathbf{r};\mathbf{r}') = \sum_i \frac{1}{2\pi}\int_{-\pi}^{\pi} d\kappa\, n_i(\kappa)\psi_i^*(\mathbf{r}';\kappa)\psi_i(\mathbf{r};k)$$

$$= \sum_{jj'}\sum_m P_{j'j}^m \sum_{m'} \chi_{j'}^{m+m'}(\mathbf{r}')\chi_j^{m'}(\mathbf{r}), \qquad (11)$$

where $n_i(\kappa)$ are the occupation numbers of the one-electron states, χ_j^m denotes $\mathcal{S}^m\chi_j(\mathbf{r})$, and P_{ij}^m are the coefficients of the real lattice expansion of the density matrix given by

$$P_{j'j}^m = \sum_i \frac{1}{2\pi}\int_{-\pi}^{\pi} d\kappa\, n_i(\kappa)c_{j'i}^*(\kappa)c_{ji}(\kappa)e^{i\kappa m} \qquad (12)$$

The total energy for the nanotube is then given by

$$E = \sum_{ij}\sum_m P_{ij}^m\left\{-\frac{1}{2}\left\langle\chi_i^m|\nabla^2|\chi_j^0\right\rangle + \left\langle\chi_i^m|\varepsilon_{xc}[\rho(\mathbf{r})]|\chi_j^0\right\rangle\right\}$$
$$+\frac{1}{2}\sum_{m''}\left\{\left(\sum_n{}'\sum_{n'}\frac{Z_nZ_{n'}}{|\mathbf{R}_n^0 - \mathbf{R}_{n'}^{m''}|}\right)\right.$$
$$+\sum_{ij}\sum_m P_{ij}^m\left(\sum_{i'j'}\sum_{m'}P_{i'j'}^{m'}[\chi_i^m\chi_j^0|\chi_{i'}^{m'+m''}\chi_{j'}^{m''}]\right.$$
$$\left.\left.-2\sum_n\left\langle\chi_i^m\left|\frac{Z_n}{|\mathbf{r}-\mathbf{R}_n^{m''}|}\right|\chi_j^0\right\rangle\right)\right\} \qquad (13)$$

where Z_n and $\mathbf{R}_n$ denote the nuclear charges and coordinates within a single unit cell, $\mathbf{R}_n^m$ denotes the nuclear coordinates in unit cell m ($\mathbf{R}_n^m \equiv \mathcal{S}^m\mathbf{R}_n$), and $[\rho_1|\rho_2]$ denotes an electrostatic interaction integral,

$$[\rho_1|\rho_2] \equiv \int d^3r_1\int d^3r_2\,\frac{\rho_1(\mathbf{r}_1)\rho_2(\mathbf{r}_2)}{r_{12}} \qquad (14)$$

Rather than solve for the total energy directly as expressed in eqn (13), we follow the suggestion of earlier workers to fit the exchange-correlation potential and the charge density (in the Coulomb potential) to a linear combination of Gaussian-type functions.

We have carried out a series of geometry optimizations on nanotubes with diameters less than 2 nm. We will present some results for a selected subset of the moderate band gap nanotubes, and then focus on results for an example chiral systems: the chiral [9,2] nanotube with a diameter of 0.8 nm. This nanotube has been chosen because its diameter corresponds to those found in relatively large amounts by Iijima[7] after the synthesis of single-walled nanotubes.

How the structural properties of the fullerene nanotubes change with conformation is one of the most important questions to be answered about these new materials. In particular, two properties are most apt for study with the LDF approach: how does the band gap change with nanotube diameter, and how does the strain energy change with nanotube diameter? We have studied these questions extensively using empirical methods[10–14,23], and are currently working on a comprehensive study of the band gaps, strain energies, and other properties using the LDF approach. We expect from eqn (7) and the previous analysis of the graphene sheet model that the moderate band gap nanotubes should have band gaps that vary roughly inversely proportional to the nanotube radius. In Fig. 6 we depict representative results for some moderate band gap nanotubes. In the figure, we see that not only do our first-principles band gaps decrease in an inverse relationship to the nanotube radius, but that the band gaps are well described with a reasonable value of $V_{pp\pi} \sim -2.5$ eV.

Second, we expect that the strain energy per carbon should increase inversely proportional to the square of the nanotube radius[23]. Based on a continuum elastic model, Tibbetts[4] derived a strain energy for a thin graphitic nanotube of the general form:

$$\sigma = \frac{\pi E L a^3}{12R} \tag{16}$$

where E is the elastic modulus, R is the radius of curvature, L is the length of the cylinder, and a is a representative thickness of the order of the graphite interplanar spacing (3.35 Å). Assuming that the total number of carbons is given by $N = 2\pi RL/\Omega$, where Ω is the area per carbon, we find that the strain energy per carbon is expected to be

$$\frac{\sigma}{N} = \frac{Ea^3}{24}\frac{\Omega}{R^2} . \tag{17}$$

In earlier work, we found this relationship was well observed, using empirical bond-order potentials for all nanotubes with radii less than 0.9 nm, and for a range of serpentine nanotubes using the LDF method. As part of our studies, we have carried out first-principles LDF calculations on a representative sample of chiral nanotubes. Iijima and Ichihashi[7] have recently reported the synthesis of single-shell fullerene nanotubes with diameters of about 1 nm, using the gas-phase product of a carbon-arc synthesis with iron vapor present. After plotting the frequency of single-shell nanotubes they observed versus nanotube diameter, they found enhanced abundances for diameters of roughly 0.8 nm and 1.1 nm. For first-principles simulations then to be useful, they should be capable of calculations for nanotubes with diameters from about 0.6–2.0 nm. In Fig. 7 we depict the calculated total energy per carbon, shifted relative to an extrapolated value for an infinite radius nanotube, for a representative sample of nanotubes over this range using the LDF approach. In this figure, the open squares denote results for unoptimized nanotubes, where the nano-

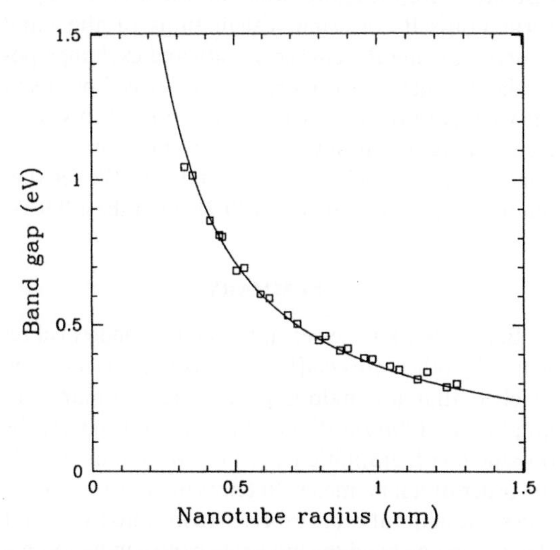

Fig. 6. Band gap as a function of nanotube radius using first-principles LDF method. Solid line shows estimates using graphene sheet model with $V_{pp\pi} = -2.5$ eV.

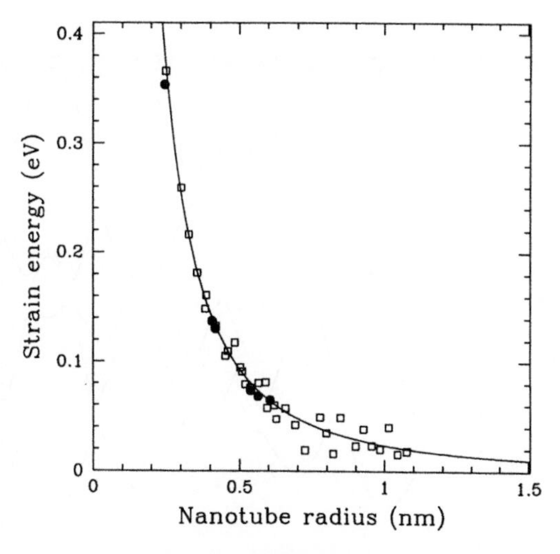

Fig. 7. Strain energy per carbon (total energy minus total energy extrapolated for the graphite sheet) as a function of nanotube radius calculated for unoptimized nanotube structures (open squares) and optimized nanotube structures (solid circles). Solid line depicts inverse square relationship drawn through point at smallest radius.

Table 1. Relaxation energy per carbon (eV) obtained by geometry optimization of nanotubes relative to extrapolated value for graphene

Nanotube	Radius (nm)	Unrelaxed energy (eV)	Relaxed energy (eV)
[12,5]	0.6050	0.067	0.064
[13,2]	0.5630	0.071	0.068
[11,4]	0.5370	0.076	0.073
[7,5]	0.4170	0.133	0.130
[9,2]	0.4060	0.140	0.137
[4,3]	0.2460	0.366	0.354

tube has been directly constructed from a conformal mapping of graphene with a carbon-carbon nearest-neighbor distance of 0.144 nm. We have carried out a limited number of complete geometry optimizations within the LDF approach. All of the nanotubes selected fall in the range of radii from 0.4–0.6 nm representative of the relatively high abundances found by Iijima and Ichihashi[7]. For the nanotubes studied, the change in energy resulting from relaxation of the lattice is minimal, as depicted by the solid circles in Fig. 7, and presented in Table 1. We see the inverse square relationship is still well observed, although at the larger nanotube radii we begin to observe some scatter in the results that arises from the numerical roundoff errors.

As an example of a nanotube representative of the diameters experimentally found in abundance, we have calculated the electronic structure of the [9,2] nanotube, which has a diameter of 0.8 nm. Figure 8 depicts the valance band structure for the [9,2] nanotube. This band structure was calculated using an unoptimized nanotube structure generated from a conformal mapping of the graphite sheet with a 0.144 nm bond distance. We used 72 evenly-spaced points in the one-dimensional Brillouin zone ($-\pi < \kappa \leq \pi$) and a carbon 7s3p Gaussian basis set. Our earlier analysis of the graphene sheet model predicts that each band of the graphene sheet transforms to a single band of the [9,2] nanotube. From our earlier work[11–14] we expect a highly oscillatory band structure (using the helical symmetry) with about 12 oscillations in each band over the Brillouin zone. In Fig. 8, we see basically six local maxima in each band in the half Brillouin zone depicted. Further, we see the band gap at about one third of the way from the left-hand side of Fig. 8. The highly oscillatory nature of the band structure in these nanotubes makes accurate measurement of the band gap difficult. In our current studies of the band gap of these materials, we typically perform the self-consistent portion of the calculations with a modest number of evenly-spaced points in the Brillouin zone (usually 72 for the entire Brillouin zone). We then perform non-self-consistent calculations of the band structure, using the charge density and exchange potential calculated, for more densely spaced points in the vicinity of the band gap. An example of this is depicted in Fig. 9, which shows the direct gap of the [9,2] nanotube in the vicinity of the band gap. We see that our predicted band gap for the [9,2] nanotube is 0.9 eV.

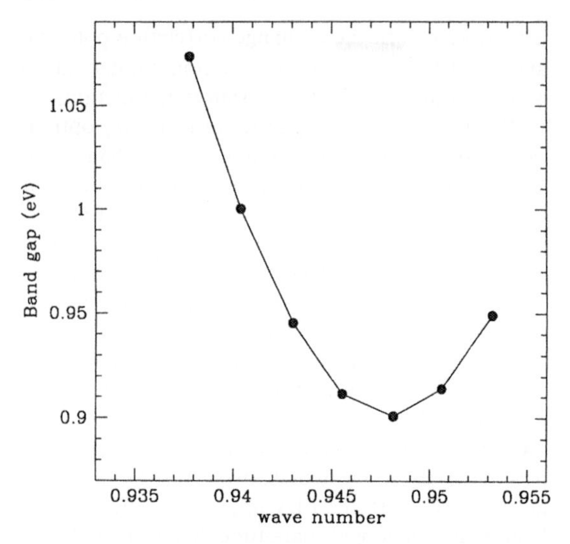

Fig. 9. Direct band gap for [9,2] nanotube in vicinity of band gap. Wave number is dimensionless coordinate κ, with one-dimensional Brillouin zone for κ defined $-\pi \leq \kappa \leq \pi$.

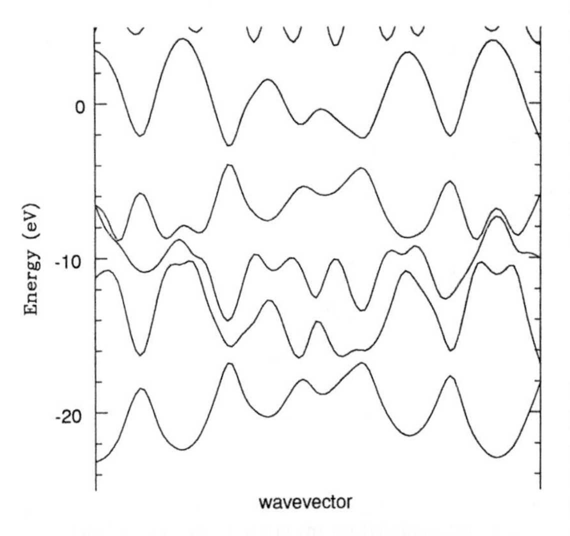

Fig. 8. LDF valence band structure of [9,2] chiral nanotube. The Fermi level lies at midgap at −3.3 eV. Dimensionless wavenumber coordinate κ ranges from 0 to π.

4. SUMMARY

Both in work reviewed herein and in theoretical research by other workers[10–30], a consensus has been reached that anomalous properties (compared to graphite and "*normal*" graphitic nanotubes) can be obtained with graphitic nanotubes with diameters of the order of a nanometer. In terms of electronic properties, the nanotubes are expected to fall into two major classes: on one hand the moderate band gap nanotubes that do not satisfy the $n_1 - n_2 = 3\,m$ condition in the graphene sheet, and on the other hand the small band gap and truly metallic serpentine conformation nanotubes that do satisfy this condition. We have earlier demonstrated[10–12] for the serpentine nanotubes that, for diameters under a nanometer, we expect that the density of states at the Fermi level is comparable to metallic densities and that the nanotubes should not Peierls-distort at normal temperatures. Independent of helicity, we find that the larger-diameter moderate band gap members of the family of moderate band gap nanotubes ($n_1 - n_2 \neq 3\ m$) have bandgaps given approximately by $E_g = |V_{pp\pi}|(r_{CC}/R_T)$ and hence do not have bandgaps approaching k_BT at room temperature until their diameters exceed approximately about 30 nm[10].

We have also examined the energetics and elastic properties of small-diameter graphitic nanotubes using both first-principles and empirical potentials[23]. We find that the strain energy per carbon relative to an unstrained graphite sheet goes as $1/R_T^2$ (where R_T is the nanotube radius) and is insensitive to other aspects of the lattice structure, indicating that relationships derivable from continuum elastic theory persist well into the small radius limit. In general, we find that the elastic properties are those expected by directly extrapolating the behavior of larger graphitic fibers to a small cross-section.

The recent advances in synthesis of single-shell nanotubes should stimulate a wealth of new experimental and theoretical studies of these promising materials aimed at determining their structural, electronic, and mechanical properties. Many questions remain to be further investigated. How can they be terminated[29,43] and can they be connected[30]? What are their electrical properties? For those that are semiconductors, can they be successfully doped[44]? What are the mechanical properties of these nanotubes? How will they respond under compression and stress[23–28]? Do they have the high strengths and rigidity that their graphitic and tubular structure implies? How are these nanotubes formed[43,45,46]? Can techniques be devised that optimize the growth and allow the extraction of macroscopic amounts of selected nanotubes[7,8]? These questions all deserve immediate attention, and the promise of these novel all-carbon materials justify this as a major research area for the current decade.

Acknowledgements—This work was supported by the Office of Naval Research (ONR) through the Naval Research Laboratory and directly through the ONR (Chemistry-Physics and Materials Divisions. We thank D. H. Robertson, D. W. Brenner, and B. I. Dunlap for many useful discussions.

REFERENCES

1. S. Iijima, *Nature* (London) **354**, 56 (1991).
2. W. Krätschmer, L. D. Lamb, K. Fostiropoulos, and D. R. Huffman, *Chem. Phys. Lett.* **170**, 167 (1990). *Nature* **347**, 354 (1990).
3. W. E. Billups and M. A. Ciufolini, eds. *Buckminsterfullerenes*. VCH, New York (1993).
4. G. G. Tibbetts, *J. Crystal Growth* **66**, 632 (1983).
5. J. S. Speck, M. Endo, and M. S. Dresselhaus, *J. Crystal Growth* **94**, 834 (1989).
6. T. W. Ebbesen and P. M. Ajayan, *Nature* (London) **358**, 220 (1992).
7. S. Iijima and T. Ichihashi, *Nature* (London) **363**, 603 (1993).
8. D. S. Bethune, C. H. Klang, M. S. DeVries, G. Gorman, R. Savoy, J. Vazquez, and R. Beyers, *Nature* (London) **363**, 605 (1993).
9. M. Endo, K. Takeuchi, S. Igarashi, K. Kobori, M. Shiraishi, and H. W. Kroto, *J. Phys. Chem. Solids* **54**, 1841 (1993).
10. J. W. Mintmire, B. I. Dunlap, and C. T. White, *Phys. Rev. Lett.* **68**, 631 (1992).
11. C. T. White, J. W. Mintmire, R. C. Mowrey, D. W. Brenner, D. H. Robertson, J. A. Harrison, and B. I. Dunlap, In *Buckminsterfullerenes* (Edited by W. E. Billups and M. A. Ciufolini) pp. 125–184. VCH, New York, (1993).
12. J. W. Mintmire, D. H. Robertson, B. I. Dunlap, R. C. Mowrey, D. W. Brenner, and C. T. White, *Electrical, Optical, and Magnetic Properties of Organic Solid State Materials* (Edited by L. Y. Chiang, A. F. Garito, and D. J. Sandman) p. 339. MRS Symposia Proceedings No. 247. Materials Research Society, Pittsburgh (1992).
13. C. T. White, D. H. Robertson, and J. W. Mintmire, *Phys. Rev. B* **47**, 5485 (1993).
14. J. W. Mintmire, D. H. Robertson, and C. T. White, *J. Phys. Chem. Solids* **54**, 1835 (1993).
15. N. Hamada, S. Sawada, and A. Oshiyamu, *Phys. Rev. Lett.* **68**, 1579 (1992).
16. R. Saito, M. Fujita, G. Dresselhaus, M. S. Dresselhaus, *Phys. Rev. B* **46**, 1804 (1992). *Mater. Res. Soc. Sym. Proc.* **247**, 333 (1992); *Appl. Phys. Lett.* **60**, 2204 (1992).
17. R. Saito, G. Dresselhaus, and M. S. Dresselhaus, *J. Appl. Phys.* **73**, 494 (1993).
18. H. Ajiki and T. Ando, *J. Phys. Soc. Japan* **62**, 1255 (1993). *J. Phys. Soc. Japan* **62**, 2470 (1993).
19. P.-J. Lin-Chung and A. K. Rajagopal, *J. Phys. C* **6**, 3697 (1994). *Phys. Rev. B* **49**, 8454 (1994).
20. X. Blase, L. X. Benedict, E. L. Shirley, and S. G. Louie, *Phys. Rev. Lett.* **72**, 1878 (1994).
21. D. J. Klein, W. A. Seitz, and T. G. Schmalz, *J. Phys. Chem.* **97**, 1231 (1993).
22. K. Harigaya, *Phys. Rev. B* **45**, 12071 (1992).
23. D. H. Robertson, D. W. Brenner, and J. W. Mintmire, *Phys. Rev. B* **45**, 12592 (1992).
24. A. A. Lucas, P. H. Lambin, and R. E. Smalley, *J. Phys. Chem. Solids* **54**, 587 (1993).
25. J.-C. Charlier and J.-P. Michenaud, *Phys. Rev. Lett.* **70**, 1858 (1993).
26. G. Overney, W. Zhong, and D. Tománek, *Z. Phys. D* **27**, 93 (1993).
27. R. S. Ruoff, J. Tersoff, D. C. Lorents, S. Subramoney, and B. Chan, *Nature* **364**, 514 (1993).
28. J. Tersoff and R. S. Ruoff, *Phys. Rev. Lett.* **73**, 676 (1994).
29. M. Fujita, R. Saito, G. Dresselhaus, M. S. Dresselhaus, *Phys. Rev. B* **45**, 13834 (1992).
30. B. I. Dunlap, *Phys. Rev. B* **46**, 1933 (1992).
31. J. C. Slater and G. F. Koster, *Phys. Rev.* **94**, 1498 (1954).
32. M. L. Elert, J. W. Mintmire, and C. T. White, *J. Phys. (Paris), Colloq.* **44**, C3–451 (1983).
33. M. L. Elert, C. T. White, and J. W. Mintmire, *Mol. Cryst. Liq. Cryst.* **125**, 329 (1985). C. T. White, D. H. Robertson, and J. W. Mintmire, unpublished.
34. J. W. Mintmire and C. T. White, *Phys. Rev. Lett.* **50**, 101 (1983). *Phys. Rev. B* **28**, 3283 (1983).
35. J. W. Mintmire, In *Density Functional Methods in Chemistry* (Edited by J. Labanowski and J. Andzelm) p. 125. Springer-Verlag, New York (1991).
36. B. I. Dunlap, J. W. D. Connolly, and J. R. Sabin, *J.*

Chem. Phys. **71**, 3396 (1979). *J. Chem. Phys.* **71**, 4993 (1979).
37. J. W. Mintmire, *Int. J. Quantum Chem. Symp.* **13**, 163 (1979).
38. J. W. Mintmire and J. R. Sabin, *Chem. Phys.* **50**, 91 (1980).
39. J. W. Mintmire and B. I. Dunlap, *Phys. Rev. A* **25**, 88 (1982).
40. J. W. Mintmire, *Int. J. Quantum Chem. Symp.* **24**, 851 (1990).
41. J. W. Mintmire and J. R. Sabin, *Int. J. Quantum Chem. Symp.* **14**, 707 (1980).
42. J. W. Mintmire, J. R. Sabin, and S. B. Trickey, *Phys. Rev. B* **26**, 1743 (1982).
43. S. Iijima, *Mater. Sci. Eng.* **B19**, 172 (1993).
44. J.-Y. Yi and J. Bernholc, *Phys. Rev. B* **47**, 1703 (1993).
45. M. Endo and H. W. Kroto, *J. Phys. Chem.* **96**, 6941 (1992).
46. S. Iijima, P. M. Ajayan, and T. Ichihashi, *Phys. Rev. Lett.* **69**, 3100 (1992).

CARBON NANOTUBES WITH SINGLE-LAYER WALLS

CHING-HWA KIANG,[1,2] WILLIAM A. GODDARD III,[2] ROBERT BEYERS,[1]
and DONALD S. BETHUNE[1]
[1]IBM Research Division, Almaden Research Center, 650 Harry Road,
San Jose, California 95120-6099, U.S.A.
[2]Materials and Molecular Simulation Center, Beckman Institute, Division of Chemistry and Chemical Engineering, California Institute of Technology, Pasadena, California 91125, U.S.A.

(*Received* 1 *November* 1994; *accepted* 10 *February* 1995)

Abstract—Macroscopic quantities of single-layer carbon nanotubes have recently been synthesized by co-condensing atomic carbon and iron group or lanthanide metal vapors in an inert gas atmosphere. The nanotubes consist solely of carbon, sp^2-bonded as in graphene strips rolled to form closed cylinders. The structure of the nanotubes has been studied using high-resolution transmission electron microscopy. Iron group catalysts, such as Co, Fe, and Ni, produce single-layer nanotubes with diameters typically between 1 and 2 nm and lengths on the order of micrometers. Groups of shorter nanotubes with similar diameters can grow radially from the surfaces of lanthanide carbide nanoparticles that condense from the gas phase. If the elements S, Bi, or Pb (which by themselves do not catalyze nanotube production) are used together with Co, the yield of nanotubes is greatly increased and tubules with diameters as large as 6 nm are produced. Single-layer nanotubes are anticipated to have novel mechanical and electrical properties, including very high tensile strength and one-dimensional conductivity. Theoretical calculations indicate that the properties of single-layer tubes will depend sensitively on their detailed structure. Other novel structures, including metallic crystallites encapsulated in graphitic polyhedra, are produced under the conditions that lead to nanotube growth.

Key Words—Carbon, nanotubes, fiber, cobalt, catalysis, fullerenes, TEM.

1. INTRODUCTION

The discovery of carbon nanotubes by Iijima in 1991[1] created much excitement and stimulated extensive research into the properties of nanometer-scale cylindrical carbon networks. These multilayered nanotubes were found in the cathode tip deposits that form when a DC arc is sustained between the graphite electrodes of a fullerene generator. They are typically composed of 2 to 50 concentric cylindrical shells, with outer diameters typically a few tens of nm and lengths on the order of μm. Each shell has the structure of a rolled up graphene sheet—with the sp^2 carbons forming a hexagonal lattice. Theoretical studies of nanotubes have predicted that they will have unusual mechanical, electrical, and magnetic properties of fundamental scientific interest and possibly of technological importance. Potential applications for them as one-dimensional conductors, reinforcing fibers in super-strong carbon composite materials, and sorption material for gases such as hydrogen have been suggested. Much of the theoretical work has focussed on single-layer carbon tubules as model systems. Methods to experimentally synthesize single-layer nanotubes were first discovered in 1993, when two groups independently found ways to produce them in macroscopic quantities[2,3]. These methods both involved co-vaporizing carbon and a transition metal catalyst and produced single-layer nanotubes approximately 1 nm in diameter and up to several microns long. In one case, Iijima and Ichihashi produced single-layer nanotubes by vaporizing graphite and Fe in an Ar/CH_4 atmosphere. The tubes were found in the deposited soot[2]. Bethune *et al.*, on the other hand, vaporized Co and graphite under helium buffer gas, and found single-layer nanotubes in both the soot and in web-like material attached to the chamber walls[3].

2. SYNTHESIS OF SINGLE-LAYER CARBON NANOTUBES

In a typical experiment to produce single-layer nanotubes, an electric arc is used to vaporize a hollow graphite anode packed with a mixture of metal or metal compound and graphite powder. Two families of metals have been tried most extensively to date: transition metals such as Fe, Co, Ni, and Cu, and lanthanides, notably Gd, Nd, La, and Y. While these two metal groups both catalyze the formation of single-layer nanotubes, the results differ in significant ways. The iron group metals have been found to produce high yields of single-layer nanotubes in the gas phase, with length-to-diameter ratios as high as several thousand[2–11]. To date, no association between the nanotubes and metal-containing particles has been clearly demonstrated. In contrast, the tubes formed with lanthanide catalysts, such as Gd, Nd, and Y, are shorter and grow radially from the surface of 10–100 nm diameter particles of metal carbide[8,10,12–15], giving rise to what have been dubbed "sea urchin" particles[12]. These particles are generally found in the soot deposited on the chamber walls.

Some other results fall in between or outside these main groups. In the case of nickel, in addition to long, straight nanotubes in the soot, shorter single-layer tubes growing radially (urchin style) for *fcc*-Ni or NiC_3 particles in the rubbery collar that forms around the cathode have also been found[16,17]. Second despite the fact that copper reportedly does not catalyze single-layer tube growth in the gas phase[11], Lin *et al.* found that numerous short, single-layer tube structures form on the cathode tip when a Cu-containing anode is used[18]. Finally, the growth of single-layer tubules on a graphite substrate by pyrolyzing a hydrogen/benzene mixture in a gas-phase flow-reactor at 1000°C was recently reported[19]. That experimental result is unique among those described here, in that no metal atoms are involved. An overview of some of the experimental results on single-layer nanotubes is presented in Table 1.

In the arc-production of nanotubes, experiments to date have been carried out in generally similar fashion. An arc is typically run with a supply voltage of 20–30 V and a DC current of 50–200 A (depending on the electrode diameters, which range from 5–20 mm). Usually He buffer gas is used, at a pressure in the range 50–600 Torr and flowing at 0–15 ml/min. The anode is hollowed out and packed with a mixture of a metal and powdered graphite. In addition to pure iron group metals, mixtures of these metals[7,8,10] and metal compounds (oxides, carbides, and sulfides) [5] have been successfully used as source materials for the catalytic metals in nanotube synthesis. The ratio of metal to graphite is set to achieve the desired metal concentration, typically a few atomic percent.

Parameter studies have shown that single-layer nanotubes can be produced by the arc method under a wide range of conditions, with large variations in variables such as the buffer gas pressure (100–500 Torr), gas flow rate, and metal concentration in the anode[4,5]. These parameters are found to change the yield of nanotubes, but not the tube characteristics such as the diameter distribution. In contrast, the presence of certain additional elements, although they do not catalyze nanotube growth when used alone, can greatly modify both the amount of nanotube production and the characteristics of the nanotubes. For example, sulfur[5], bismuth, and lead[6] all increase the yield and produce single-layer nanotubes with diameters as large as 6 nm, much larger than those formed with Co catalyst alone. Sulfur also appears to promote the encapsulation of Co-containing crystallites into graphitic polyhedra. Lambert *et al.* recently reported that a platinum/cobalt 1:1 mixture also significantly increased the yield of nanotubes[11], even though Pt alone also has not produced nanotubes[9,11].

Different product morphologies have been found in different regions of the arc-reactor chamber. On the cold walls, a primary soot is deposited. In normal fullerene production, this soot has a crumbly, flocculent character. However, under conditions that lead to abundant nanotube growth, the density of tubes in this soot can be high enough to give it a rubbery character, allowing it to be peeled off the chamber wall in sheets. This rubbery character may be caused by either chemical or physical cross-linking between the nanotubes and the soot. We note that fullerenes in amounts comparable to those obtained without a metal present can be extracted from the rubbery soots using the normal solvents. Second, a hard slag is deposited on the cathode tip. This cathode tip contains high densities of multilayer nanotubes and polyhedral particles[20,21]. The fact that the transition-metal-catalyzed single-layer nanotubes are distributed throughout the soot and rarely in the slag deposit leads to the conclusion

Table 1. Results on single-layer nanotubes*

Elements	D (nm)	d_{mp} (nm)	Crystallites	Ref.	Note
Fe	0.7–1.6	0.80, 1.05	Fe_3C	[2]	a
Fe	0.6–1.3	0.7–0.8	—	[8]	b
Co	0.9–2.4	1.3, 1.5	*fcc*-Co	[3,5]	c
Co	1–2	1.2–1.3	Co wrapped with graphene layers	[4]	
Co	0.6–1.8	—	—	[8]	b
Ni	1.2–1.5	—	*fcc*-Ni in polyhedra in cathode deposit	[16]	
Ni	0.6–1.3	0.7–0.8	—	[8]	b
Fe+Ni	0.9–3.1	1.7	—	[8]	d
Fe+Ni	>0.6	1.3–1.8	—	[8]	b
Co+Ni	>0.6	1.2–1.3	—		
Co+S	1.0–6.0	1.3, 1.5	Co(C) in polyhedra and *fcc*-Co	[5]	e
Co+Bi	0.8–5	1.2, 1.5	—	[6]	c
Co+Pb	0.7–4	—	—	[6]	c
Co+Pt	≈2	—	CoPt	[11]	
Y	1.1–1.7	—	YC_2 in polyhedra	[8]	f
Cu	1–4	—	Cu in polyhedra	[18]	g
no metal	>2	—	graphite substrate	[19]	h

*Unless specified, samples were from soot deposited on the chamber wall and the buffer gas was helium. Elements are those incorporated in the graphite anode, D is the nanotube diameter range, D_{mp} is the most abundant nanotube diameter, and Crystallites refers to metal-containing particles generated by the arc process and found in the soot.

[a]Statistics from 60 tubes; [b]from 40 tubes; [c]from over 100 tubes; [d]from 70 tubes; [e]from over 300 tubes; [f]Nanotubes grew radially out of YC_2 crystals, 15–100 nm long; [g]Nanotubes found in the cathode deposit, 3–40 nm long; [h]Nanotubes formed by C_6H_6 pyrolysis on graphite substrate.

that, in that case, the nanotubes form in the gas phase. Third, a soft rubbery blanket or collar builds up around the cathode when iron group metals are used. This material has been found to contain graphitic polyhedral particles, metals or metal carbides encapsulated in polyhedral particles, string of beads structures[8,16], and single-layer nanotubes[4,8,16]. Finally, with some catalysts, notably Co, mixtures of Co with Fe, Ni, Pt, S, Bi, and Pb, and Fe/Ni mixtures, web-like materials form inside the chamber when the arc is running [3,6,8,11].

Figure 1 is a scanning electron micrograph (SEM) of a sample of the web-like material obtained by vaporizing Co and C under 400 Torr He[3]. The threads and bundles of carbon nanotubes, often partly clad with a layer of non-crystalline carbon and fullerenes. The threads connect rounded particles with typical diameters of a few tens of nanometers. Figure 2a is a transmission electron microscope (TEM) image of the nanotube bundles. The sample was prepared by sonicating some soot in ethanol for a few minutes and placing a drop of the liquid on a holey-carbon-coated copper TEM grid. Shown in the micrograph is a region where a gap in the holey-carbon film was formed after the soot was put on the grid. Bundled and individual nanotubes bridge the ≈ 0.25 μm gap. The soot particles themselves consist of non-crystalline carbon containing dark spots that have been identified by Energy Dispersive X-ray Spectroscopy (EDS) and electron diffraction to be *fcc*-Co particles[3]. Figure 2b, taken at higher magnification, shows a region where a high density of tubes span a gap in the soot. The process of pulling the tubules out of the soot mass has aligned them to a striking degree. A high resolution TEM (HRTEM) image of a group of nanotubes (Fig. 2c) demonstrates their tendency to aggregate into bundles. The aggregation process is presumably driven by van der Waals attraction, which has been shown experimentally to give rise to significant forces between adjacent multilayer nanotubes[22], and is predicted to give rise to ordering of bundled single-layer nanotubes into crystalline arrays[23]. A micrograph showing a bundle of Ni-catalyzed nanotubes end-on lends some support to this idea[17].

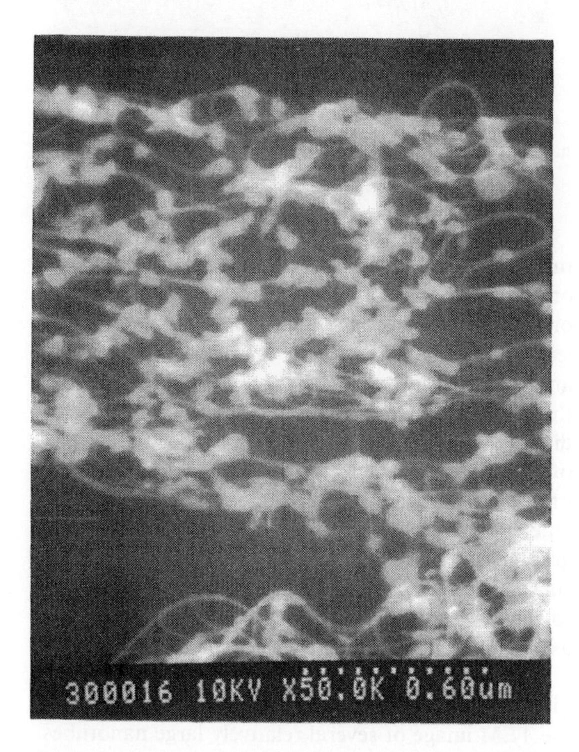

Fig. 1. Scanning electron micrograph of the soot taken from the chamber wall; the threads are nanotube bundles.

The metals Y and Gd have been found to facilitate the growth of urchin particles—consisting of bundles of relatively short single-layer nanotubes rooted on and extending radially outward from metal carbide particles, such as Gd_xC_y[12,15] and YC_2[8,14,17]. These tubules have diameters of 1 to 2 nm, similar to the longer tubules produced by the iron group metals, but have lengths of only 10 to 100 nm. These structures have been found in the primary soot, suggesting that they form in the gas phase. However, the similar structures reported for the case of Ni were found in the rubbery blanket surrounding the cathode[16,17]. In that case, the nanotubes radiated from metal particles that were identified by electron diffraction to be crystalline *fcc*-Ni or Ni_3C. The Ni-containing particles were typically encased in several graphitic carbon layers, and the free ends of the short, radial single-walled tubes were generally observed to be capped.

In the experiment of Lin *et al.*, Cu was used in the anode and single-layer nanotubes formed in the center region of the cathode deposit[18]. These tubes had lengths of a few tens of nanometers and diameters of 1–4 nm. Unlike tubes produced using transition metals or lanthanides, these nanotubes usually had irregular shapes, with diameters varying along the tube axes. From this Lin *et al.* infer that the nanotube structures contain relatively high densities of pentagonal and heptagonal defects. The tubes were not found to be associated with Cu-containing particles. Copper crystallites loosely wrapped in graphitic carbon were occasionally found in the cathode deposit.

Recently, a non-arc method leading to single-layer nanotube production was reported. Endo *et al.* demonstrated that sections of single-layer nanotubes form at early times when a benzene/hydrogen mixture is pyrolyzed at 1000°C over a graphite substrate[19]. In this work, primary nanotubes quite similar to arc-produced carbon nanotubes form, in some cases with only single-layer graphene walls and diameters as small as 2–3 nm. At later times, these primary pyrolytic carbon nanotubes (or PCNTs) accrete additional amorphous pyrolytic carbon and grow into fibers with μm diameters and cm lengths. High-temperature annealing can then be used to increase the crystallinity of the fibers. The process to make PCNTs is distinguished from that used to make vapor-grown carbon fibers (VCGCF)[24,25] by the fact that VGCF is produced by thermally decomposing hydrocarbon vapor in the presence of a transition metal catalyst.

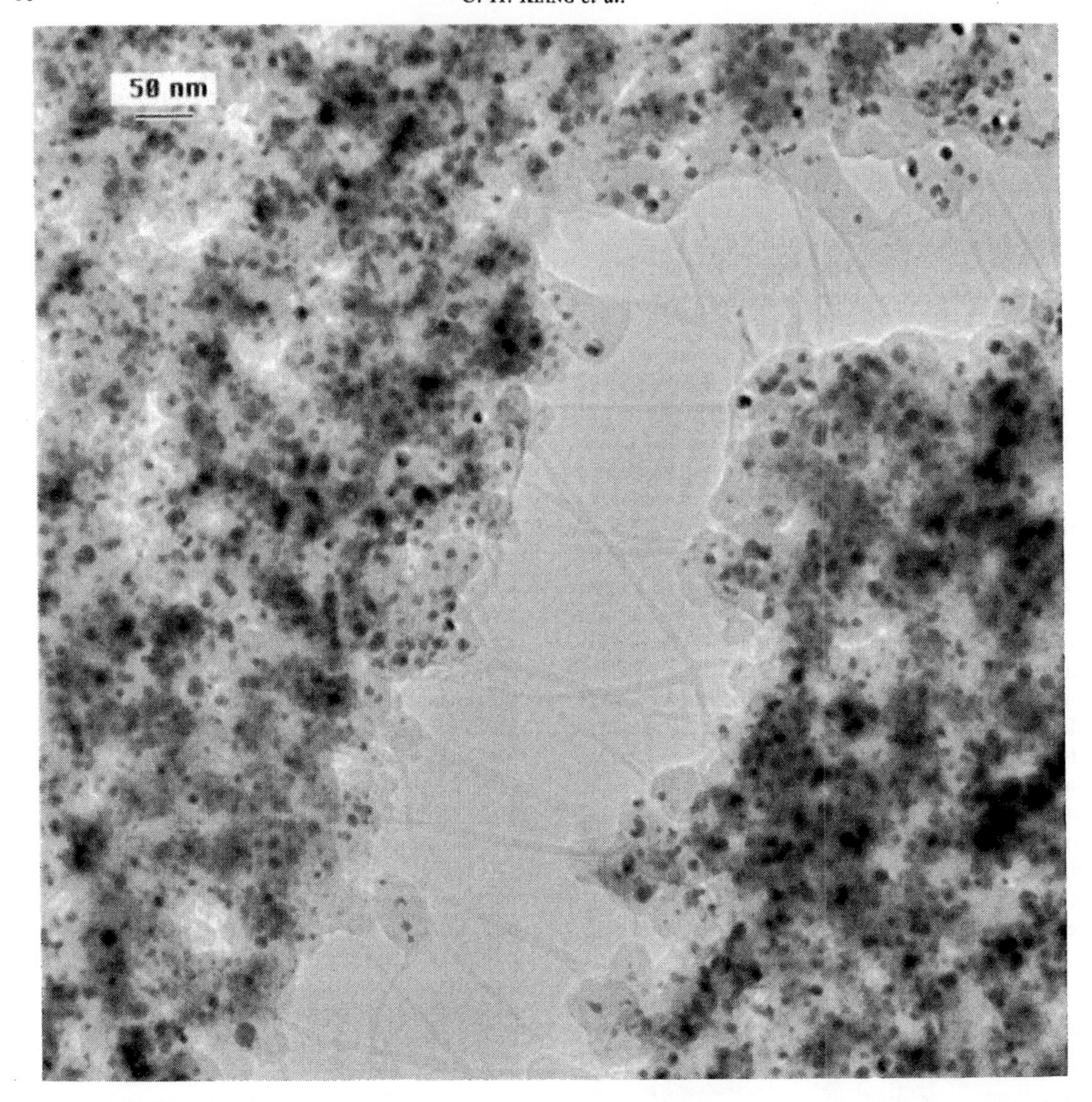

Fig. 2a. Bundles and individual single-layer carbon nanotubes bridge across a gap in a carbon film.

3. STRUCTURE OF SINGLE-LAYER CARBON NANOTUBES

The structure of an ideal straight, infinitely long, single-layer nanotube can be specified by only two parameters: its diameter, D, and its helicity angle, α, which take on discrete values with small increments[2,26]. These atomic scale structural parameters can in principle be determined from selected area diffraction patterns taken from a single tubule. Although for multilayer tubes numerous electron diffraction studies confirming their graphitic structure have been published, the very weak scattering from nm diameter single-layer tubes and their susceptibility to damage by a 100–200 keV electron beam make it very difficult to make such measurements. Iijima was able to show by diffraction that a single-layer tube indeed had a cylindrical graphene sheet structure[2]. Saito reported a similar conclusion based on diffraction from a bundle of several single-layer tubes[17].

On the molecular scale, single-layer carbon nanotubes can be viewed either as one-dimensional crystals or as all-carbon semi-flexible polymers. Alternatively, one can think of capped nanotubes as extended fullerenes[27]. For example, one can take a C_{60} molecule and add a belt of carbon to form a C_{70}. By repeating the process, one can make a long tubule of 0.7 nm diameter with zero helicity[26]. Likewise, joining belts of 75 edge-sharing benzene rings generates a nanotube of about 6 nm diameter. Nanotubes typically have diameters smaller than multilayer tubes. TEM micrographs show that single-layer tubules with diameters smaller than 2 nm are quite flexible, and often are seen to be bent, as in Fig. 3a. Bends with radii of curvature as small as ten nm can be observed.

Tubes with diameters larger than 2 nm usually exhibit defects, kinks, and twists. This is illustrated in the TEM image of several relatively large nanotubes shown in Fig. 3b. The diameter of the tubes seems to vary slightly along the tube axis due to radial defor-

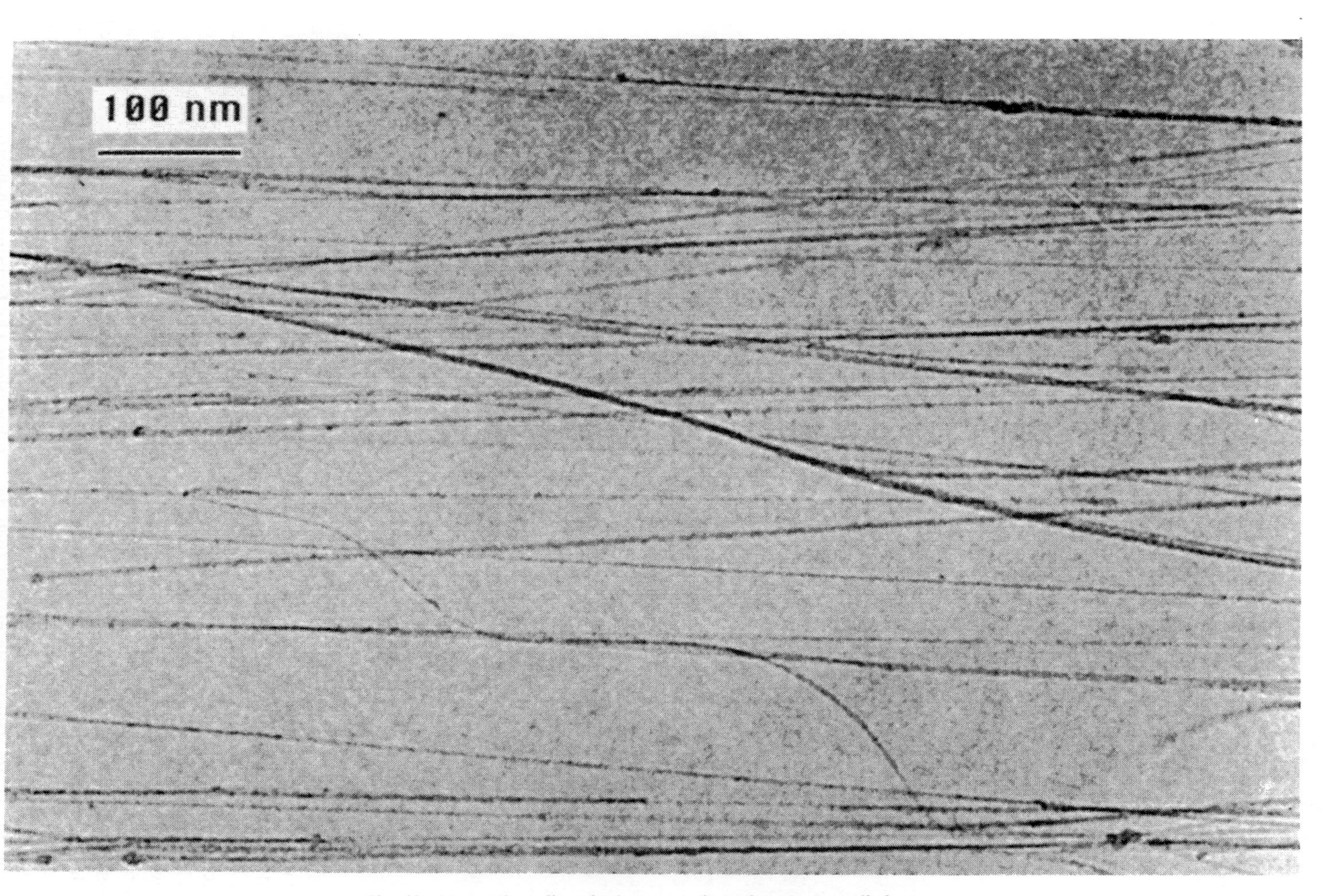

Fig. 2b. Nanotubes aligned when a portion of soot was pulled apart.

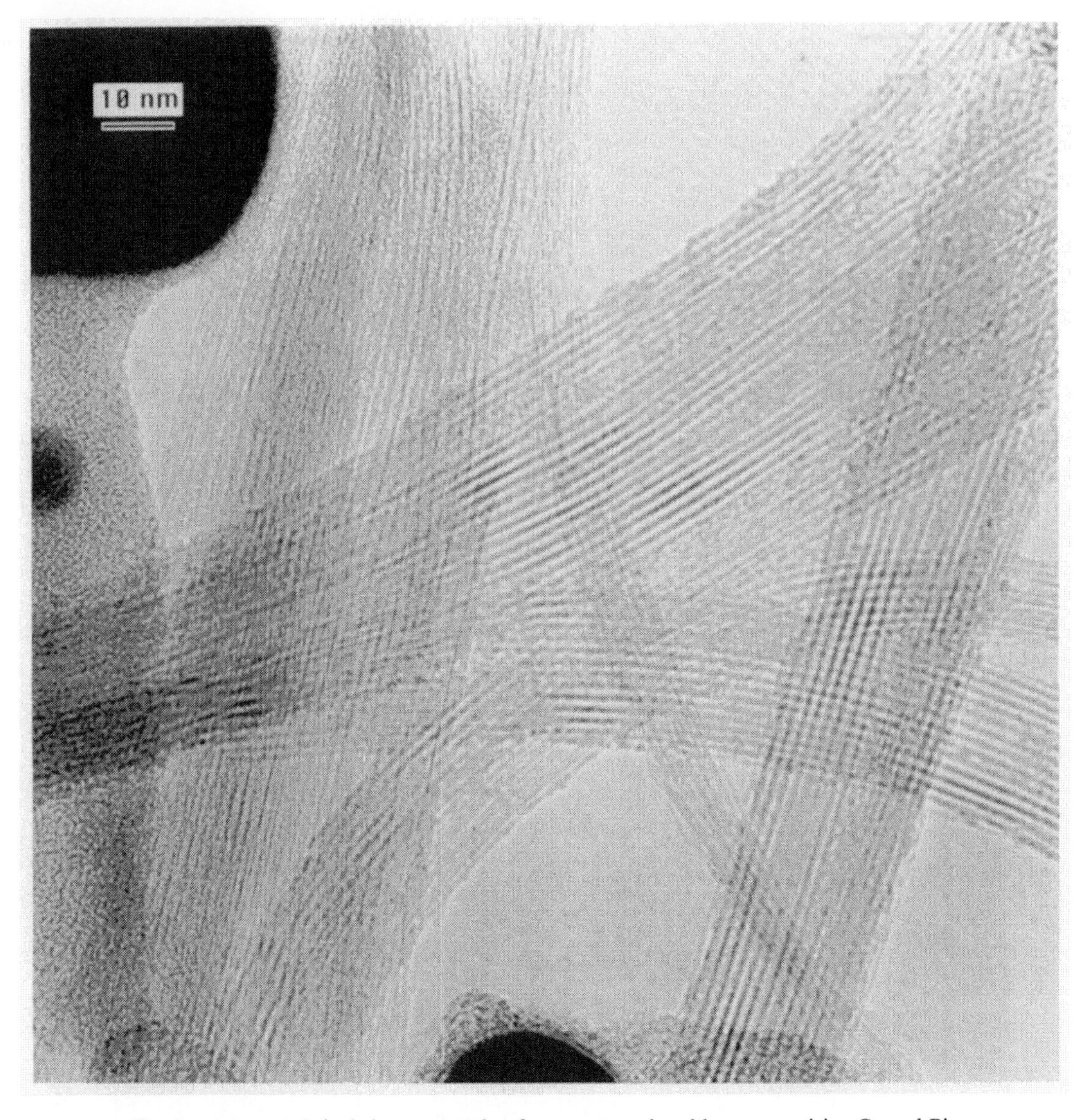

Fig. 2c. Aggregated single-layer nanotubes from soot produced by co-vaporizing Co and Bi.

mation. Classical mechanical calculations show that the tubes with diameters greater than 2 nm will deform radially when packed into a crystal[23], as indicated by TEM images presented by Ruoff *et al.*[22]. The stability of nanotubes is predicted to be lowered by the defects[28]. This may account for the observation that smaller tubes often appear to be more perfect because, with their higher degree of intrinsic strain, smaller tubes may not survive if defective. The most likely defects involve the occurrence of 5- and 7-fold rings. These defects have discrete, specific effects on the tube morphology, and can give changes in tube diameter, bends at specific angles, or tube closure, for example[29,30,31]. By deliberately placing such defects in specific locations, it would be possible in principle to create various branches and joints, and thus to connect nanotubes together into elaborate 3-D networks[32,33].

Despite a wealth of theoretical work on the electronic structure [26,34–41], and vibrational properties [38,42,43] of single-layer nanotubes, very little characterization beyond TEM microscopy and diffraction has been possible to date, due to the difficulty in separating them from the myriad of other carbon structures and metal particles produced by the relatively primitive synthetic methods so far employed. Recently, it was reported that the Raman spectrum of a sample containing Co-catalyzed nanotubes showed striking features that could be correlated with theoretical predictions of the vibrational properties of single-layer tubules[44]. Kuzuo *et al.*[45] were able to use transmission electron energy loss spectroscopy to study the electronic structure of a bundle of single-layer nanotubes selected by focusing the electron beam to a 100-nm diameter circular area. Features of the spectra obtained were shifted and broadened compared to the corresponding features for graphite and multilayer nanotubes. These changes were tentatively interpreted to be effects of the strong curvature of the nanotube wall.

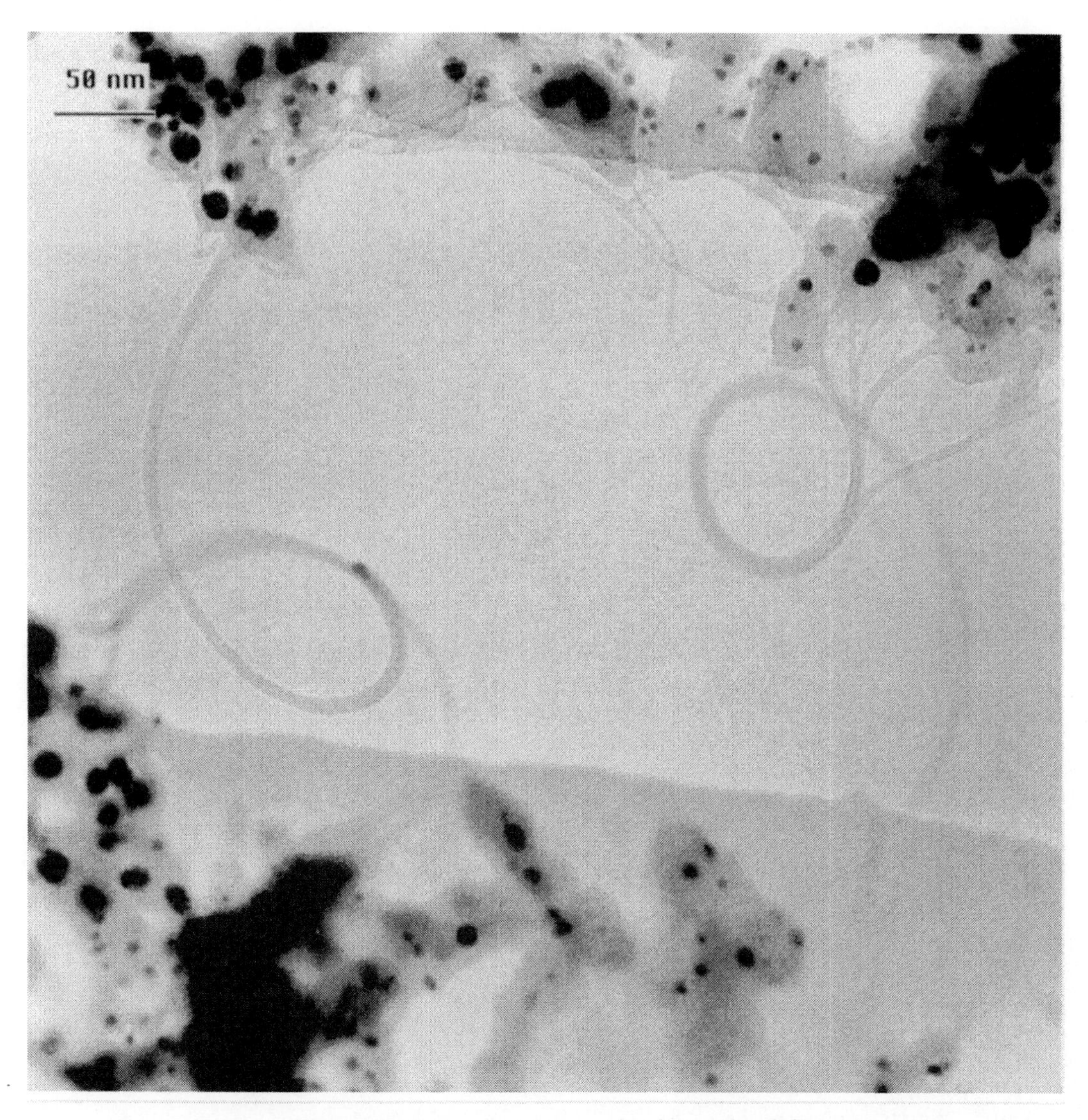

Fig. 3a. Small-diameter tubes are often bent and curled.

4. THE METAL PARTICLES

The encapsulated ferromagnetic particles produced by this process may eventually be of technological interest, for example, in the field of magnetic storage media. Some work characterizing the magnetic properties of the encapsulated Co particles produced by arc co-evaporation with carbon has been recently reported[46]. The phase and composition of the metal-containing particles may also provide information on the growth conditions in the reactor. The temporal and spatial profiles of temperature, metal and carbon densities, and reaction rates all affect the growth of both these particles and the nanotubes. The composition of the metal-containing particles in the soot deposited at regions away from the electrode is not the same as for those found in the cathode deposit. For iron group metals, pure metallic particle as well as cementite phase (Fe_3C, Co_3C, and Ni_3C) exist in the outer surface of the cathode deposit[16]. These particles appear spherical and are wrapped with layers of graphene sheet with no gaps. The low-temperature phases, α-Fe and α-*hcp* Co, form abundantly, whereas the high-temperature phases, β-Fe and β-*fcc* Co, comprise less than 10% of the metal particle. In contrast, the metal particles found in the soot on the chamber wall contain mostly high-temperature phases, such as Fe_3C[2], *fcc*-Co[3,47], and *fcc*-Ni[16], and not all of the particles are wrapped in graphitic layers. These findings show that as the particles move away from the arc their temperature is rapidly quenched. The relatively fast time scale for reaction that this implies may be crucial for the growth of single-layer carbon nanotubes and, in particular, it may preclude the growth of additional layers of carbon on the single-layer tubules.

The presence of sulfur is found to enhance the formation of graphitic carbon shells around cobalt-containing particles, so that cobalt or cobalt carbide particles encapsulated in graphitic polyhedra are found throughout the soot along with the single-layer nano-

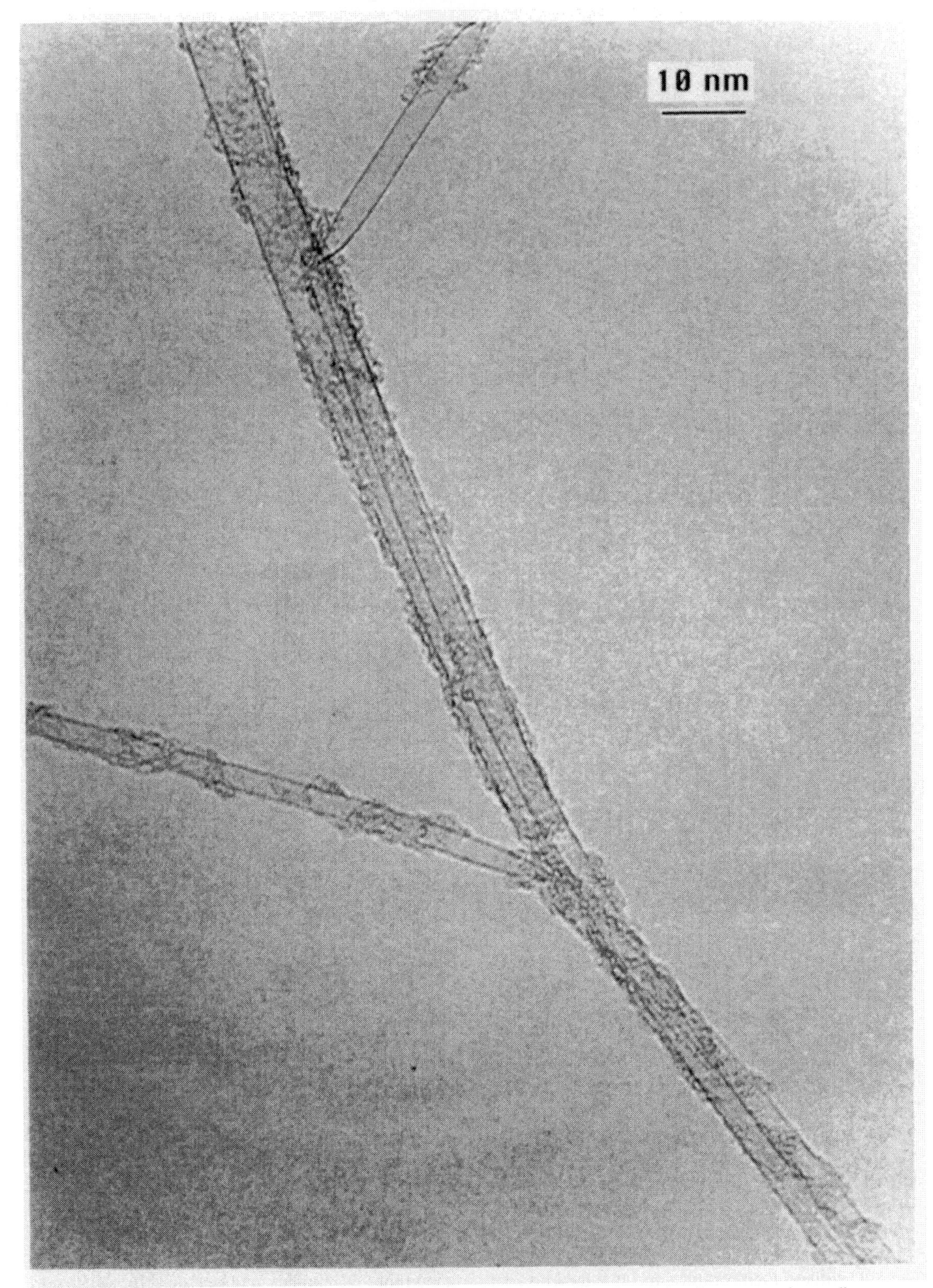

Fig. 3b. Large-diameter tubes produced with Co and S present; the tubes shown have approximate diameters of 5.7, 3.1, and 2.6 nm.

tubes. Figure 4 is a high-resolution image of some encapsulated Co particles, which have structures reminiscent of those observed for LaC_2 and YC_2 particles found in cathode deposits[16,48–50]. Crystallites encapsulated in graphitic polyhedra constitute about 30% of the total Co-containing particles. The role of sulfur in the formation of these filled polyhedra is not clear. Sulfur is known to assist the graphitization of

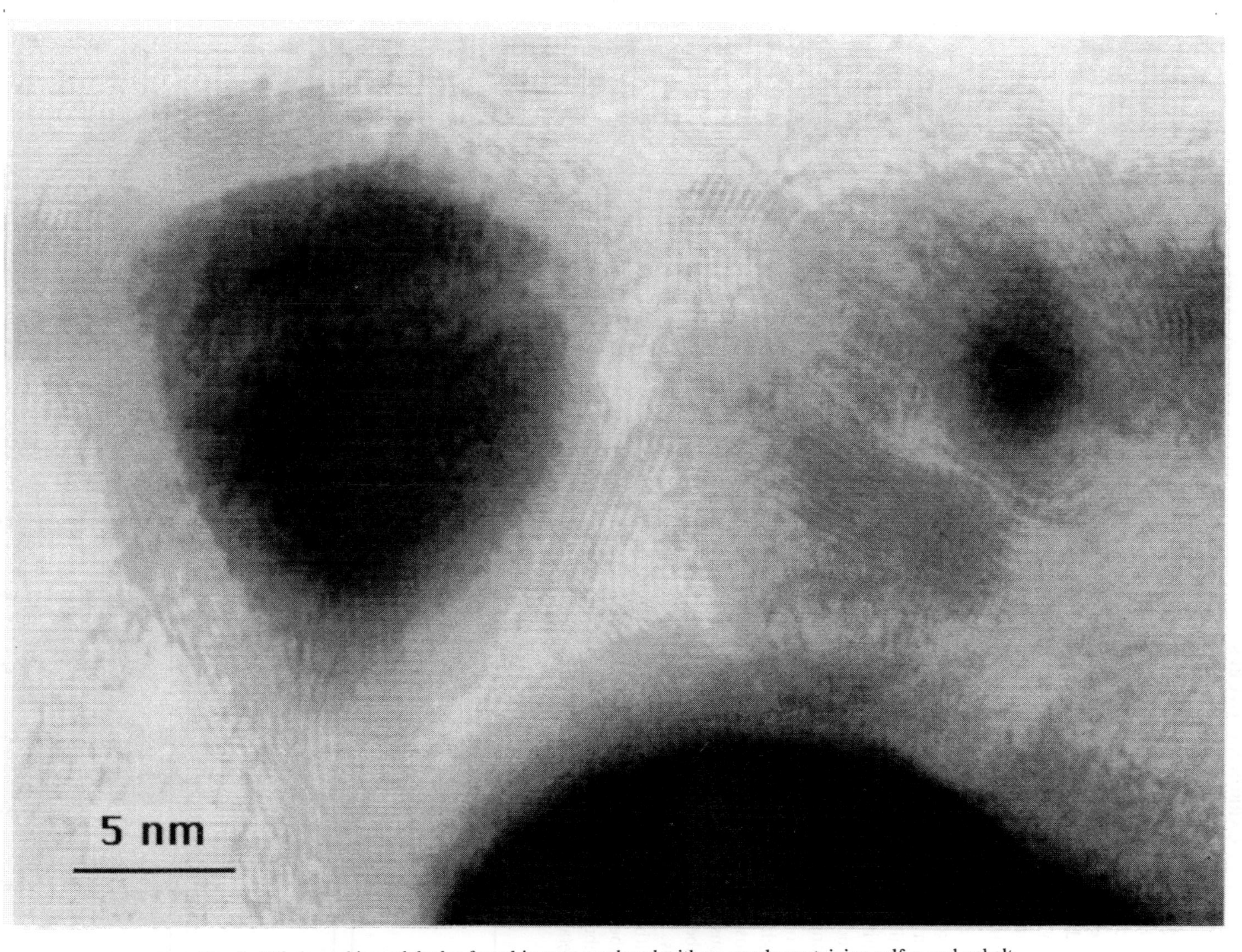

Fig. 4. Filled graphite polyhedra found in soot produced with an anode containing sulfur and cobalt.

vapor-grown carbon fibers, but the detailed process is not yet understood[51].

5. GROWTH OF SINGLE-LAYER CARBON NANOTUBES

There remains a major puzzle as to what controls the growth of these nanotubes, and how it precludes the formation of additional layers. The reaction conditions in the electric arc environment used for nanotube production to date are not ideal for mechanistic studies, since the plasma composition near the arc is very complex and inhomogeneous, making individual variables impossible to isolate. So far, we can only examine the product composition to extract clues about the growth mechanism. One feature that can be analyzed is the diameter distribution of single-layer carbon nanotubes formed. Table 1 summarizes the data available. This should be considered to be only a qualitative description, given the non-systematic sampling procedures, statistical uncertainties, and wide variations in the growth conditions used in various laboratories. The nanotube diameters were obtained from high-resolution TEM images. At a gross level, the most interesting aspect of the accumulated data is the consistency of the production of 1–2 nm diameter tubes by the various metals and combinations of metals. The exceptional cases are the combinations of Co with S, Pb, or Bi, which produce considerably large tubes. Even in those cases, the main peak in the distribution occurs between 1 and 2 nm. Figure 5 presents detailed histograms of the abundance of different diameter nanotubes produced with Fe, Co, and Co/S, adapted from earlier reports[2,5]. In comparing the diameter distributions produced using Co and Co/S, there is striking correlation of both the overall maxima and even the fine structures exhibited by the distributions (Figs. 5b and 5c, respectively). For the cases where large diameter tubes (> 3 nm) are produced by adding S, Pb, or Bi to the cobalt, the tubes are still exclusively single-layered. We observed only one double-layer nanotube out of over a thousand tubes observed. This suggests that nucleation of additional layers must be strongly inhibited. The stability of nanotubes as a function of their diameter has been investigated theoretically via classical mechanical calculations[52,53]. The tube energies vary smoothly with diameter, with larger diameter tubes more stable than smaller ones. The narrow diameter distributions and occurrence of only single-layer tubes both point to the importance of growth kinetics rather than energetic considerations in the nanotube formation process.

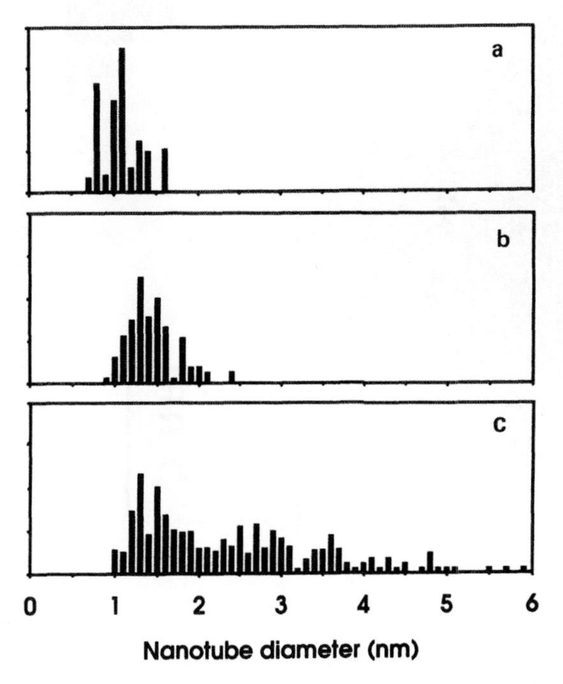

Fig. 5. Diameter distributions of nanotubes produced via different methods: (a) Fe catalyst in an Ar/CH_4 atmosphere, adapted from Ref. 2; (b) Co catalyst in He atmosphere, adapted from Ref. 5; (c) Co catalyst with sulfur, about 4 at.% each, adapted from Ref. 5.

S, Pb, and Bi affect the Co-catalyzed production of single-layer nanotubes by greatly increasing the yield and the maximum size of the nanotubes. The formation of web-like material in the chamber is very dramatically enhanced. As noted above, these elements do not produce nanotubes without a transition metal present. How these effects arise and whether they involve a common mechanism is not known. In the production of VGCF, sulfur was found to be an effective scavenger for removing blocking groups at graphite basal edges[51]. The added elements may assist the transport of carbon species crucial for the growth of nanotubes in the vicinity of the arc. Or they could act as co-catalysts interacting with Co to catalyze the reaction, or as promoters helping to stabilize the reactants, or simply as scavengers that remove blocking groups that inhibit tube growth.

Growth models for vapor-grown carbon fibers (VGCF) have been proposed[24,25]. Those fibers, produced by hydrocarbon decomposition at temperatures around 1200°C, are believed to grow from the surface of a catalyst particle, with carbon deposited on the particle by decomposition of the hydrocarbon migrating by diffusion through the particle, or over its surface, to the site where the fiber is growing. The fiber size is comparable to the size of the catalytic particle, but can thicken if additional pyrolytic carbon is deposited onto the fiber surface. It is tempting to think that single-wall nanotubes may also grow at the surfaces of transition metal particles, but particles much smaller than those typical in VGCF production. To date however, the long single-layer nanotubes found in the soot have not been definitely associated with metal particles. Thus, how the metal exerts its catalytic influence, and even what the catalytic species are, remain open questions. The urchin particles produced by lanthanide or Ni catalysts do show an association between the single-layer nanotubes and catalyst particles. In this case, the particles are 10 to 100 times larger than the tube diameters. In the case of single-layer tubes produced by Cu in the cathode deposit, growth occurs under extreme conditions of temperature and carbon density. The nanotubes produced also have very different characteristics. Therefore, we expect that their formation mechanism will be quite different. It is possible that instead of growth occurring at a metal particle interface, as has been proposed for VGCF, urchin particles[8,17], and long single-walled nanotubes[5], a mechanism more akin to those proposed for the growth of multilayer nanotubes on the cathode tip in an all-carbon environment may be involved[21,26,29,54]. In that case, it has been suggested that growth occurs at the free end of the nanotube, which protrudes out into the carbon plasma.

Some features of the arc process are known and are relevant to growth models for single-layer nanotubes. Earlier isotope labelling analyses of fullerene formation shows that fullerenes formed in the arc are built up from atomic carbon[55–57]. Also, the production of nanotubes does not seem to depend on whether metal oxide or pure metal is used in the graphite anode. These results imply that both the metal and the carbon are completely atomized under the arc conditions, and that both the catalytic species and nanotubes must be built up from atoms or atomic ions. This fact, together with the consistency of the diameters of the single-layer nanotubes produced in the gas phase by transition metal catalysts, suggests a model where small catalytic particles rapidly assemble in a region of high carbon density. Single-layer tubules nucleate and grow very rapidly on these particles as soon as they reach a critical size, leading to the relatively narrow diameter distributions observed. If nucleation of additional layers is slow, the rapid drops in temperature and carbon density as the tubes move away from the arc could turn off the growth processes before multilayers can form.

6. FUTURE DIRECTIONS

Experimental research on single-layer nanotubes is still in a very early stage. Understanding the growth mechanism of these nanotubes remains a great challenge for scientists working in this area. Not even the nature of the catalytically active species has been established to date. Developing better controlled systems than standard arc reactors will be necessary to allow the dependence of tube growth on the various important parameters to be isolated. The temperature and the carbon and metal densities are obvious examples of such parameters. In the arc plasma, they are highly coupled and extremely inhomogeneous. Knowledge of the growth mechanism will possibly allow us to optimize the fabrication scheme and the characteristics of the nanotubes.

A second key problem is to devise means to separate the tubes from the soot and metal particles, either chemically or mechanically. This is an essential step towards manipulation and thorough characterization of these materials. Recently, it was reported that a significant fraction of the metal particles could be removed from the sample by vacuum annealing it at high (1600°C) temperature[11]. The amorphous carbon soot particles, however, are difficult to remove, and the oxidative approach used with some success to isolate multilayer tubes[58] seems to destroy the single-layer tubes. Measurement of the mechanical, optical, electrical, and magnetic properties requires a clean sample to interpret the data unambiguously. Tests of the dependence of electrical conductivity and mechanical strength on the tube diameter should be done, and may soon be feasible with the availability of nanotubes with a wide range of diameters.

The unique properties of single-layer carbon nanotubes will continue to inspire scientists in diverse fields to explore their properties and possible applications. Defect-free nanotubes are predicted to have very high tensile strength. A theoretical calculation of the elastic constant for single-layer nanotubes[52] gives a result consistent with a simple estimate based on the c_{11} elastic constant of graphite (c_{11} = 1.06 TPa). Using this constant, one finds a force constant of 350 Nt/(m of edge) for graphene sheet. Multiplying this value by the circumference of a 1.3 nm diameter nanotube gives an elastic constant of 1.45×10^{-6} Nt for such a tube. Macroscopically, a bundle of these tubes 25 μms in diameter would support a 1-kg weight at a strain of 3%. In comparison, a steel wire of that diameter would break under a load of about 50 gm. When nanotubes are assembled into crystalline bundles, the elastic modulus does not decrease linearly with tube diameter but, rather, it remains constant for tube diameters between 3 and 6 nm, suggesting the strength-to-weight ratio of the crystal increases as the tube diameter increases [23]. The anisotropy inherent in the extreme aspect ratios characteristic of these fibers is an important feature, particularly if they can be aligned. *Ab initio* calculations show that these nanotubes could be one-dimensional electric conductors or semiconductors, depending on their diameter and helicity[36,39,59]. Other applications of carbon nanotubes have been proposed in areas that range widely, from physics, chemistry, and materials to biology. Examples, such as hydrogen storage media, nanowire templates, scanning tunneling microscopy tips, catalyst supports, seeds for growing carbon fibers, batteries materials, reinforcing fillings in concrete, *etc.* provide ample motivation for further research on this pseudo-one-dimensional form of carbon.

Acknowledgement—This research is partially supported by the NSF (ASC-9217368) and by the Materials and Molecular Simulation Center. We thank J. Vazquez for help with the SEM imaging of nanotubes, G. Gorman and R. Savoy for X-ray analysis, and M. S. de Vries for mass spectrometry.

REFERENCES

1. S. Iijima, *Nature* **354**, 56 (1991).
2. S. Iijima and T. Ichihashi, *Nature* **363**, 603 (1993).
3. D. S. Bethune, C. H. Kiang, M. S. de Vries, G. Gorman, R. Savoy, J. Vazquez and R. Beyers, *Nature* **363**, 605 (1993).
4. P. M. Ajayan, J. M Lambert, P. Bernier, L. Barbedette,

C. Colliex, and J. M. Planeix, *Chem. Phys. Lett.* **215**, 509 (1993).
5. C.-H. Kiang, W. A. Goddard III, R. Beyers, J. R. Salem, and D. S. Bethune, *J. Phys. Chem.* **98**, 6612 (1994).
6. C.-H. Kiang, P. H. M. van Loosdrecht, R. Beyers, J. R. Salem, D. S. Bethune, W. A. Goddard III, H. C. Dorn, P. Burbank, and S. Stevenson, *Proceedings of the Seventh International Symposium on Small Particles and Inorganic Clusters*, Kobe, Japan, *Surf. Rev. Lett.* (in press).
7. S. Seraphin and D. Zhou, *Appl. Phys. Lett.* **64**, 2087 (1994).
8. S. Seraphin, *J. Electrochem. Soc.* (USA) **142**, 290 (1995).
9. S. Seraphin, D. Zhou, J. Jiao, M. S. Minke, S. Wang, T. Yadav, and J. C. Withers, *Chem. Phys. Lett.* **217**, 191 (1994).
10. Y. Tapesh, J. C. Withers, S. Seraphin, S. Wang, and D. Zhou, In *Novel Forms of Carbon*, MRS Proc. 349, pp. 275–281. Mat. Res. Soc., Pittsburgh (1994).
11. J. M. Lambert, P. M. Ajayan, P. Bernier, J. M. Planeix, V. Brotons, B. Cog, and J. Castaing, *Chem. Phys. Lett.* **226**, 364 (1994).
12. S. Subramoney, R. S. Ruoff, D. C. Lorents, and R. Malhotra, *Nature* **366**, 637 (1994).
13. S. Wang and D. Zhou, *Chem. Phys. Lett.* **225**, 165 (1994).
14. D. Zhou, S. Seraphin, and S. Wang, *Appl. Phys. Lett.* **65**, 1593 (1994).
15. S. Subramoney, P. van Kavelaar, R. S. Ruoff, D. C. Lorents, and A. J. Kazmer, In *Recent Advances in the Chemistry and Physics of Fullerenes and Related Materials* (Edited by K. M. Kadish and R. S. Ruoff). The Electrochemical Society, Pennington, NJ (1994).
16. Y. Saito, T. Yoshikawa, M. Okuda, N. Fujimoto, K. Sumiyama, K. Suzuki, A. Kasuya, and Y. Nishina, *J. Phys. Chem. Solids* **54**, 1849 (1993).
17. Y. Saito, M. Okuda, N. Fujimoto, T. Yoshikawa, M. Tomita, and T. Hayashi, *Jpn. J. Appl. Phys. 2 Lett.* (Japan) **33**, L526 (1994).
18. X. Lin, X. K. Wang, V. P. Dravid, R. P. H. Chang, and J. B. Ketterson, *Appl. Phys. Lett.* **64**, 181 (1994).
19. M. Endo, K. Takeuchi, S. Igarashi, K. Kobori, M. Shiraishi, and H. W. Kroto, *J. Phys. Chem. Solids* **54**, 1841 (1993).
20. D. Ugarte, *Nature* **359**, 707 (1992).
21. Y. Saito, T. Yoshikawa, M. Inagaki, M. Tomita, and T. Hayashi, *Chem. Phys. Lett.* **204**, 277 (1993).
22. R. S. Ruoff, J. Tersoff, D. C. Lorents, S. Subramoney, and B. Chan, *Nature* **364**, 514 (1994).
23. J. Tersoff and R. S. Ruoff, *Phys. Rev. Lett.* **73**, 676 (1994).
24. M. Endo, *Chemtech* **18**, 568 (1988).
25. N. M. Rodriguez, *J. Mater. Res.* **8**, 3233 (1993).
26. R. Saito, M. Fujita, G. Dresselhaus, and M. S. Dresselhaus, *Matls. Sci. Eng.* **B19**, 185 (1993).
27. P. W. Fowler, *J. Phys. Chem. Solids* (UK) **54**, 1825 (1993).
28. R. Saito, G. Dresselhaus, and M. S. Dresselhaus, *Chem. Phys. Lett.* **195**, 537 (1992).
29. S. Iijima, *Matls. Sci. Engin.* **B19**, 172 (1993).
30. B. I. Dunlap, *Phys. Rev. B* **49**, 5643 (1994).
31. B. I. Dunlap, *Phys. Rev. B* **50**, 8134 (1994).
32. B. I. Dunlap, *Phys. Rev. B* **46**, 1933 (1992).
33. L. A. Chernozatonskii, *Physics Lett. A* (Netherlands) **172**, 173 (1992).
34. J. W. Mintmire, D. H. Robertson, and C. T. White, *J. Phys. Chem. Solids* **54**, 1835 (1993).
35. R. Saito, M. Fujita, G. Dresselhaus, and M. S. Dresselhaus, *Phys. Rev. B***46**, 1804 (1992).
36. R. A. Jishi, M. S. Dresselhaus, and G. Dresselhaus, *Phys. Rev. B***48**, 11385 (1993).
37. M. S. Dresselhaus, G. Dresselhaus, and R. Saito, *Solid State Commun.* (USA) **84**, 201 (1992).
38. R. A. Jishi, D. Inomata, K. Nakao, M. S. Dresselhaus, and G. Dresselhaus, *J. Phys. Soc. Jpn.* (Japan) **63**, 2252 (1994).
39. N. Hamada, S. I. Sawada, and A. Oshiyama, *Phys. Rev. Lett.* **68**, 1579 (1992).
40. C. J. Mei and V. H. Smith Jr., *Physica C* **213**, 157 (1993).
41. M. Springborg and S. Satpathy, *Chem. Phys. Lett.* **225**, 454 (1994).
42. R. A. Jishi and M. S. Dresselhaus, *Phys. Rev. B***45**, 11305 (1992).
43. R. A Jishi, L. Venkataraman, M. S. Dresselhaus, and G. Dresselhaus, *Chem. Phys. Lett.* **209**, 77 (1993).
44. J. M. Holden, P. Zhou, X-x Bi, P. C. Eklund, S. Bandow, R. A. Jishi, K. Das Chowdhury, G. Dresselhaus, and M. S. Dresselhaus, *Chem. Phys. Lett.* **220**, 186 (1994).
45. R. Kuzuo, M. Terauchi, M. Tanaka, and Y. Saito, *Jpn. J. Appl. Phys. 2, Lett.* **33**, L1316 (1994).
46. M. E. McHenry, S. A. Majetich, J. O. Artman, M. DeGraef, and S. W. Staley, *Phys. Rev. B* **49**, 11358 (1994).
47. S. A. Majetich, J. O. Artman, M. E. McHenry, N. T. Nuhfer, and S. W Staley, *Phys. Rev. B* **48**, 16845 (1993).
48. R. S. Ruoff, D. C. Lorents, B. Chan, R. Malhotra, and S. Subramoney, *Science* **259**, 346 (1993).
49. Y. Saito, T. Yoshikawa, M. Okuda, M. Ohkohchi, M. Inagaki, Y. Ando, A. Kasuya, and Y. Nishina, *Chem. Phys. Lett.* **209**, 72 (1993).
50. S. Seraphin, D. Zhou, J. Jiao, J. C. Withers and R. Loutfy, *Appl. Phys. Lett.* (USA) **63**, 2073 (1993).
51. H. Katsuki, K. Matsunaga, M. Egashira, and S. Kawasumi, *Carbon* **19**, 148 (1981).
52. D. H. Robertson, D. W. Brenner, and J. W. Mintmire, *Phys. Rev. B* **45**, 12592 (1992).
53. A. A. Lucas, Ph. Lambin, and R. E. Smalley, *J. Phys. Chem. Solids* **54**, 587 (1993).
54. M. Endo and H. W. Kroto, *J. Phys. Chem.* **96**, 6941 (1992).
55. G. Meijer and D. S. Bethune, *J. Chem. Phys.* **93**, 7800 (1990).
56. R. D. Johnson, C. S. Yannoni, J. R. Salem, and D. S. Bethune, In *Clusters and Cluster-Assembled Materials*, MRS Proc. 26 (Edited by R. S. Averbach, J. Bernholc, and D. L. Nelson), pp. 715–720. Materials Research Society, Pittsburgh, PA (1991).
57. T. W. Ebbesen, J. Tabuchi, and K. Tanagaki, *Chem. Phys. Lett.* **191**, 336 (1992).
58. T. W. Ebbesen, P. M Ajayan, H. Hiura, and K. Tanigaki, *Nature* **367**, 519 (1994).
59. R. Saito, M. Fujita, G. Dresselhaus, and M. S. Dresselhaus, *Appl. Phys. Lett.* **60**, 2204 (1992).

CARBON NANOTUBES: I. GEOMETRICAL CONSIDERATIONS

R. SETTON
Centre de Recherche sur la Matière Divisée, CNRS, 1 B rue de la Férollerie, F 45071 Orléans Cedex 2, France

(*Received* 22 *August* 1994; *accepted* 15 *September* 1994)

Abstract—The geometrical conditions pertaining to closure, helicity, and interlayer distance between successive layers with circular cross-sections in carbon tubules (nanotubes) have been examined. Both the intralayer length of the C—C bonds and the interlayer distance between successive layers must vary with the radius of the layers. The division into groups of the sheets in nanotubes is found to be due to the reciprocal interaction of the interlayer distance variations and of the conditions required to maintain constancy of the pitch angle.

Key Words—Carbon nanotubes, pitch angle, helix angle, interlayer distance, carbon-carbon intralayer distance.

1. INTRODUCTION

Carbon tubules (or nanotubes) are a new form of elemental carbon recently isolated from the soot obtained during the arc-discharge synthesis of fullerenes[1]. High-resolution electron micrographs do not favor a scroll-like helical structure, but rather concentric tubular shells of 2 to 50 layers, with tips closed by curved, cone-shaped, or even polygonal caps. Later work[2] has shown the possibility of obtaining single-shell seamless nanotubes.

Recently, the structure of some helical carbon nanotubes was examined[3], and the present work is an attempt at completing the geometrical approach to the structural problems encountered in the case of tubules with circular cross-sections. However, most of the conclusions in the present work are applicable to nanotubes with polygonal cross-sections that have also been shown to exist.

2. THEORY

Leaving aside the complications engendered by the presence of end-caps and considering, therefore, only the cylindrical part of the tubules, one must differentiate between two types of helicity.

2.1 *Scroll helicity*

The presence of scroll helicity replaces a set of concentric cylinders by a single sheet rolled upon itself (Fig. 1). Assuming that the distance between the successive rolls of the scroll is constant, its cross-section can be conveniently represented in polar coordinates by the Archimedean spiral:

$$\rho = a_0 + (\Theta/2\pi)d, \tag{1}$$

where a_0 is the initial radius of the innermost fold and d is the (constant) distance between successive folds. The sign of Θ determines the helicity of the scroll, counterclockwise ($\Theta > 0$) or clockwise ($\Theta < 0$).

The consequence of the presence of scroll helicity in a tubule is expected to be that any increase (decrease) of the intralayer C—C distance G will increase (decrease) the local length of the spiral, but not necessarily the mean interlayer distance, since the scroll can easily adapt its radius of curvature to minimize, if necessary, any energetic strain due to a stress in the local bond lengths.

2.2 *Screw helicity*

The second type of helicity can affect both scrolls and individual cylinders. It will be present when neither the **a** nor the **b** unit vector of the basic two-dimensional graphite lattice of the unfolded scroll or cylinder is at an angle of 0° or at a multiple of 30° with the direction of the cylinder axis (Fig. 2). The situation is complicated by the fact that whenever screw helicity affects a seamless cylindrical layer, any change in the value of G must necessarily affect the radius r of the cylinder. It is therefore highly likely that the interlayer distance d between two successive cylinders is not strictly constant in any multilayered tubule with a circular cross-section, since the adverse effect of the elastic strain on the sp^2 orbitals and the resulting distribution of charges—which undoubtedly affect the intralayer C—C distance—progressively decreases as the circumference and the radius of curvature of the cylinders increase. As justified later, one is therefore led to introduce a parameter δd or δr to characterize the small variations of r with respect to a hypothetical nominal value likely to be encountered even if, strictly speaking, the variation δG should also be taken into account.

2.3 *Symmetric non-helical tubules and cylinders*

In the absence of both types of helicity, the non-helical or *symmetric tubule* consists of a set of cylin-

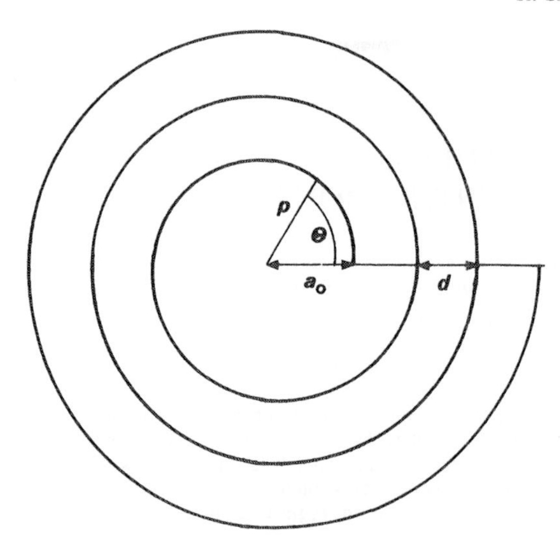

Fig. 1. Approximate cross section of a tubule with scroll helicity. The distance d between successive rolls of the spiral is assumed to be constant, and a_0 is the initial radius of the innermost fold.

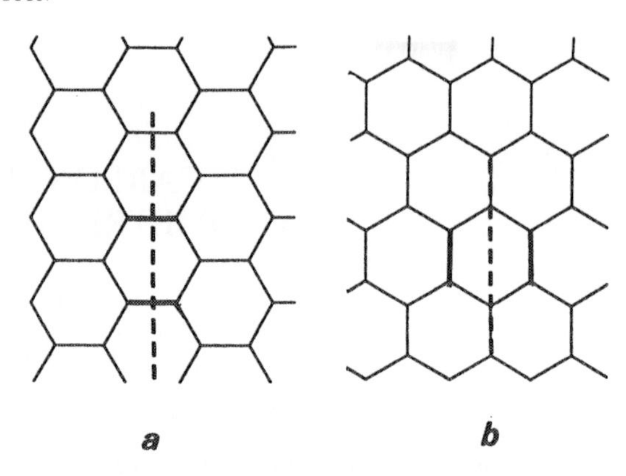

Fig. 3. The two possible orientations of the hexagons in non-helical sheets, relative to the cylinder axis (broken line); (a) case I; (b) case II (see text).

drical sheets which, for simplicity, will be assumed to have a common axis and a circular cross-section. The hexagons of the unrolled cylinders all have one of their three sets of parallel opposite sides respectively perpendicular (case I) or parallel (case II) to the common axis of the cylinders (Fig. 3).

The orientation of the hexagons with respect to the tube axis of the unrolled ith cylinder as well as the radius r_i of the cylinder can conveniently be described in terms of two orthogonal vectors[4], $\mathbf{P_i}$ and $\mathbf{Q_i}$ (Fig. 4), respectively integral multiples of the unit vectors $\mathbf{x}$ and $\mathbf{y}$ with moduli $|\mathbf{x}| = (3/2)G$ and $|\mathbf{y}| = (\sqrt{3}/2)G$. For the ith layer, we can write:

$$\mathbf{R_i} = \mathbf{P_i} + \mathbf{Q_i} \qquad (2)$$

or, for the scalars R_i, P_i, Q_i,

$$|\mathbf{R_i}| = [(P_i 3G/2)^2 + (Q_i \sqrt{3}G/2)^2]^{1/2} \qquad (3)$$

where P_i, Q_i are integers of identical parity, and

$$R_i = 2\pi r_i. \qquad (4)$$

The pitch angle α_i characteristic of this orientation of the hexagons is:

$$\alpha_i = \arctan(Q_i/\sqrt{3}P_i), \qquad (5)$$

but one can also distinguish an *apparent angle of pitch* β_i between the helical rows of unbroken hexagons and the plane perpendicular to the cylinder axis, with

$$\beta_i = 30° - \alpha_i. \qquad (6)$$

Although the limits of α are:

$$0° \leq \alpha < 60°, \qquad (7)$$

this angle can be limited to 30°, since every (P_i, Q_i) doublet is associated with a corresponding doublet $(P_i, -Q_i)$, symmetric of (P_i, Q_i) with respect to $\alpha = 0°$, yielding the same value of the radius (Fig. 5) and characterizing the stereoisomer of the chiral ith cylindrical sheet. The doublet (P_i, Q_i) is also equivalent to the doublet (P_j, Q_j), with $P_j = (P_i + Q_i)/2$ and $Q_j = (3P_i - 2Q_i)/2$, with the pitch angle $\alpha_j = 60° - \alpha_i$, which also yields the same value of r_i.

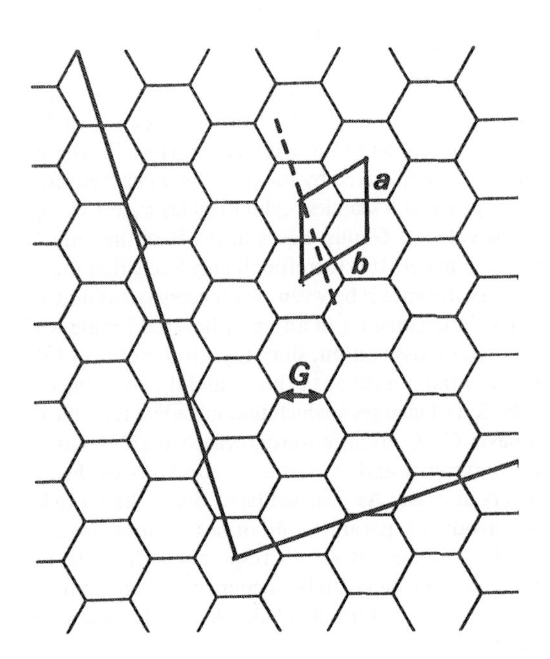

Fig. 2. A portion of an unrolled cylinder with screw helicity. The broken line is parallel to the cylinder axis, and the cylindrical sheet has been cut along a generatrix (full line parallel to the cylinder axis). **a** and **b** are the unit vectors of the two-dimensional carbon layer in sp^2 hybridization.

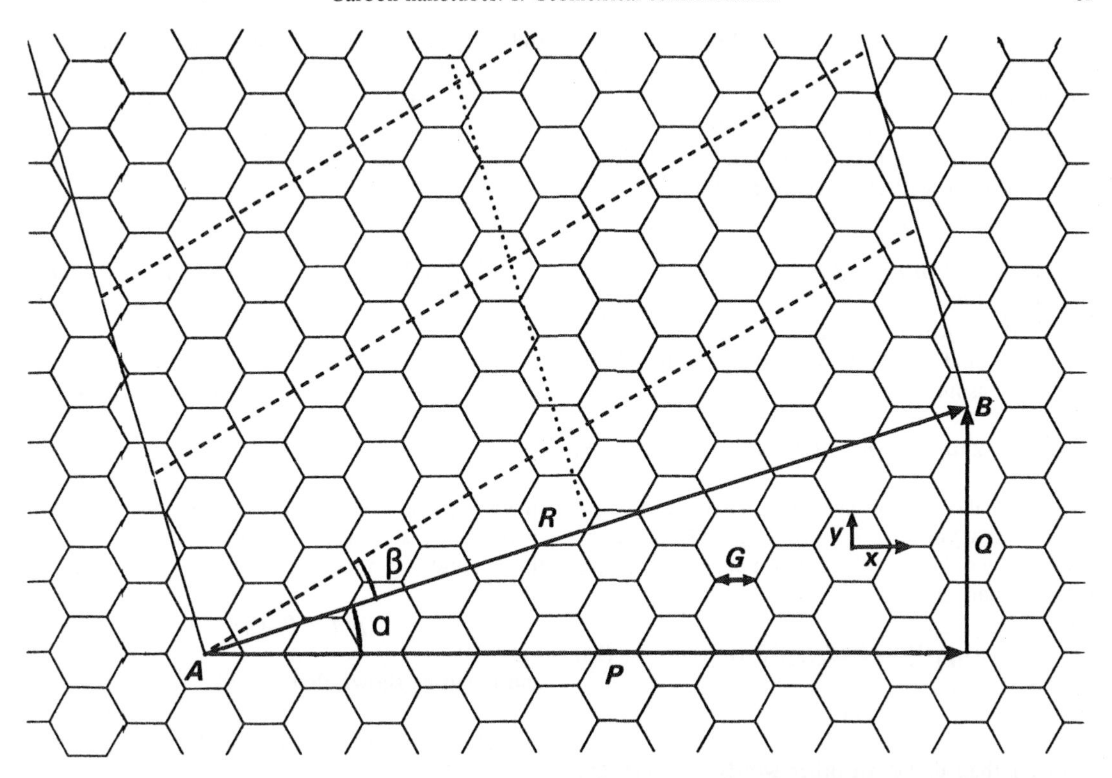

Fig. 4. Screw helicity: the system of (P, Q) coordinates used to describe the orientation of the two-dimensional sp^2 carbon layer in an unrolled cylindrical sheet whose edges are shown by the slanted unlabelled full lines. Closure of the cylinder is obtained by rolling the sheet around the direction of the cylinder axis given by the dotted line and superimposing hexagons A and B. The slanted dashed lines correspond to a continuous line of unbroken hexagons of the cylinder, and indicate the apparent angle of pitch β.

It is therefore possible to impose the limits:

$$1 \leq P_i \leq \text{int}[2\pi r_i/(3G/2)]$$

$$(\text{int} \equiv \text{integral part of}) \quad (8)$$

$$0 \leq Q_i \leq P_i \quad (9)$$

in which $Q_i = 0$ and $Q_i = P_i$ correspond to cases I and II respectively, and for which eqn (4) becomes:

$$2\pi r_i(\text{I}) = P_i 3G/2 \quad (10)$$

and

$$2\pi r_i(\text{II}) = P_i\sqrt{3}G. \quad (11)$$

Referring now to the symmetric tubule, the interlayer distance $d_{i,i+n}$ between two layers of radii r_i and r_{i+n} is:

$$d_{i,i+n} = r_{i+n} - r_i \quad (12)$$

and, for two consecutive layers, $d_{i,i+1} \approx 0.339$ nm[5]. For the configuration of hexagons in case I, the increase in length of the circumference of the $(i + 1)$th layer with respect to the circumference of the ith layer must be equal to an integral and even multiple of $|\mathbf{x}|$; hence, assuming that $d_{i,i+1}$ is practically constant and writing d for $d_{i,i+1}$,

$$2\pi d = n_{i,i+1}(\text{I})3G/2 \quad (13)$$

$$n_{i,i+1}(\text{I}) = 10 \quad (14)$$

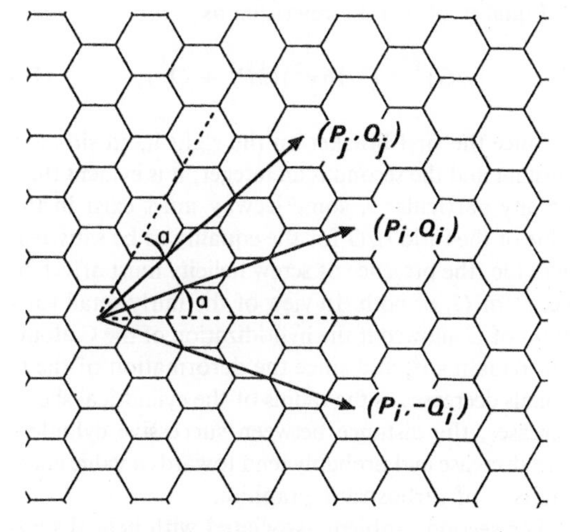

Fig. 5. Screw helicity: the three vectors corresponding to $|\mathbf{R}| = G\sqrt{26}$, with $P_i = 5$, $Q_i = 3$ and $P_j = 4$, $Q_j = 6$ (see text). The vector $(\mathbf{P_i}, -\mathbf{Q_i})$ characterizes the optical isomer of $(\mathbf{P_i}, \mathbf{Q_i})$.

to better than 1×10^{-5} if $d = 0.339$ nm and $G = 0.142$ nm. This is well illustrated by the sequence formed from the values $r_1 = 0.678$ nm, $G = 0.142$ nm, $d = 0.339$ nm, and $\delta r/r = \pm 1.5\%$, which give more than 30 consecutive non-helical ($\alpha = 0°$) symmetric cylinders, with characteristics $r_i\,(\text{nm}) \approx 0.678 + (i-1)d$ and $P_i = (i+1)10$.

2.4 *Symmetrical helical cylinders with $\alpha = 30°$*

For the configuration of hexagons in case II, $n_{i,i+1}(\text{II})$ should now be an integral and even multiple of $|\mathbf{y}|$ or, using eqn (11),

$$2\pi d_{i,i+1} = n_{i,i+1}(\text{II})\sqrt{3}G. \qquad (15)$$

This equality is not even approximately satisfied for small integral values of $n_{i,i+1}(\text{II})$. There is, however, a possibility offered by the distance $d_{i,i+3}$, that is, for 1.02 nm, namely:

$$n_{i,i+3}(\text{II}) = 2\pi d_{i,i+3}/\sqrt{3}G \qquad (16)$$

$$\approx 26$$

to better than 0.2%. In other words, if a (P_i, Q_i) doublet yields $\alpha = 30°$, which is the pitch angle corresponding to case II, the chances are high that this angle will be repeated for the $(i+3)$th cylinder. A good example of this is given by the series described by the values $r_1 = 0.71$ nm, $G = 0.1421$ nm, $d = 0.339$ nm, and $\delta d/d = 2\%$, which yield $\alpha = 30°$ for $i = 1, 4, 10, 19, 22, 27, 30, 33, 36, 39, \ldots$, with most differences between successive values of i either equal to 3 or to a multiple of 3, and the values $P_i = Q_i = 18, 44, 96, 174, 200, 243, 269, 295, 321, 347, \ldots$, which show the expected differences.

2.5 *Symmetric helical tubes and cylinder with α different from 0° or 30°*

Equality (4) can be rewritten as:

$$(r_i/G)^2 = (3/16\pi^2)(3P_i^2 + Q_i^2). \qquad (17)$$

Since the first bracket on the right-hand side is a constant and the second is an integer, it is evident that, for any particular i, some leeway must exist in the value of the ratio r_i/G for the equality to be satisfied. Here too, the presence of screw helicity must affect either r_i, or G, or both. In view of the fairly small variations of G allowed if the hybridization of the C atoms is to remain sp^2, and since the deformation of the C orbitals decreases as the radius of the cylindrical sheets increases, the distance between successive cylinders must decrease and probably tend towards a value characteristic of turbostratic graphite.

The second problem associated with helical symmetric cylindrical sheets addresses the possibility of finding the same helix angle or pitch in two neighbouring sheets. Apart from cases similar to the one described above (case I, section 2.3) in which it is possible to find long series of successive cylindrical sheets with $\alpha = 0°$, it is clear that one should have, for two successive cylinders with identical helicity:

$$Q_{i+1}/P_{i+1} = Q_i/P_i \qquad (18)$$

with

$$(r_{i+1}/G)^2 = (3/16\pi^2)(3P_{i+1}^2 + Q_{i+1}^2) \qquad (19)$$

or

$$[(r_i + d)/G]^2 = (3/16\pi^2)[3(P_i + p)^2 + (Q_i + q)^2] \qquad (20)$$

in which p and q are, respectively, integral increments of P_i and Q_i. For the pitch angles of the two cylinders to be equal, we must have:

$$p/q = P_i/Q_i, \qquad (21)$$

and it can be shown that

$$p = (2\pi d/G)\{2/[9 + 3(Q_i/P_i)^2]^{1/2}\}. \qquad (22)$$

In view of the remarks in § 2.3, it is legitimate to substitute the value 15 for the round bracket in the r.h.s. of eqn (22). Taking then into account the fact that the maximum and minimum possible values of p are obtained when $q = 0$ and $q = p$, respectively, eqn (22) yields:

$$8.66 \leq p \leq 10. \qquad (23)$$

The only doublets consistent with this inequality, with the necessity of identical parity for p and q, and with values of r_1 close to 0.34 nm, are given in Table 1 (from which the doublet (10,0) can be excluded since it is characteristic of a symmetrical, non-helical sheet). Hence, if $\delta r = 0$, *the necessary conditions for two successive helical cylindrical sheets to have strictly identical pitch angles are*:

$$q/p = Q_i/P_i, \qquad (24)$$

$$q{:}p = 7{:}9, \text{ or } 2{:}10, \text{ or } 4{:}10, \qquad (25)$$

since all other values would give either $d < 0.334$ nm (which is already less than d of graphite and, there-

Table 1. (p,q) increments for obtaining identical successive pitch angles

(p,q)	Interlayer distance d (nm)	Pitch angle $\alpha°$
(9,7)	0.334	24.2
(10,0)	0.339	0
(10,2)	0.341	6.6
(10,4)	0.348	13.0

fore, highly unlikely) or $d > 0.348$ nm (i.e., at least 4% above the value of graphite and also unlikely). If both conditions are met, the pitch angle is 24.18°, or 6.59°, or 13.00°. If only condition (25) above is operative, the successive values of α will be *nearly* but not *strictly* equal, and will be different from these three values. Two examples illustrate these points:

(a) the unbroken series of 21 cylinders with $\alpha = 6.59°$, $r_1 = 0.6825$ nm, $G = 0.142$ nm, $d = 0.341$ nm, $\delta d/d = \pm 1.5\%$, with $Q_i:P_i = 2:10$, $P_1 = 20$, $P_{21} = 220$, and $q:p = 2:10$ in all cases;

(b) the five successive cylindrical sheets with $r_1 = 10.22$ nm, $G = 0.142$ nm, $d = 0.34$ nm, $\delta d/d = \pm 1.5\%$, and the characteristics shown in Table 2. A "single-helix angle of about 3°" was observed in the selected area electron diffraction pattern of a four-sheet symmetric helical tube[3], but the small variations of α shown in Table 2 would obviously not have been visible in the diffractogram shown in the reference.

2.6 *Sheet groups*

It has also been stated[3] that "the helix angle (of a multiple-sheet helical tube) changes about every three to five sheets" so that the gross structure of the tube is constituted by successive groups of cylindrical sheets, with the helix angles increasing by a constant value from each group to the next. In light of the arguments developed in the preceding paragraphs, and except for the specific cases $\alpha = 0°$, 6.59°, 13.00°, 24.18°, the pitch angle of the sheets within a group is more likely to be nearly constant rather than strictly constant. This is illustrated by the values in Table 3 showing the characteristics of the 28 sheets in a tube whose inner and outer radii are, respectively, equal to 0.475 and 9.66 nm, calculated on the basis of $r_1 = 0.476$ nm, $d = 0.34$ nm, $G = 0.142$ nm, $\delta d/d = \pm 0.8\%$. Since α is not strictly constant within each group, it is more appropriate to consider a mean value $\langle\alpha\rangle$ of the pitch angle for each group, as shown in Table 4. Only in the first group of sheets (i = 1 to 2) is α truly constant ($\alpha = 0°$). For all the other groups, the increments (p,q) of the (P_i, Q_i) doublets are constant and equal to (10,2). In fact, it is because these increments are identical that the group pitch angles are nearly constant when the values of P_i and Q_i are sufficiently large, which is not yet the case in the second group.

The above series also illustrates an experimentally established fact[3]: Except for the change of mean values between groups 2 and 3, and 5 and 6, the successive values of $\langle\alpha\rangle$ increase by a nearly constant amount, namely, about 3 to 4°. The question then arises as to the reason causing the modification of α between two successive groups, such as the one occurring between $i = 11$ and $i = 12$. If (P_{12}, Q_{12}) had been (124,18) instead of (123,31), r_{12} would have been 4.218 nm with $\alpha = 4.78°$ (i.e., within $\delta d/d = \pm 1.0\%$ instead of $\pm 0.8\%$ imposed for the values in Table 3). Hence, during the actual synthesis of the tube, *the end of each group must be determined by the tolerance in the G values accepted by the sp² orbitals compatible with the need to close the cylindrical sheet.*

Although the physical growth mechanism of the cylindrical sheets is not yet fully known, the fact that

Table 2. Computed characteristics of a five-sheet symmetric tubule with a nearly constant pitch angle

i	r_i (nm)	(P_i, Q_i)	$\alpha°$
1	10.22	(301,29)	3.18
2	10.56	(311,31)	3.29
3	10.90	(321,31)	3.19
4	11.24	(331,33)	3.29
5	11.58	(341,33)	3.20

Table 3. Computed characteristics of a 28-sheet tubule showing the division into groups of sheets

i	(P_i, Q_i)	r_i (nm)	$\alpha°$
1	(14,0)	0.475	0
2	(24,0)	0.814	0
3	(34,2)	1.153	1.94
4	(44,4)	1.494	3.00
5	(54,6)	1.834	3.67
6	(64,8)	2.175	4.13
7	(74,10)	2.516	4.46
8	(84,12)	2.857	4.72
9	(94,12)	3.195	4.22
10	(104,14)	3.566	4.44
11	(114,16)	3.877	4.63
12	(123,31)	4.214	8.28
13	(133,33)	4.555	8.15
14	(143,35)	4.896	8.04
15	(153,37)	5.237	7.95
16	(163,39)	5.578	7.87
17	(171,61)	5.919	11.64
18	(181,63)	6.259	11.36
19	(191,65)	6.599	11.12
20	(202,56)	6.935	9.09
21	(212,58)	7.276	8.98
22	(222,60)	7.617	8.87
23	(232,62)	7.958	8.77
24	(239,91)	8.296	12.40
25	(249,93)	8.635	12.17
26	(259,95)	8.975	11.96
27	(269,97)	9.315	11.76
28	(279,99)	9.655	11.56

Table 4. Group characteristics of the 28-sheet nanotube described in Table 3

Group	Range of i	p,q	$\langle\alpha°\rangle$
1	1 to 2	10,0	0
2	3 to 8	10,2	3.65
3	9 to 11	10,2	4.43
4	12 to 16	10,2	8.06
5	17 to 19	10,2	11.37
6	20 to 23	10,2	8.93
7	24 to 28	10,2	11.97

successive sheets have nearly constant helicity suggests the following mechanism: once the ith (topmost) sheet is formed, C_n species in the gas phase ($n = 3,1,2\ldots$, in order of decreasing abundance[6]) condense onto the outer surface of the tube, probably at a distance slightly larger than 0.34 nm[7], in registry with the underlying substrate and in sp^2 hybridization, since this configuration is the one that best minimizes the energy of the system[8]. The ith sheet thus constitutes the template for the construction of the $(i+1)$th sheet[9], and the helicity of the as-yet-unclosed cylinder during its formation is the same as that of the underlying substrate. This process continues until the C atoms on either side of the cut form the C—C bonds that close the cylinder. If the geometric characteristics of these bonds are too different from 120° and $G \approx 0.142$ nm, the pitch angle will change to the immediately higher or lower value, thus ensuring closure. Once the cylinder is closed, and even if closure only involved a single row of hexagons, the following rows of the cylindrical sheet will necessarily have the same pitch angle, at least as long as the sheet is growing over an existing substrate.

2.7 *Smallest inner tube diameter*

Liu and Cowley[3] have reported the observation of innermost cylinders with diameters as small as 0.7 or 1.3 nm. The first value is only slightly larger than 0.68 nm, the well known diameter of the C_{60} molecule[10], while the second is in fair agreement with 1.36 nm, the calculated diameter C_{240}. As shown in Table 5, there are 5 (P_1, Q_1) doublets which, with $G = 0.142$ nm, give $r_1 = 0.35$ nm $\pm 3\%$, and 14 (P_1, Q_1) doublets (not all are given in Table 4) for the larger of the two values within $\pm 5\%$. It is interesting to find that there is, in both cases, the possibility of forming a non-helical cylindrical sheet with $\alpha = 0°$, namely $(P_1, Q_1) = (10,0)$, $r_1 = 0.339$ nm, and $(P_1, Q_1) = (20,0)$, $r_1 = 0.678$ nm, with the first row of C atoms corresponding to one of the equatorial cuts of C_{60} or of C_{240}; but ignorance of the actual mechanism of formation of the first cylindrical sheet forbids an objective choice among the various possibilities.

Table 5. Some possible characteristics of symmetric sheets with the smallest observed radii

(P_i, Q_i)	r_i (nm)	$\alpha°$
(10,0)	0.339	0
(10,2)	0.341	6.59
(10,4)	0.348	13.00
(9,9)	0.352	30
(10,6)	0.359	19.11
(16,16)	0.626	30
(19,1)	0.644	1.74
(17,17)	0.665	30
(20,0)	0.678	0
(20,2)	0.679	3.30

2.8 *Influence of the innermost cylinder*

If the template effect of Endo and Kroto[9] does indeed exist and the growth of the tubule occurs outwards, a knowledge of the characteristics of the innermost tube is then of paramount importance in any attempt at modelizing the structure of a symmetric nanotube[11]. By reversing eqn (17), possible values for P_1 and Q_1 can be obtained from

$$3P_1^2 + Q_1^2 = (16\pi^2/3)\,[(r_1/G)^2 \pm \delta r/r], \qquad (26)$$

and the inequalities (8) and (9), provided r_1 has been determined (say, by HRTEM—high resolution transmission electron microscopy), and G is arbitrarily fixed, say, at $G = 0.142$ nm. Although the task may become formidable if r_1 and/or $\delta r/r$ are large, the number of possibilities is strictly limited, especially since selected area electron diffraction (SAED) can provide narrower limits by giving a first approximation of the pitch angle[3].

3. CONCLUSION

It is clear that a large number of parameters influence the formation of the sheets in a tubule and that their relative importance is still unknown, as is also the cause of the occurrence of the defects responsible for the eventual polygonization of the sheets. Although the model presented here highlights the necessity of including, as one of the parameters, the uncertainty δr or δd on the separation of successive cylindrical sheets, *it is impossible to predict with absolute certainty the final characteristics of any of these sheets, symmetric or not, on the basis of the characteristics of the previous one*. Nevertheless, a number of features of their structure, such as the presence or absence of helicity, and the presence of groups of sheets with nearly the same angle of pitch, can be explained and quantified.

REFERENCES

1. S. Iijima, *Nature* **354**, 56 (1991).
2. X. Lin, X. K. Wang, V. P. Dravid, R. P. H. Chang, and J. B. Ketterson, *Appl. Phys. Letters* **64**, 181 (1994).
3. M. Liu and J. M. Cowley, *Carbon* **32**, 393 (1994).
4. A. Setton and R. Setton, *Synth. Met.* **4**, 59 (1981).
5. T. W. Ebbessen and P. M. Ajayan, *Nature* **358**, 220 (1992).
6. D. R. Stull and G. C. Sinke, *Thermodynamic properties of the elements* (No. 18, Advances in Chemistry Series), pp. 66–69. American Chemical Society, Washington (1956).
7. L. A. Girifalco and R. A. Lad, *J. Chem. Phys.* **25**, 693 (1956).
8. J.-C. Charlier and J.-P. Michenaud, *Phys. Rev. Lett.* **70**, 1858 (1993).
9. M. Endo and H. W. Kroto, *J. Phys. Chem.* **96**, 6941 (1992).
10. P. A. Heiney, J. E. Fischer, A. R. McGhie, W. J. Romanow, A. M. Denenstein, J. P. McCauley, Jr., A. P. Smith III, and D. E. Cox, *Phys. Rev. Lett.* **66**, 291 (1991).
11. V. A. Drits and C. Tchoubar, *X-Ray diffraction by disordered lamellar structures*. Springer-Verlag, Berlin (1990).

SCANNING TUNNELING MICROSCOPY OF CARBON NANOTUBES AND NANOCONES

KLAUS SATTLER
University of Hawaii, Department of Physics and Astronomy, 2505 Correa Road, Honolulu, HI 96822, U.S.A.

(*Received* 18 *July* 1994; *accepted* 10 *February* 1995)

Abstract—Tubular and conical carbon shell structures can be synthesized in the vapor phase. Very hot carbon vapor, after being deposited onto highly oriented pyrolytic graphite (HOPG), forms a variety of nanostructures, in particular single-shell tubes, multishell tubes, bundles of tubes, and cones. The structures were analyzed by scanning tunneling microscopy (STM) in UHV. Atomic resolution images show directly the surface atomic structures of the tubes and their helicities. A growth pathway is proposed for fullerenes, tubes, and cones.

Key Words—Carbon nanotubes, fullerenes, STM, fibers, nanostructures, vapor growth.

1. INTRODUCTION

Hollow carbon nanostructures are exciting new systems for research and for the design of potential nanoelectronic devices. Their atomic structures are closely related to their outer shapes and are described by hexagonal/pentagonal network configurations. The surfaces of such structures are atomically smooth and perfect. The most prominent of these objects are fullerenes and nanotubes[1]. Other such novel structures are carbon onions[2] and nanocones[3].

Various techniques have been used to image nanotubes: scanning electron microscopy (SEM)[1], scanning tunneling microscopy (STM)[4–7], and atomic force microscopy (AFM)[8]. Scanning probe microscopes are proximity probes. They can provide three-dimensional topographic images and, in addition, can give the atomic structure of the surface net. They can also be used to measure the electronic (STM) and elastic (AFM) properties of small structures. STM is restricted to electrically conducting objects, but AFM does not have this constraint.

STM and AFM images give directly the three-dimensional morphology of tubes and are consistent with the structures inferred from SEM. In addition, atomically resolved STM images make direct helicity determinations possible[4]. They give information about the nature of stacking of concentric carbon layers within the nanotubes via modifications of their surface density of states. STM is sensitive to such small lateral local density of states variations. Contours taken from the ends of the tubes show that some of them are open and others are closed. Many images indicate that the closed tubes have hemispherical caps. Such terminations can be modeled by fullerene hemispheres with 5/6 networks.

It is not easy to determine detailed properties of the tube terminations using STM or AFM. These microscopes cannot image undercut surfaces and the tip shape is convoluted with the cap shape of the nanotube. However, the tips may have very sharp edges yielding quite realistic three-dimensional images from tube terminations. Also, besides a difference in morphology, open and closed ends show a difference in electronic structure. Open ends appear with 'highlighted' edges in STM images, which is due to an enhanced dangling bond electron state density. Closed ends do not have such highlighted edges.

The growth pathway of various fullerene- and graphene-type nano-objects may be related. They are synthesized in the vapor phase and often appear simultaneously on the same sample. A common growth mechanism with similar nucleation seeds may, therefore, lead to these different structures.

2. EXPERIMENTAL

Graphite was used as substrate for the deposition of carbon vapor. Prior to the tube and cone studies, this substrate was studied by us carefully by STM because it may exhibit anomalous behavior with unusual periodic surface structures[9,10]. In particular, the cluster-substrate interaction was investigated[11]. At low submonolayer coverages, small clusters and islands are observed. These tend to have linear structures[12]. Much higher coverages are required for the synthesis of nanotubes and nanocones. In addition, the carbon vapor has to be very hot, typically >3000°C. We note that the production of nanotubes by arc discharge occurs also at an intense heat (of the plasma in the arc) of >3000°C.

The graphite (grade-A HOPG) was freshly cleaved in UHV and carefully examined by STM before the deposition. The HOPG surface was determined to be atomically flat and defect-free over micrometer dimensions. Any defect or adatom would have easily been detected. The graphite was cooled to −30°C during evaporation. The carbon vapor was produced by resistively heating a 99.99% purity carbon foil (0.5 mm thick) in UHV (base pressure 2×10^{-8} Torr). The deposition rate of 0.5 Å/s was controlled by a quartz crystal film thickness monitor. After deposition, the samples were transferred to a STM operated at 2×10^{-10} Torr, without breaking vacuum. Our evaporation and condensation process leads to the formation of various nanostructures, with 70% nanotubes on the average, of the overall products. In some areas, which may be as large as one square micron, we find 100% nanotubes. The yield for single-wall tubes varied from experiment to experiment from a few percent to 80–90%. Bundles of multiwall tubes were found in some areas, but were usually less abundant than isolated tubes. Individual nanocones were observed together with tubes, but were quite seldom. The microscope was operated in the constant current and in the constant height mode. Atomic resolution images were recorded in the constant current mode, in which the tip-to-sample distance is kept constant by means of an electronic feedback control. Bias voltages of 100 to 800 mV (both positive and negative) and tunneling currents of 0.5 to 3.0 nA were applied. A mechanically shaped Pt/Ir tip was used.

We did not observe any voltage dependent variation of the tube images. Also, the measured heights of the tubes were comparable with their diameters. Both of these observations indicate that the tubes have rather metallic than semiconducting properties.

The tubes were stable over long periods of time. After several months of being stored in UHV we still observed the same features as shortly after their preparation. Some of the samples were transferred to an STM operated in air. Again, we observed similar structures as seen in UHV. This shows the high stability of the tubes. It appears that the vapor-phase growth technique produces defect-free tubes, with dangling bonds at the tube edges often being saturated by cap terminations.

3. SINGLE-SHELL TUBES

Single-shell tubes are formed from a single layer of graphite. The surface of the cylinders has a honeycomb-lattice pattern, just as in a two-dimensional graphite plane. From a theoretical point of view they are interesting as the embodiment of a one-dimensional (1-D) periodic structure along the tube axis. In the circumferential direction, periodic boundary conditions apply to the enlarged unit cell. In addition to the chiral structures, there exist two nonchiral configurations, zigzag and armchair[13].

Part of a 15-nm long, 10 Å tube, is given in Fig. 1. Its surface atomic structure is displayed[14]. A periodic lattice is clearly seen. The cross-sectional profile was also taken, showing the atomically resolved curved surface of the tube (inset in Fig. 1). Asymmetry variations in the unit cell and other distortions in the image are attributed to electronic or mechanical tip-surface interactions[15,16]. From the helical arrangement of the tube, we find that it has zigzag configuration.

The zigzag and armchair tubes can be closed by hemispherical C_{60} caps, with 3-fold and 5-fold symmetry, respectively. Both caps contain six pentagons being equally distributed. We note that most of the nanotubes that we analyzed showed hemispherical terminations. Therefore, we might assume that the tubes start to grow from an incomplete fullerene cap and that the C_{60} hemisphere is the nucleation seed for the growth of the 10 Å tube. After the C_{60} hemisphere is formed, growth may continue as an all-hexagon network, forming a tube, rather than continuing as an alternating hexagon/pentagon network leading to the C_{60} sphere. The two caps, on both sides of the 10 Å zigzag tube (C_{60+18j})[17,18] are identical, with a total number of 12 pentagons, following Euler's theorem. The two caps for the 10 Å armchair tube (C_{60+10j}) are 36° rotated relative to each other.

It is interesting that we find the zigzag configuration for the tube network. The zigzag tube (Fig. 2) is the only nonhelical one among all the possible tube configurations. A cut normal to the C_{60+18j} tube axis leaves 18 dangling bonds, compared to 10 dangling bonds for the C_{60+10j} tube. For the armchair tube, it may be easy to incorporate pentagonal defects leading to an early closure because only one additional atom is required to form a pentagon at the growth periphery. For the zigzag tube, however, two atoms are required to form a pentagon and the structure might rather continue as a hexagonal network. Therefore, the zigzag 10 Å single-shell tubes might have a higher probability for growth.

4. MULTI-SHELL TUBES

There is an infinite number of possible atomic structures of graphene tubules. Each structure is characterized by its diameter and the helical arrangement of the carbon hexagons. Presumably, only single-shell tubes with small diameters of about 10 Å are formed and tubes with larger diameters are multishell tubes.

We produced multilayer tubes with diameters between 20 Å and 70 Å and up to 2000 Å in length[4]. An STM image of such tubes is shown in Fig. 3. The cylindrical shapes are well displayed.

We observed in some cases coaxial arrangement of the outermost and an inner tube. The outer tube may be terminated and the adjacent inner one is imaged simultaneously[4]. We measure an interlayer spacing of 3.4 Å, which is about the graphite interlayer distance (3.35 Å).

We find that the tubes are placed almost horizontally on the substrate. Irregular nanostructures were also formed, as displayed in the images. However, the high occurrence of tubes clearly shows that carbon prefers to condense to tubular structures, as opposed to other nanostructures, under our preparation conditions.

In Fig. 4 we show an atomic resolution image of a carbon tube. The structure imaged at the upper right corner of the picture comes from another tube. Both of them were ~1000 Å long. A perfect honeycomb surface structure is observed. By taking into account the curvature of the tube surface and the STM imaging profile, we find the same lattice parameter as that of graphite (1.42 Å). This directly proves that the tubu-

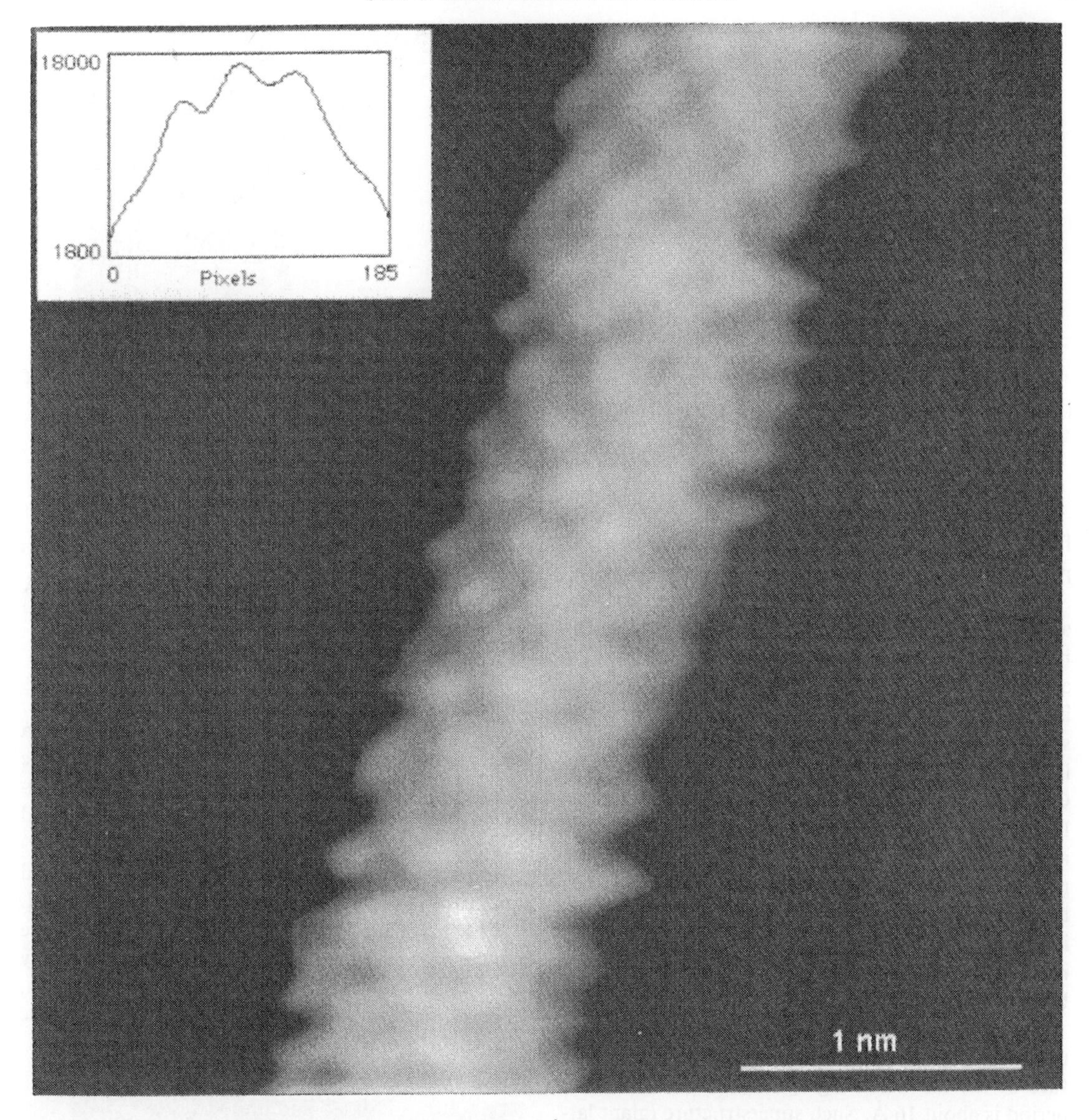

Fig. 1. Atomic resolution STM image of a 10 Å nanotube of carbon with cross-sectional plot.

lar surface is a graphitic network. The honeycomb surface net is arranged in a helical way, with a chiral angle of 5.0 ± 0.5°.

In a common method for the production of tubular carbon fibers, the growth is initiated by submicrometer size catalytic metal particles[19]. Tube growth out of a graphite rod during arc-discharge might also be related to nanoparticle-like seeds present at the substrate. In vapor-phase growth without a catalyst, the tubes may grow out of a hollow fullerene cap instead of a compact spherical particle.

This view is supported by our observation of hemispherically capped single-wall and multi-wall tubes on the same samples. It suggests that the C_{60}-derived tube could be the core of possible multilayer concentric graphitic tubes. After the single-shell tube has been

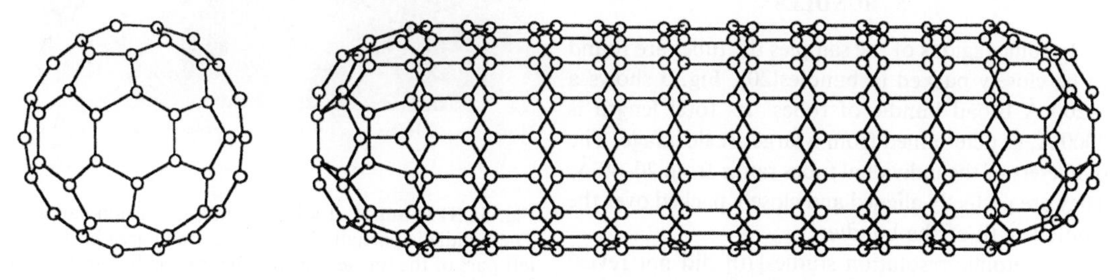

Fig. 2. Model of the C_{60}-based 10 Å tube with zigzag configuration (C_{60+18j}).

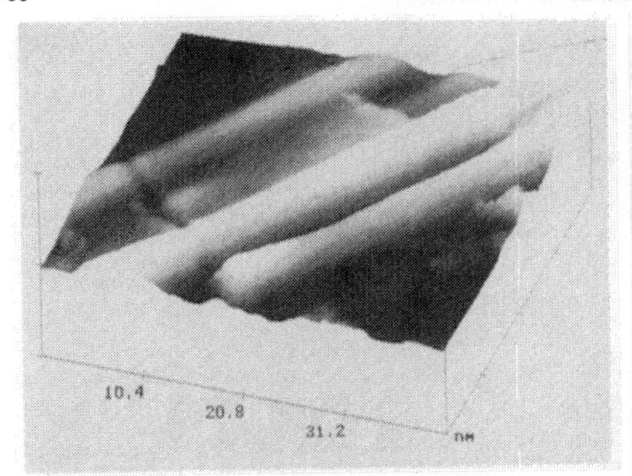

Fig. 3. STM images of fullerene tubes on a graphite substrate.

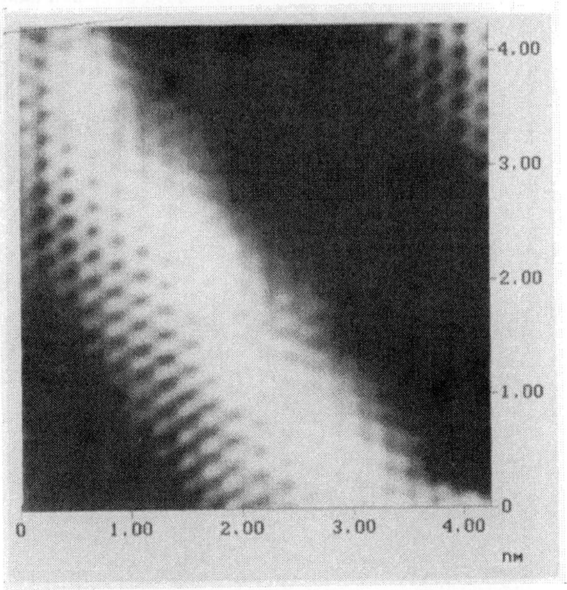

Fig. 4. Atomic resolution STM image of a carbon tube, 35 Å in diameter. In addition to the atomic structure, a zigzag superpattern along the tube axis can be seen.

formed, further concentric shells can be added by graphitic cylindrical layer growth.

The C_{60+10j} (j = 1,2,3, . . .) tube has an outer diameter of 9.6 Å[18]. In its armchair configuration, the hexagonal rings are arranged in a helical fashion with a chiral angle of 30 degrees. In this case, the tube axis is the five-fold symmetric axis of the C_{60}-cap. The single-shell tube can be treated as a rolled-up graphite sheet that matches perfectly at the closure line. Choosing the cylinder joint in different directions leads to different helicities. One single helicity gives a set of discrete diameters. To obtain the diameter that matches exactly the required interlayer spacing, the tube layers need to adjust their helicities. Therefore, in general, different helicities for different layers in a multilayer tube are expected. In fact, this is confirmed by our experiment.

In Fig. 4 we observe, in addition to the atomic lattice, a zigzag superpattern along the axis at the surface of the tube. The zigzag angle is 120° and the period is about 16 Å. Such superstructure (giant lattice) was found earlier for plane graphite[10]. It is a Moiré electron state density lattice produced by different helicities of top shell and second shell of a tube. The measured period of 16 Å reveals that the second layer of the tube is rotated relative to the first layer by 9°. The first and the second cylindrical layers, therefore, have chiral angles of 5° and −4°, respectively. This proves that the tubes are, indeed, composed of at least two coaxial graphitic cylinders with different helicities.

5. BUNDLES

In some regions of the samples the tubes are found to be closely packed in bundles[20]. Fig. 5 shows a ~200 Å broad bundle of tubes. Its total length is 2000 Å, as determined from a larger scale image. The diameters of the individual tubes range from 20–40 Å. They are perfectly aligned and closely packed over the whole length of the bundle.

Our atomic-resolution studies[10] did not reveal any steps or edges, which shows that the tubes are perfect graphitic cylinders. The bundle is disturbed in a small region in the upper left part of the image. In the closer view in Fig. 6, we recognize six tubes at the bundle surface. The outer shell of each tube is broken, and an inner tube is exposed. We measure again an intertube spacing of ~3.4 Å. This shows that the exposed inner tube is the adjacent concentric graphene shell.

The fact that all the outer shells of the tubes in the bundles are broken suggests that the tubes are strongly coupled through the outer shells. The inner tubes, however, were not disturbed, which indicates that the

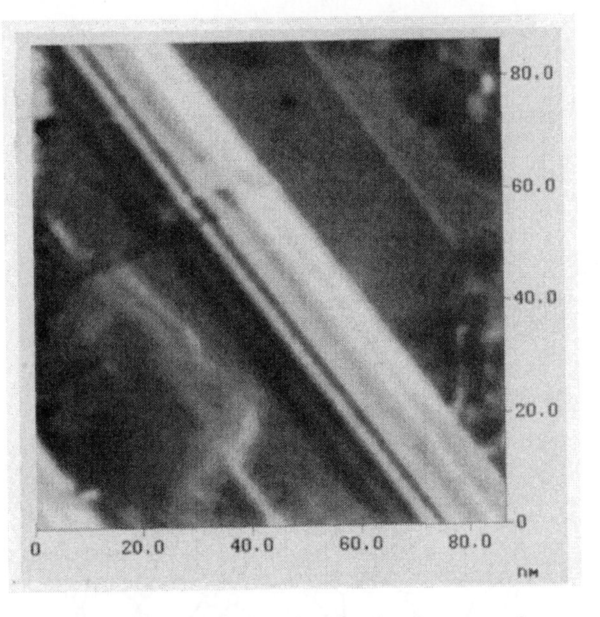

Fig. 5. STM image of a long bundle of carbon nanotubes. The bundle is partially broken in a small area in the upper left part of the image. Single tubes on the flat graphite surface are also displayed.

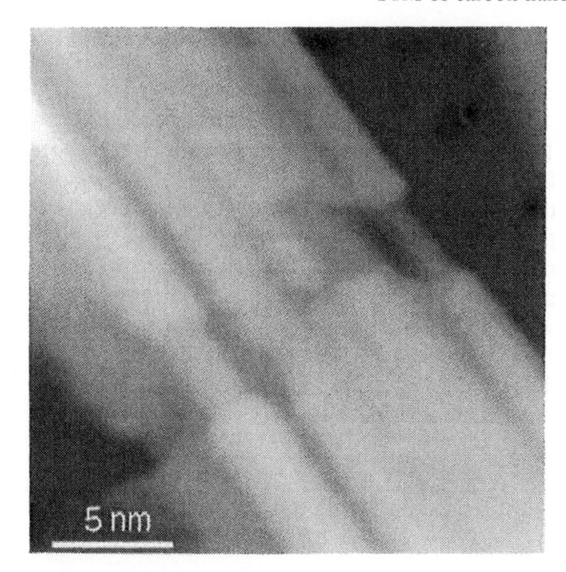

Fig. 6. A closer view of the disturbed area of the bundle in Fig. 5; the concentric nature of the tubes is shown. The outermost tubes are broken and the adjacent inner tubes are complete.

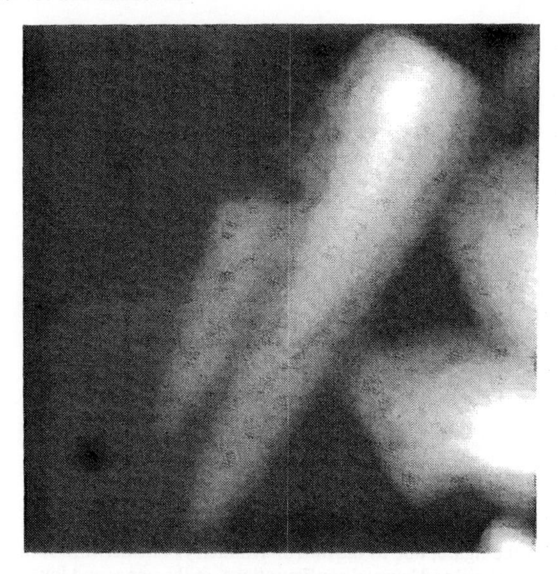

Fig. 7. A (244 Å) STM image of two fullerene cones.

intertube interaction is weaker than the intratube interaction. This might be the reason for bundle formation in the vapor phase. After a certain diameter is reached for a single tube, growth of adjacent tubes might be energetically favorable over the addition of further concentric graphene shells, leading to the generation of bundles.

6. CONES

Nanocones of carbon are found[3] in some areas on the substrate together with tubes and other mesoscopic structures. In Fig. 7 two carbon cones are displayed. For both cones we measure opening angles of $19.0 \pm 0.5°$. The cones are 240 Å and 130 Å long. Strikingly, all the observed cones (as many as 10 in a $(800 \text{ Å})^2$ area) have nearly identical cone angles $\sim 19°$.

At the cone bases, flat or rounded terminations were found. The large cone in Fig. 7 shows a sharp edge at the base, which suggests that it is open. The small cone in this image appears closed by a spherical-shaped cap.

We can model a cone by rolling a sector of a sheet around its apex and joining the two open sides. If the sheet is periodically textured, matching the structure at the closure line is required to form a complete network, leading to a set of discrete opening angles. The higher the symmetry of the network, the larger the set. In the case of a honeycomb structure, the sectors with angles of $n \times (2p/6)$ ($n = 1,2,3,4,5$) can satisfy perfect matching. Each cone angle is determined by the corresponding sector. The possible cone angles are 19.2°, 38.9°, 60°, 86.6°, and 123.6°, as illustrated in Fig. 8. Only the 19.2° angle was observed for all the cones in our experiment. A ball-and-stick model of the 19.2° fullerene cone is shown in Fig. 9. The body part is a hexagon network, while the apex contains five pentagons. The 19.2° cone has mirror symmetry through a plane which bisects the 'armchair' and 'zigzag' hexagon rows.

It is interesting that both carbon tubules and cones have graphene networks. A honeycomb lattice without inclusion of pentagons forms both structures. However, their surface nets are configured differently. The graphitic tubule is characterized by its diameter and its helicity, and the graphitic cone is entirely characterized by its cone angle. Helicity is not defined for the graphitic cone. The hexagon rows are rather arranged in helical-like fashion locally. Such 'local helicity' varies monotonously along the axis direction of the cone, as the curvature gradually changes. One can

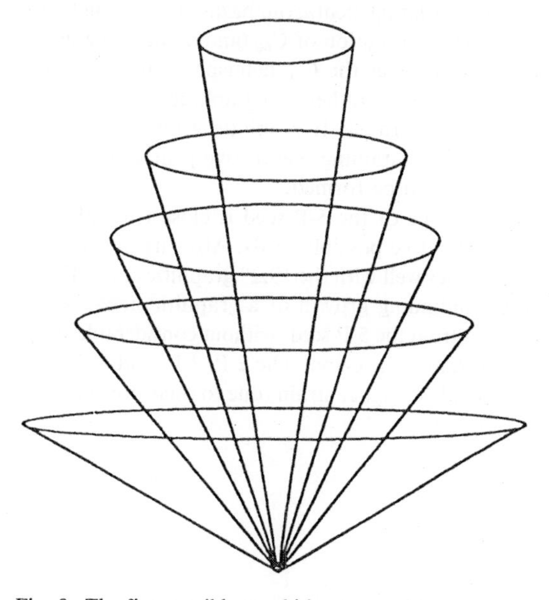

Fig. 8. The five possible graphitic cones, with cone angles of 19.2°, 38.9°, 60°, 86.6°, and 123.6°.

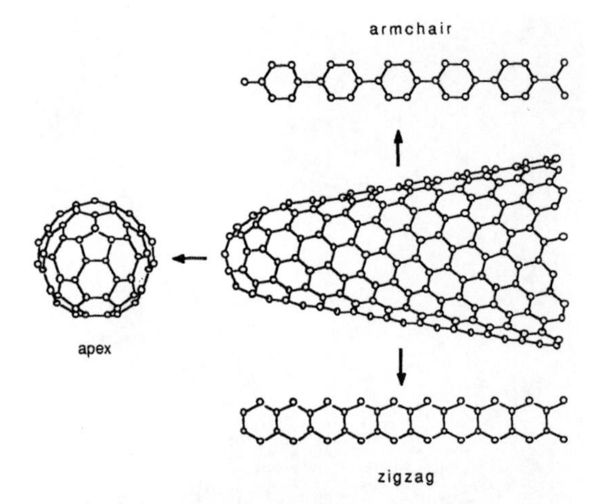

Fig. 9. Ball-and-stick model for a 19.2° fullerene cone. The back part of the cone is identical to the front part displayed in the figure, due to the mirror symmetry. The network is in 'armchair' and 'zigzag' configurations, at the upper and lower sides, respectively. The apex of the cone is a fullerene-type cap containing five pentagons.

easily show that moving a 'pitch' (the distance between two equivalent sites in the network) along any closure line of the network leads to another identical cone. For the 19.2° cone, a hexagon row changes its 'local helical' direction at half a turn around the cone axis and comes back after a full turn, due to its mirror symmetry in respect to its axis. The other four cones, with larger opening angles, have D_{nd} ($n = 2,3,4,5$) symmetry along the axis. The 'local helicity' changes its direction at each of their symmetry planes.

As fullerenes, tubes, and cones are produced in the vapor phase we consider all three structures being originated by a similar-type nucleation seed, a small curved carbon sheet composed of hexagons and pentagons. The number of pentagons (m) in this fullerene-type (m-P) seed determines its shape. Continuing growth of an alternating pentagon/hexagon (5/6) network leads to the formation of C_{60} (and higher fullerenes). If however, after the C_{60} hemisphere is completed, growth continues rather as a graphitic (6/6) network, a tubule is formed. If graphitic growth progresses from seeds containing one to five pentagons, fullerene cones can be formed.

The shape of the 5-P seed is closest to spherical among the five possible seeds. Also, its opening angle matches well with the 19.2° graphitic cone. Therefore, continuing growth of a graphitic network can proceed from the 5-P seed, without considerable strain in the transition region. The 2-P, 3-P, and 4-P seeds would induce higher strain (due to their nonspherical shapes) to match their corresponding cones and are unlikely to form. This explains why only the 19.2° cones have been observed in our experiment.

Carbon cones are peculiar mesoscopic objects. They are characterized by a continuous transition from fullerene to graphite through a tubular-like intermedium. The dimensionality changes gradually as the cone opens. It resembles a 0-D cluster at the apex, then proceeds to a 1-D 'pipe' and finally approaches a 2-D layer. The carbon cones may have complex band structures and fascinating charge transport properties, from insulating at the apex to metallic at the base. They might be used as building units in future nanoscale electronics devices.

Acknowledgements—Financial support from the National Science Foundation, Grant No. DMR-9106374, is gratefully acknowledged.

REFERENCES

1. S. Iijima, *Nature* **354**, 56 (1991). T. W. Ebbesen and P. M. Ajayan, *Nature* **358**, 220 (1992).
2. D. Ugarte, *Nature* **359**, 707 (1992).
3. M. Ge and K. Sattler, *Chem. Phys. Lett.* **220**, 192 (1994).
4. M. Ge and K. Sattler, *Science* **260**, 515 (1993).
5. T. W. Ebbesen, H. Hiura, J. Fijita, Y. Ochiai, S. Matsui, and K. Tanigaki, *Chem. Phys. Lett.* **209**, 83 (1993).
6. M. J. Gallagher, D. Chen, B. P. Jakobsen, D. Sarid, L. D. Lamb, F. A. Tinker, J. Jiao, D. R. Huffman, S. Seraphin, and D. Zhou, *Surf. Sci. Lett.* **281**, L335 (1993).
7. Z. Zhang and Ch. M. Lieber, *Appl. Phys. Lett.* **62**, 2792 (1993).
8. R. Hoeper, R. K. Workman, D. Chen, D. Sarid, T. Yadav, J. C. Withers, and R. O. Loutfy, *Surf. Sci.* **311**, L371 (1994).
9. K. Sattler, *Int. J. Mod. Phys. B* **6**, 3603 (1992).
10. J. Xhie, K. Sattler, N. Venkateswaran, and M. Ge, *Phys. Rev. B* **47**, 15835 (1993).
11. J. Xhie, K. Sattler, U. Mueller, G. Raina, and N. Venkateswaran, *Phys. Rev. B* **43**, 8917 (1991).
12. M. Ge, K. Sattler, J. Xhie, and N. Venkateswaran, In *Novel forms of carbon* (Edited by C. L. Renschler, J. Pouch, and D. Cox) *Mat. Res. Soc. Proc.* **270**, 109 (1992).
13. R. Saito, M. Fujita, G. Dresselhaus, and M. S. Dresselhaus, *Appl. Phys. Lett.* **60**, 2204 (1992).
14. M. Ge and K. Sattler, *Appl. Phys. Lett.* **65**, 2284 (1994).
15. D. Tomanek, S. G. Louie, H. J. Mamin, D. W. Abraham, R. E. Thomson, E. Ganz, and J. Clarke, *Phys. Rev. B* **35**, 7790 (1987).
16. J. M. Soler, A. M. Baro, N. Garcia, and H. Rohrer, *Phys. Rev. Lett.* **57**, 444 (1986).
17. N. Hamada, S. Samada, and A. Oshiyama, *Phys. Rev. Lett.* **68**, 1579 (1992).
18. M. S. Dresselhaus, G. Dresselhaus, and R. Saito, *Phys. Rev. B* **45**, 6234 (1992).
19. G. Tibbetts, *J. Cryst. Growth* **66**, 632 (1984).
20. M. Ge and K. Sattler, *Appl. Phys. Lett.* **64**, 710 (1994).

TOPOLOGICAL AND SP³ DEFECT STRUCTURES IN NANOTUBES

T. W. EBBESEN[1] and T. TAKADA[2]
[1]NEC Research Institute, 4 Independence Way, Princeton, NJ 08540, U.S.A.
[2]Fundamental Research Laboratories, NEC Corporation, 34 Miyukigaoka, Tsukuba 305, Japan

(*Received* 25 *November* 1994; *accepted* 10 *February* 1995)

Abstract—Evidence is accumulating that carbon nanotubes are rarely as perfect as they were once thought to be. Possible defect structures can be classified into three groups: topological, rehybridization, and incomplete bonding defects. The presence and significance of these defects in carbon nanotubes are discussed. It is clear that some nanotube properties, such as their conductivity and band gap, will be strongly affected by such defects and that the interpretation of experimental data must be done with great caution.

Key Words—Defects, topology, nanotubes, rehybridization.

1. INTRODUCTION

Carbon nanotubes were first thought of as perfect seamless cylindrical graphene sheets—a defect-free structure. However, with time and as more studies have been undertaken, it is clear that nanotubes are not necessarily that perfect; this issue is not simple because of a variety of seemingly contradictory observations. The issue is further complicated by the fact that the quality of a nanotube sample depends very much on the type of machine used to prepare it[1]. Although nanotubes have been available in large quantities since 1992[2], it is only recently that a purification method was found[3]. So, it is now possible to undertake various accurate property measurements of nanotubes. However, for those measurements to be meaningful, the presence and role of defects must be clearly understood.

The question which then arises is: What do we call a defect in a nanotube? To answer this question, we need to define what would be a perfect nanotube. Nanotubes are microcrystals whose properties are mainly defined by the hexagonal network that forms the central cylindrical part of the tube. After all, with an aspect ratio (length over diameter) of 100 to 1000, the tip structure will be a small perturbation except near the ends. This is clear from Raman studies[4] and is also the basis for calculations on nanotube properties[5–7]. So, a perfect nanotube would be a cylindrical graphene sheet composed only of hexagons having a minimum of defects at the tips to form a closed seamless structure.

Needless to say, the issue of defects in nanotubes is strongly related to the issue of defects in graphene. Although earlier studies of graphite help us understand nanotubes, the concepts derived from fullerenes has given us a new insight into traditional carbon materials. So, the discussion that follows, although aimed at nanotubes, is relevant to all graphitic materials. First, different types of possible defects are described. Then, recent evidences for defects in nanotubes and their implications are discussed.

2. CLASSES OF DEFECTS

Figure 1 show examples of nanotubes that are far from perfect upon close inspection. They reveal some of the types of defects that can occur, and will be discussed below. Having defined a perfect nanotube as a cylindrical sheet of graphene with the minimum number of pentagons at each tip to form a seamless structure, we can classify the defects into three groups: 1) topological defects, 2) rehybridization defects and 3) incomplete bonding and other defects. Some defects will belong to more than one of these groups, as will be indicated.

2.1 *Topological defects*

The introduction of ring sizes other than hexagons, such as pentagons and heptagons, in the graphene sheet creates topological changes that can be treated as local defects. Examples of the effect of pentagons and heptagons on the nanotube structure is shown in Fig. 1 (a). The resulting three dimensional topology follows Euler's theorem[8] in the approximation that we assume that all the individual rings in the sheet are flat. In other words, it is assumed that all the atoms of a given cycle form a plane, although there might be angles between the planes formed in each cycle. In reality, the strain induced by the three-dimensional geometry on the graphitic sheet can lead to deformation of the rings, complicating the ideal picture, as we shall see below.

From Euler's theorem, one can derive the following simple relation between the number and type of cycles n_i (where the subscript i stands for the number of sides to the ring) necessary to close the hexagonal network of a graphene sheet:

$$3n_3 + 2n_4 + n_5 - n_7 - 2n_8 - 3n_9 = 12$$

where 12 corresponds to a total disclination of 4π (i.e., a sphere). For example, in the absence of other cycles one needs 12 pentagons (n_5) in the hexagonal net-

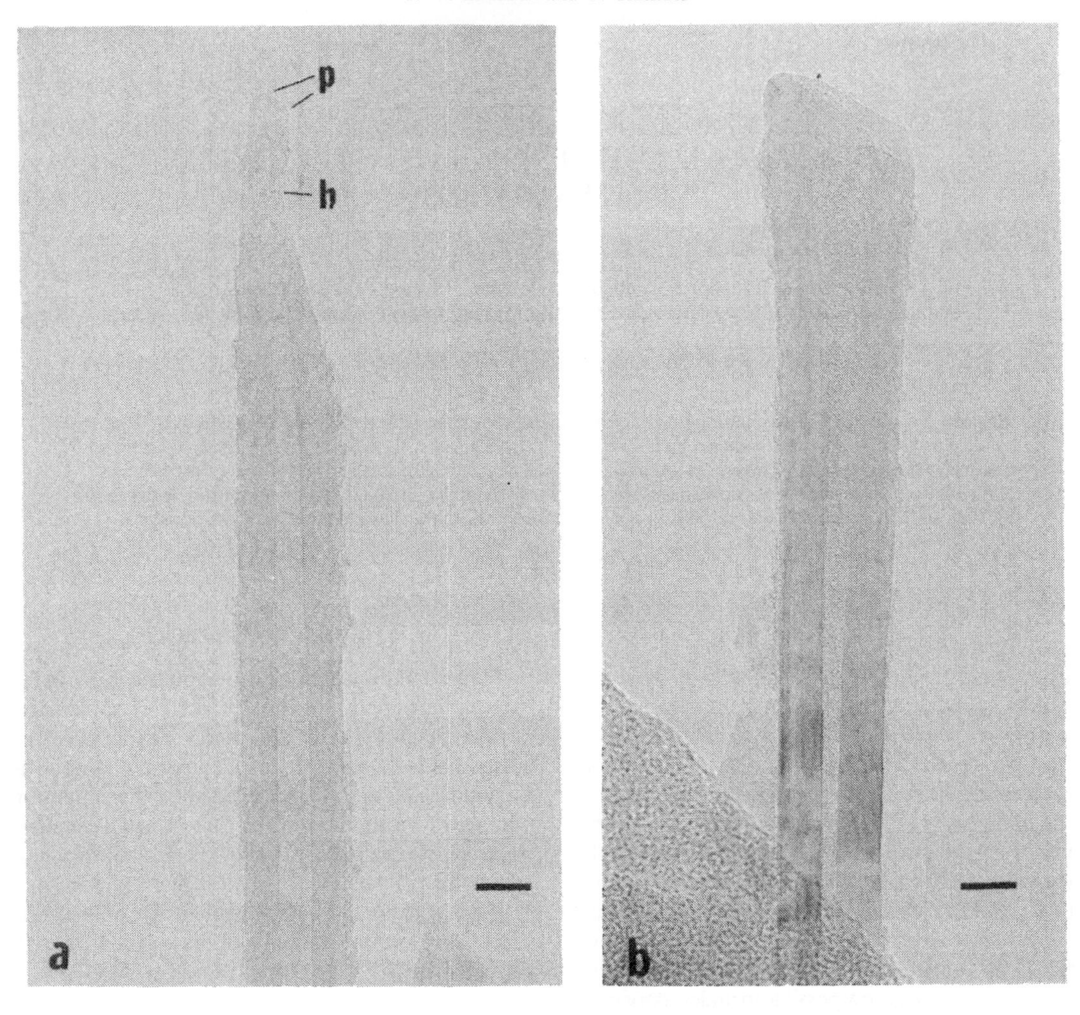

Fig. 1. Five examples of nanotubes showing evidence of defects in their structure (p: pentagon, h: heptagon, d: dislocation); see text (the scale bars equal 10 nm).

work to close the structure. The addition of one heptagon (n_7) to the nanotube will require the presence of 13 pentagons to close the structure (and so forth) because they induce opposite 60° disclinations in the surface. Although the presence of pentagons (n_5) and heptagons (n_7) in nanotubes[9,10] is clear from the disclinations observed in their structures (Fig. 1a), we are not aware of any evidence for larger or smaller cycles (probably because the strain would be too great).

A single heptagon or pentagon can be thought of as point defects and their properties have been calculated[11]. Typical nanotubes don't have large numbers of these defects, except close to the tips. However, the point defects polygonize the tip of the nanotubes, as shown in Fig. 2. This might also favor the polygonalization of the entire length of the nanotube as illustrated by the dotted lines in Fig. 2. Liu and Cowley have shown that a large fraction of nanotubes are polygonized in the core[12,13]. This will undoubtedly have significant effects on their properties due to local rehybridization, as will be discussed in the next section. The nanotube in Fig. 1 (e) appears to be polygonized (notice the different spacing between the layers on the left and right-hand side of the nanotube).

Another common defect appears to be the aniline structure that is formed by attaching a pentagon and a heptagon to each other. Their presence is hard to detect directly because they create only a small local deformation in the width of the nanotube. However, from time to time, when a very large number of them are accidentally aligned, the nanotube becomes gradually thicker and thicker, as shown in Fig. 1 (b). The existence of such tubes indicates that such pairs are probably much more common in nanotubes, but that they normally go undetected because they cancel each other out (random alignment). The frequency of occurrence of these aligned 5/7 pairs can be estimated to be about 1 per 3 nm from the change in the diameter of the tube. Randomly aligned 5/7 pairs should be present at even higher frequencies, seriously affecting the nanotube properties. Various aspects of such pairs have been discussed from a theoretical point of view in the literature[14,15]. In particular, it has been pointed out by Saito *et al.*[14] that such defect pairs

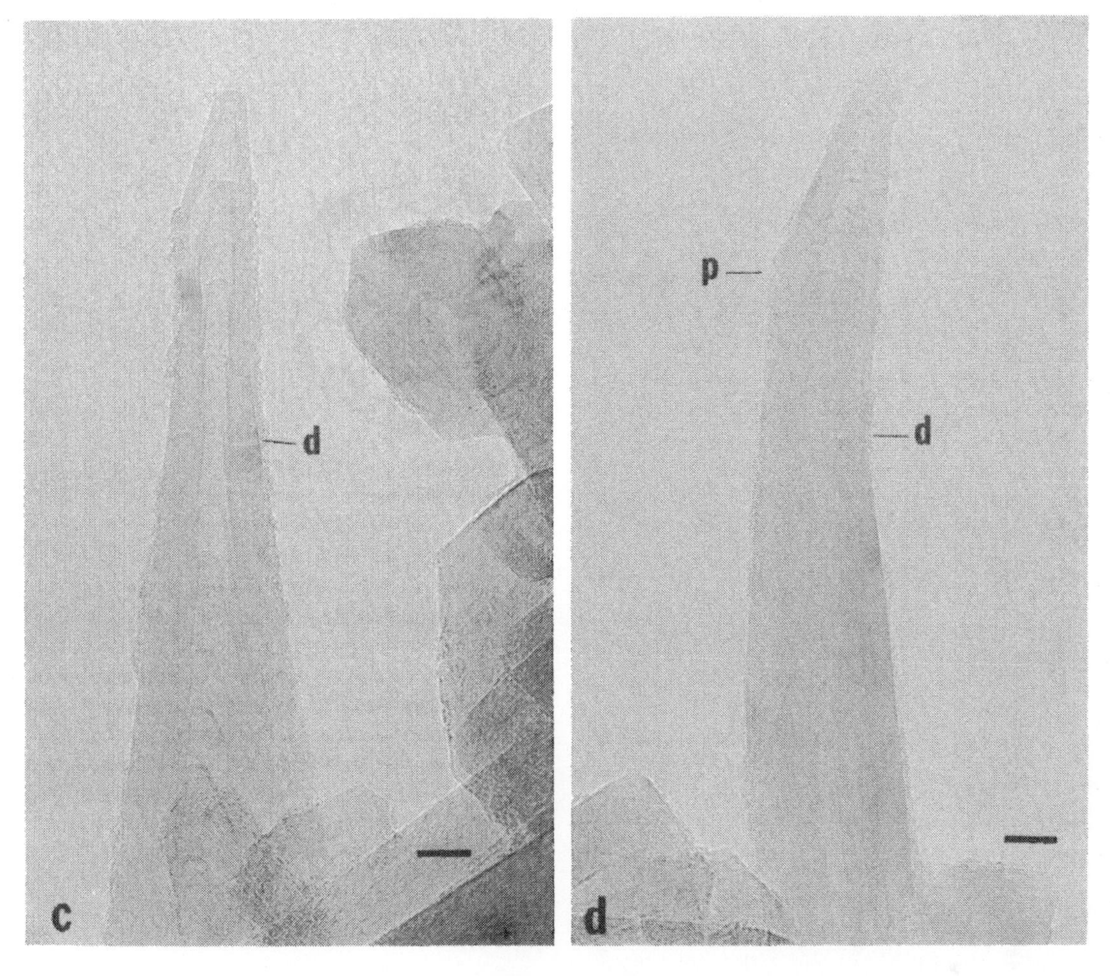

Fig. 1 continued.

can annihilate, which would be relevant to the annealing away of such defects at high temperature as discussed in the next section.

It is not possible to exclude the presence of other unusual ring defects, such as those observed in graphitic sheets[16,17]. For example, there might be heptagon-triangle pairs in which there is one sp^3 carbon atom bonded to 4 neighboring atoms, as shown in Fig. 3. Although there must be strong local structural distortions, the graphene sheet remains flat overall[16,17]. This is a case where Euler's theorem does not apply. The possibility of sp^3 carbons in the graphene sheet brings us to the subject of rehybridization.

2.2 *Rehybridization defects*

The root of the versatility of carbon is its ability to rehybridize between *sp*, sp^2, and sp^3. While diamond and graphite are examples of pure sp^3 and sp^2 hybridized states of carbon, it must not be forgotten that many intermediate degrees of hybridization are possible. This allows for the out-of-plane flexibility of graphene, in contrast to its extreme in-plane rigidity. As the graphene sheet is bent out-of-plane, it must lose some of its sp^2 character and gain some sp^3 character or, to put it more accurately, it will have $sp^{2+\alpha}$ character. The size of α will depend on the degree of curvature of the bend. The complete folding of the graphene sheet will result in the formation of a defect line having strong sp^3 character in the fold.

We have shown elsewhere that line defects having sp^3 character form preferentially along the symmetry axes of the graphite sheet[18]. This is best understood by remembering that the change from sp^2 to sp^3 must naturally involve a pair of carbon atoms because a double bond is perturbed. In the hexagonal network of graphite shown in Fig. 4, it can be seen that there are 4 different pairs of carbon atoms along which the sp^3 type line defect can form. Two pairs each are found along the [100] and [210] symmetry axes. Furthermore, there are 2 possible conformations, "boat" and "chair," for three of these distinct line defects and a single conformation of one of them. These are illustrated in Fig. 5.

In the polygonized nanotubes observed by Liu and Cowley[12,13], the edges of the polygon must have more sp^3 character than the flat faces in between. These are defect lines in the sp^2 network. Nanotubes mechanically deformed appear to be rippled, indicat-

Fig. 1 continued.

ing the presence of ridges with sp^3 character[18]. Because the symmetry axes of graphene and the long axis of the nanotubes are not always aligned, any defect line will be discontinuous on the atomic scale as it traverses the entire length of the tube. Furthermore, in the multi-layered nanotubes, where each shell has a different helicity, the discontinuity will not be superimposable. In other words, in view of the turbostratic nature of the multi-shelled nanotubes, an edge along the tube will result in slightly different defect lines in each shell.

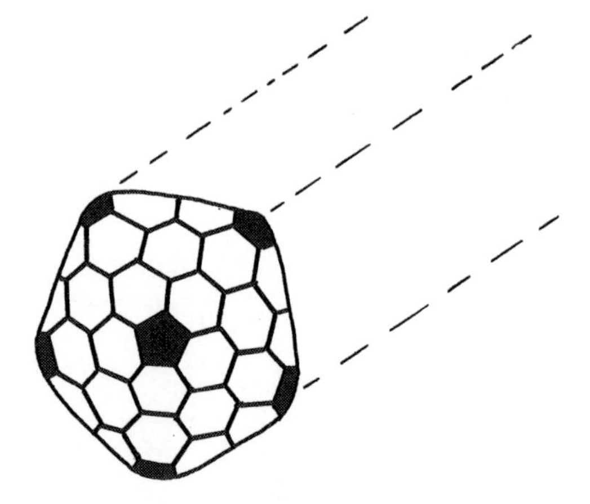

Fig. 2. Nanotube tip structure seen from the top; the presence of pentagons can clearly polygonize the tip.

2.3 *Incomplete bonding and other defects*

Defects traditionally associated with graphite might also be present in nanotubes, although there is not yet much evidence for their presence. For instance, point defects such as vacancies in the graphene sheet might be present in the nanotubes. Dislocations are occasionally observed, as can be seen in Fig. 1 (c) and (d), but they appear to be quite rare for the nanotubes formed at the high temperatures of the carbon arc. It might be quite different for catalytically grown nanotubes. In general, edges of graphitic domains and vacancies should be chemically very reactive as will be discussed below.

3. DISCUSSION

There are now clear experimental indications that nanotubes are not perfect in the sense defined in the introduction[12,13,19,20]. The first full paper dedicated to this issue was by Zhou *et al.*[19], where both pressure and intercalation experiments indicated that the particles in the sample (including nanotubes) could not be perfectly closed graphitic structures. It was pro-

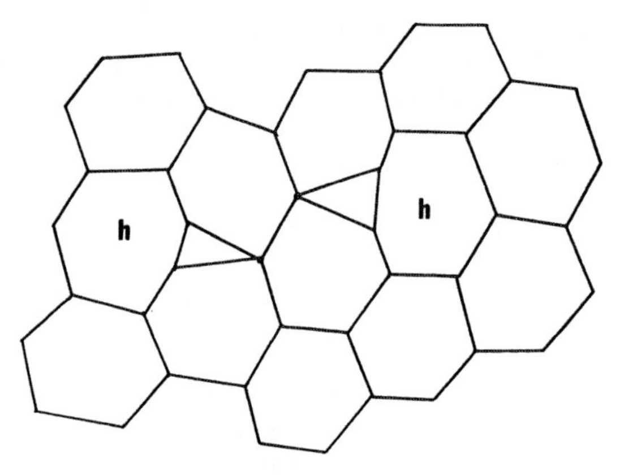

Fig. 3. Schematic diagram of heptagon-triangle defects [16,17].

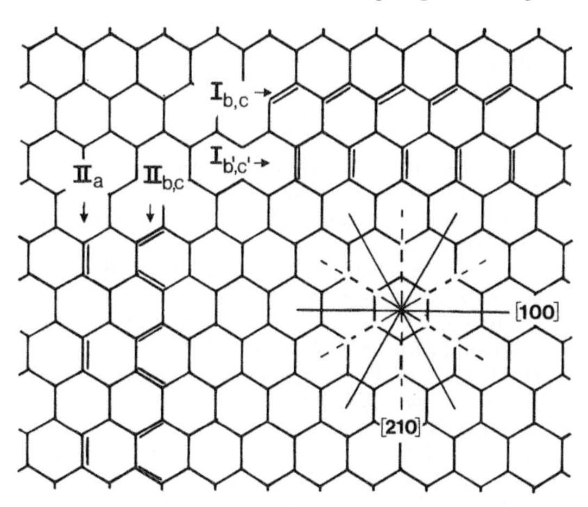

Fig. 4. Hexagonal network of graphite and the 4 different pairs of carbon atoms across which the sp^3-like defect line may form[18].

posed that nanotubes were composed of pieces of graphitic sheets stuck together in a paper-maché model. The problem with this model is that it is not consistent with two other observations. First, when nanotubes are oxidized they are consumed from the tip inwards, layer by layer[21,22]. If there were smaller domains along the cylindrical part, their edges would be expected to react very fast to oxidation, contrary to observation. Second, ESR studies[23] do not reveal any strong signal from dangling bonds and other defects, which would be expected from the numerous edges in the paper-maché model.

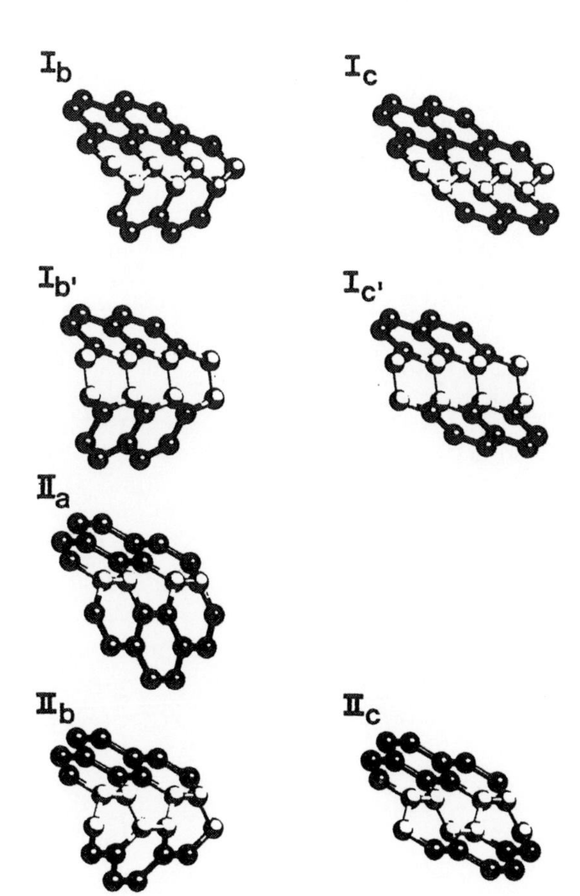

Fig. 5. Conformations of the 4 types of defect lines that can occur in the graphene sheet[18].

To try to clarify this issue, we recently analyzed crude nanotube samples and purified nanotubes before and after annealing them at high temperature[20]. It is well known that defects can be annealed away at high temperatures (ca. 2850°C). The annealing effect was very significant on the ESR properties, indicating clearly the presence of defects in the nanotubes[20]. However, our nanotubes do not fit the defect structure proposed in the paper-maché model for the reason discussed in the previous paragraph. Considering the types of possible defects (see part 2), the presence of either a large number of pentagon/heptagon pairs in the nanotubes and/or polygonal nanotubes, as observed by Liu and Cowley[12,13], could possibly account for these results. Both the 5/7 pairs and the edges of the polygon would significantly perturb the electronic properties of the nanotubes and could be annealed away at very high temperatures. The sensitivity of these defects to oxidation is unknown.

In attempting to reconcile these results with those of other studies, one is limited by the variation in sample quality from one study to another. For instance, ESR measurements undertaken on bulk samples in three different laboratories show very different results[19,23,24]. As we have pointed out elsewhere, the quantity of nanotubes (and their quality) varies from a few percent to over 60% of the crude samples, depending on the current control and the extent of cooling in the carbon arc apparatus[1]. The type and distribution of defects might also be strongly affected by the conditions during nanotube production. The effect of pressure on the spacing between the graphene sheets observed by Zhou *et al.* argues most strongly in favor of the particles in the sample having a non-closed structure[19]. Harris *et al.* actually observe that nanoparticles in these samples sometimes do not form closed structures[25]. It would be interesting to repeat the pressure study on purified nanotubes before and after annealing with samples of various origins. This should give significant information on the nature of the defects. The results taken before annealing will, no doubt, vary depending on where and how the sample was prepared. The results after sufficient annealing should be consistent and independent of sample origin.

4. CONCLUSION

The issue of defects in nanotubes is very important in interpreting the observed properties of nanotubes. For instance, electronic and magnetic properties will be significantly altered as is already clear from observation of the conduction electron spin resonance[20,23].

It would be worthwhile making theoretical calculations to evaluate the effect of defects on the nanotube properties. The chemistry might be affected, although to a lesser degree because nanotubes, like graphite, are chemically quite inert. If at all possible, nanotubes should be annealed (if not also purified) before physical measurements are made. Only then are the results likely to be consistent and unambiguous.

REFERENCES

1. T. W. Ebbesen, *Annu. Rev. Mater. Sci.* **24**, 235 (1994).
2. T. W. Ebbesen and P. M. Ajayan, *Nature* **358**, 220 (1992).
3. T. W. Ebbesen, P. M. Ajayan, H. Hiura, and K. Tanigaki, *Nature* **367**, 519 (1994).
4. H. Hiura, T. W. Ebbesen, K. Tanigaki, and H. Takahashi, *Chem. Phys. Lett.* **202**, 509 (1993).
5. J. W. Mintmire, B. I. Dunlap, and C. T. White, *Phys. Rev. Lett.* **68**, 631 (1992).
6. N. Hamada, S. Sawada, and A. Oshiyama, *Phys. Rev. Lett.* **68**, 1579 (1992).
7. R. Saito, M. Fujita, G. Dresselhaus, and M. S. Dresselhaus, *Appl. Phys. Lett.* **60**, 2204 (1992).
8. H. Terrones and A. L. Mackay, *Carbon* **30**, 1251 (1992).
9. P. M. Ajayan, T. Ichihashi, and S. Iijima, *Chem. Phys. Lett.* **202**, 384 (1993).
10. S. Iijima, T. Ichihashi, and Y. Ando, *Nature* **356**, 776 (1992).
11. R. Tamura and M. Tsukada, *Phys. Rev. B* **49**, 7697 (1994).
12. M. Liu and J. M. Cowley, *Carbon* **32**, 393 (1994).
13. M. Liu and J. M. Cowley, *Ultramicroscopy* **53**, 333 (1994).
14. R. Saito, G. Dresselhaus, and M. S. Dresselhaus, *Chem. Phys. Lett.* **195**, 537 (1992).
15. C. J. Brabec, A. Maiti, and J. Bernholc, *Chem. Phys. Lett.* **219**, 473 (1994).
16. J. C. Roux, S. Flandrois, C. Daulan, H. Saadaoui, and O. Gonzalez, *Ann. Chim. Fr.* **17**, 251 (1992).
17. B. Nysten, J. C. Roux, S. Flandrois, C. Daulan, and H. Saadaoui, *Phys. Rev. B* **48**, 12527 (1993).
18. H. Hiura, T. W. Ebbesen, J. Fujita, K. Tanigaki, and T. Takada, *Nature* **367**, 148 (1994).
19. O. Zhou, R. M. Fleming, D. W. Murphy, C. H. Chen, R. C. Haddon, A. P. Ramirez, and S. H. Glarum, *Science* **263**, 1744 (1994).
20. M. Kosaka, T. W. Ebbesen, H. Hiura, and K. Tanigaki, *Chem. Phys. Lett.* **233**, 47 (1995).
21. S. C. Tsang, P. J. F. Harris, and M. L. H. Green, *Nature* **362**, 520 (1993).
22. P. M. Ajayan, T. W. Ebbesen, T. Ichihashi, S. Iijima, K. Tanigaki, and H. Hiura, *Nature* **362**, 522 (1993).
23. M. Kosaka, T. W. Ebbesen, H. Hiura, and K. Tanigaki, *Chem. Phys. Lett.* **225**, 161 (1994).
24. K. Tanaka, T. Sato, T. Yamabe, K. Okahara, *et al.*, *Chem. Phys. Lett.* **223**, 65 (1994).
25. P. J. F. Harris, M. L. H. Green, and S. C. Tsang, *J. Chem. Soc. Faraday Trans.* **89**, 1189 (1993).

HELICALLY COILED AND TOROIDAL CAGE FORMS OF GRAPHITIC CARBON

SIGEO IHARA and SATOSHI ITOH
Central Research Laboratory, Hitachi Ltd., Kokubunji, Tokyo 185, Japan

(*Received* 22 *August* 1994; *accepted in revised form* 10 *February* 1995)

Abstract—Toroidal forms for graphitic carbon are classified into five possible prototypes by the ratios of their inner and outer diameters, and the height of the torus. Present status of research of helical and toroidal forms, which contain pentagons, hexagons, and heptagons of carbon atoms, are reviewed. By molecular-dynamics simulations, we studied the length and width dependence of the stability of the elongated toroidal structures derived from torus C_{240} and discuss their relation to nanotubes. The atomic arrangements of the structures of the helically coiled forms of the carbon cage for the single layer, which are found to be thermodynamically stable, are compared to those of the experimental helically coiled forms of single- and multi-layered graphitic forms that have recently been experimentally observed.

Key Words—Carbon, molecular dynamics, torus, helix, graphitic forms.

1. INTRODUCTION

Due, in part, to the geometrical uniqueness of their cage structure and, in part, to their potentially technological use in various fields, fullerenes have been the focus of very intense research[1]. Recently, higher numbers of fullerenes with spherical forms have been available[2]. It is generally recognized that in the fullerene, C_{60}, which consists of pentagons and hexagons formed by carbon atoms, pentagons play an essential role in creating the convex plane. This fact was used in the architecture of the geodesic dome invented by Robert Buckminster Fuller[3], and in traditional bamboo art[4] ('toke-zaiku',# for example).

By wrapping a cylinder with a sheet of graphite, we can obtain a carbon nanotube, as experimentally observed by Iijima[5]. Tight binding calculations indicate that if the wrapping is charged (i.e., the chirality of the surface changes), the electrical conductivity changes: the material can behave as a semiconductor or metal depending on tube diameter and chirality[6].

In the study of the growth of the tubes, Iijima found that heptagons, seven-fold rings of carbon atoms, appear in the negatively curved surface. Theoretically, it is possible to construct a crystal with only a negatively curved surface, which is called a minimal surface[7]. However, such surfaces of carbon atoms are yet to be synthesized. The positively curved surface is created by insertion of pentagons into a hexagonal sheet, and a negatively curved surface is created by heptagons. Combining these surfaces, one could, in principle, put forward a new form of carbon, having new features of considerable technological interest by solving the problem of tiling the surface with pentagons, heptagons, and hexagons.

The toroidal and helical forms that we consider here are created as such examples; these forms have quite interesting geometrical properties that may lead to interesting electrical and magnetic properties, as well as nonlinear optical properties. Although the method of the simulations through which we evaluate the reality of the structure we have imagined is omitted, the construction of toroidal forms and their properties, especially their thermodynamic stability, are discussed in detail. Recent experimental results on toroidal and helically coiled forms are compared with theoretical predictions.

2. TOPOLOGY OF TOROIDAL AND HELICAL FORMS

2.1 *Tiling rule for cage structure of graphitic carbon*

Because of the sp^2 bonding nature of carbon atoms, the atoms on a graphite sheet should be connected by the three bonds. Therefore, we consider how to tile the hexagons created by carbon atoms on the toroidal surfaces. Of the various bonding lengths that can be taken by carbon atoms, we can tile the toroidal surface using only hexagons. Such examples are provided by Heilbonner[8] and Miyazaki[9]. However, the side lengths of the hexagons vary substantially. If we restrict the side length to be almost constant as in graphite, we must introduce, at least, pentagons and heptagons.

Assuming that the surface consists of pentagons, hexagons, and heptagons, we apply Euler's theorem. Because the number of hexagons is eliminated by a kind of cancellation, the relation thus obtained contains only the number of pentagons and heptagons: $f_5 - f_7 = 12(1\text{-}g)$, where f_5 stands for the number of pentagons, f_7 the number of heptagons, and g is the genius (the number of topological holes) of the surface.

#At the Ooishi shrine of Ako in Japan, a geodesic dome made of bamboo with three golden balls, which was the symbol called "Umajirushi" used by a general named Mori Misaemon'nojyo Yoshinari at the battle of Okehazama in 1560, has been kept in custody. (See ref. [4]).

In the spherical forms (i.e., $g = 0$), $f_5 = f_7 + 12$. In C_{60}, for example, there are no heptagons ($f_7 = 0$), so that $f_5 = 12$. If the torus whose genius (g) is one, $f_5 = f_7$. As we mentioned in the introduction, pentagons and heptagons provide Gaussian positive and negative curvatures, respectively. Therefore, pentagons should be located at the outermost region of the torus and heptagons at the innermost.

2.2 *Classifications of tori*

Here, the topological nature of the tori will be discussed briefly. Figure 1 shows the five possible prototypes of toroidal forms that are considered to be related to fullerenes. These structures are classified by the ratios of the inner and outer diameters r_i and r_o, and the height of the torus, h. (Note that r_o is larger than r_i). As depicted in Fig. 1, if $r_i \approx r_o$, and $h \ll r_i$, and $h \approx (r_o - r_i)$ then the toroidal forms are of type (A). If $r_i < r_o$, and $r_o \sim h$, (thus $h \sim (r_o - r_i)$) then the type of the torus is of type (D). If $r_i \sim r_o \sim h$, and $h \approx (r_o - r_i)$ then the type of the torus is (B). In these tori, $h \sim (r_o - r_i)$ and we call them normal toroidal forms. However, if $h \ll (r_o - r_i)$, then the type of the torus is (C). Furthermore, If $(r_o - r_i) \ll h$, then the type of the torus is (E). These are the elongated toroidal forms, as we can see from the definition of type (C) and (E).

2.3 *Derivation of the helical forms*

In constructing a helix, the bond lengths of the hexagons substantially vary without the introduction of pentagons and/or heptagons. Thus, to make a graphitic form, it may be a good hypothesis that a helical structure will consist of pentagons, hexagons, and heptagons of carbon atoms. Therefore, a helical structure tiled by polygons was topologically constructed by cutting the torus into small pieces along the toroidal direction and replacing them, having the same toroidal direction, but slightly displaced upwards along the axis. The helix thus created contains one torus per pitch without loss of generality. Because the helix is created by the torus in our case, the properties of the helix strongly depend on the types of the torus.

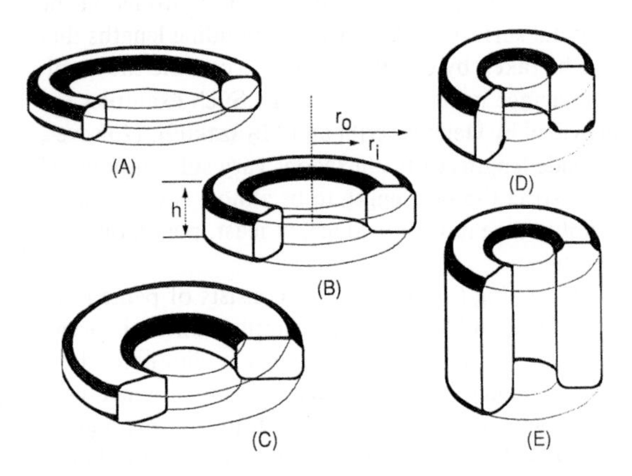

Fig. 1. Five possible simple prototypes of the toroidal forms of graphitic carbon. All cross-sections of the tube are square. Here r_o, r_i, and h are the outer and inner radii and height of the torus, respectively.

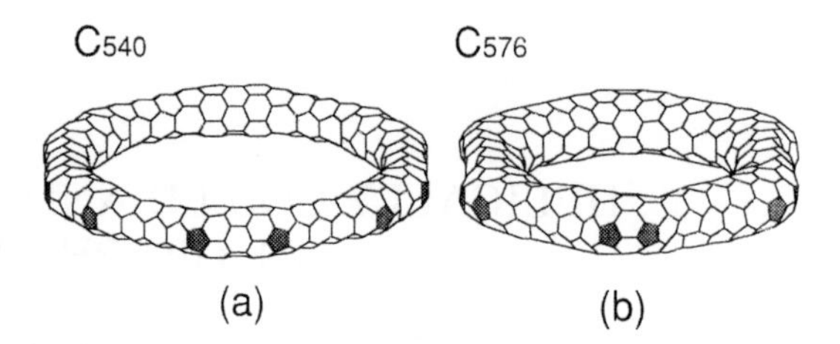

Fig. 2. Optimized toroidal structures of Dunlap's tori: (a) torus C_{540} and (b) torus C_{576}; pentagons and heptagons are shaded.

3. TOROIDAL FORMS OF GRAPHITIC CARBON

3.1 *Construction and properties of normal tori*

3.1.1 *Geometric construction of tori.* Possible constructions of tori with pentagons, heptagons, and hexagons of carbon atoms are given independently by Dunlap[10], Chernozatonskii[11], and us[12–17]. In ref. [18], the method-of-development map was used to define various structures of tori. For other tori, see ref. [19].

By connecting the sliced parts of tubes, Dunlap proposed toroidal structures C_{540} and C_{576}, both of which have six-fold rotational symmetry; both contain twelve pentagon-heptagon pairs in their equators[10] (See Fig. 2). Dunlap's construction of the tori connects carbon tubules $(2L,0)$ and (L,L) of integer L in his notation[10]. The bird's eye view of the structures of tori C_{540} and C_{576} are shown in Fig. 2. This picture is useful for understanding the difference between Dunlap's construction and ours. Dunlap's tori belong to the Type (A) according to the classification proposed in section 2.2.

Recently, we become aware that Chernozatonskii hypothetically proposed some structures of different types of toroidal forms[11], which belong to type (B) of our classification. He proposed toroidal forms by creating suitable joints between tubes. See Fig. 3 of ref. [9]. He inserts octagons or heptagons into hexagons to create a negatively curved surface, as Dunlap and we did. His tori C_{340} and C_{440} have five-fold rotational symmetry as our tori, the number of pairs of pentagons and hexagons is ten. But the pentagons (at the outer surface) and the heptagons (at the inner surface) are located in the equator of the tori as Dunlap. Chernozatonskii's tori may be in the intermediate structure between Dunlap's and ours, but two heptagons (a kind of defect) are connected to each other (which he called an Anna saddle). Since two heptagons are nearest neighbors, his torus would be energetically higher and would not be thermodynamically stable, as the placement of the pentagons follows the isolated pentagon rule. Other types of toroidal forms, such as type (C) and (E), are discussed later in section 3.2.2.

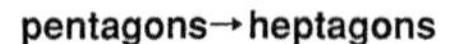

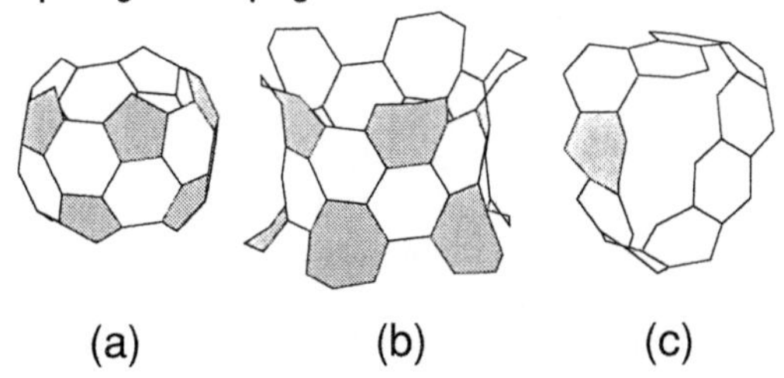

Fig. 3. Pentagon-heptagon transformation: (a) five-fold rotational surface of C_{60}; (b) negatively curved surface created by pentagon-heptagon transformation; (c) a part of the remaining surface in creating the C_{360} torus.

Contrary to the previous models, our tori[12,13, 15–17] were derived from the C_{60} fullerene because the inner surface of the tori was obtained by removing the two parallel pentagons in C_{60}, and replacing the ten remaining pentagons with heptagons, as shown in Fig. 3. The inner surface thus obtained forms arcs when cut by a vertical cross-section, and the outer surface of the torus was constructed by extending the arc until the arc became closed. Because the great circle of C_{60} consists of ten polygons, the arc of the torus was also closed by connecting ten polygons (which consists of a pentagon and a heptagon and eight hexagons). Finally, gaps were filled by hexagon rings. Using the guiding condition that $f_5 = f_7$, we created tori with 360 carbon atoms and with 240 carbon atoms as shown in Fig. 4 (a) and (b), respectively. The torus C_{360}[12,13] belongs to type (B) and the torus C_{240}[15] is type (D). As shown in Fig. 4, our tori belongs to the point group D_{5d}. Note that tori C_{360} turns out to be derived from tubules (8, 2) and that none of the pentagon-heptagon pairs lies on the equator. In refs. [13] and [15], larger or smaller tori were derived by using the Goldberg transformation, where hexagons are inserted into the original torus.

3.1.2 *Thermodynamic properties.* A molecular-dynamics simulation method (using a steepest decent method) with Stillinger-Weber potential is employed to optimize structures and to obtain the cohesive energies of the tori[12,20,21]. To confirm the thermodynamic stability, simulations at higher temperatures using a second-order equations-of-motion method were also performed. For details see ref. [13].

For the tori C_{360}, C_{240}, C_{540}, and C_{576}, the values of the cohesive energies per atom are −7.41, −7.33, −7.40, and −7.39 eV, respectively. Because the torus C_{240} has the highest ratio of the number of pentagons and heptagons to hexagons among them, torus C_{240} affects the distortion caused by the insertion of pentagons and heptagons. For tori C_{240} and C_{360}, the difference between them arises from the shape of the outer surface of these tori, because the inner surfaces of both are derived from the same surface of a spherical fullerene C_{60} with the same pentagon-to-heptagon replacements. As we raise the temperature up to 2000 K, tori C_{360}, C_{240}, C_{540}, and C_{576} retained their stability, indicating that they will be viable once they are formed.

3.1.3 *Rotational symmetric properties of tori.* We will study the various rotational symmetries of the tori. The k-rotational symmetric structures were prepared by cutting the k_0 symmetric torus along the radius of curvature into k_0 equal pieces, and by continuously combining the k pieces. Here k can be larger or smaller than k_0. Because torus C_{240} has five-fold symmetry ($k_0 = 5$), each piece contains 48 atoms. Thus, we generated tori C_{192}, C_{288}, C_{336}, and C_{384} for $k = 4$, 6, 7, and 8. For other tori, a similar procedure was used to generate various rotational symmetric forms[15].

The relaxed structures of the various (rotational) symmetric toroidal forms were obtained by steepest decent molecular-dynamics simulations[15]. For the elongated tori derived from torus C_{240}, the seven-fold rotational symmetry is found to be the most stable. Either five-fold or six-fold rotational symmetry is the most stable for the toroidal forms derived from tori C_{360} and C_{540}, respectively (see Fig. 5).

Because the cohesive energy of the fullerene C_{60} is −7.29 eV/atom and that of the graphite sheet is −7.44 eV/atom, the toroidal forms (except torus C_{192}) are energetically stable (see Fig. 5). Finite temperature molecular-dynamics simulations show that all tori (except torus C_{192}) are thermodynamically stable.

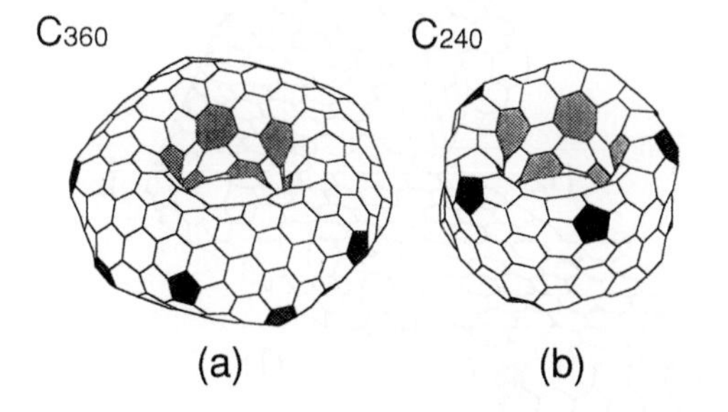

Fig. 4. Optimized toroidal structures: (a) torus C_{360} and (b) torus C_{240}; Pentagons and heptagons are shaded. The diameters of the tube of the stable torus C_{360} determined by optimization using molecular dynamics with Stillinger-Weber potential[21], is 8.8 Å. The diameter of the hole is 7.8 Å, which is quite close to the diameter of fullerence C_{60}.

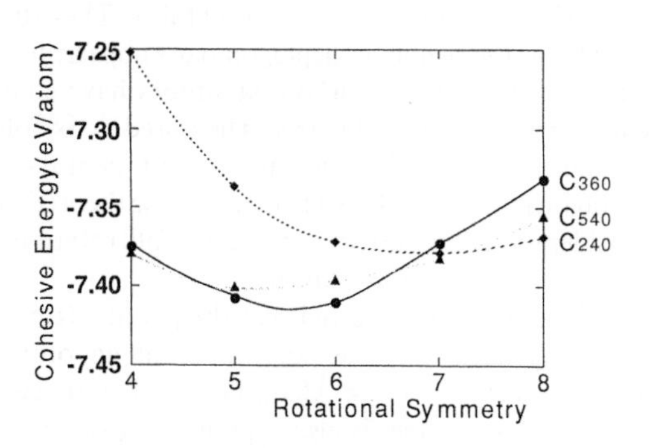

Fig. 5. Dependence of the cohesive energy of tori C_{360}, C_{240}, and C_{540} on the rotational symmetry.

3.1.4 *Recent results of electronic calculations.* Total energy calculations or molecular orbital calculations are necessary to explore electronic, optical, and chemical properties of toroidal forms. From the *ab initio* self-consistent field (SCF) calculation [22] for the torus C_{120}, the HOMO-LUMO (highest-occupied-lowest-unoccupied molecular orbitals) gap, which is responsible for the chemical stability, is 7.5 eV. This is close to that of SCF calculation for C_{60} of 7.4 eV. (If the value of the HOMO-LUMO gap is zero, the molecule is chemically active, thus unstable.) In SCF, the HOMO-LUMO is different from local density approximation. For stability, ours is consistent with the result of the all-electron local density approximation calculation where the value is 1.0 eV for the HOMO-LUMO gap[23]. Recent tight-binding calculation of the same author[24], indicates that the HOMO-LUMO gap for C_{360} is 0.3 eV. These values indicate that toroidal structures are chemically stable. The tight-binding calculation of the HOMO-LUMO gap for tori C_{540} and C_{576} gives 0.04 eV and 0.02 eV, respectively[10].

Our Hückel-type calculation for isomers of C_{240}[16] indicates that the positions and directions of the polygons change the electronic structures substantially for C_{240} or C_{250}. Because of the geometrical complexity of the torus, any simple systematics, as have been found for the band gaps of the carbon nanotubes[6], could not be derived from our calculations. But, the common characteristics of the isomers for C_{240} with large HOMO-LUMO gaps are that their inner and outer tubes have the same helicities or that the pentagons and heptagons are radially aligned. Note that the HOMO-LUMO gap of the torus C_{240}, which is shown in Fig. 3 (b), is 0.497 eV.

3.2. *Results of the experiments and elongated tori*

3.2.1 *Results of the experiments.* Several experimental groups try to offer support for the existence of the toroidal form of graphitic carbon[25]. Transmission electron microscopy (TEM) images taken by Iijima, Ajayan, and Ichihashi[26] provided experimental evidence for the existence of pairs of pentagons (outer rim) and heptagons (inner rim), which are essential in creating the toroidal structure[10–17], in the turn-over edge (or turn-around edge[26]) of carbon nanometer-sized tubes. They suggested that the pentagon-heptagon pairs appearing in the turn-over edge of carbon nanotubes have some symmetry along the tube axis. They used a six-fold symmetric case where the number of pentagon-heptagon pairs is six. This accords with the theoretical consideration that the five-, six-, seven-fold rotational symmetric tori are most stable.

Iijima *et al.* also showed that the parallel fringes appearing in the turn-over edge of carbon nanotubes have a separation of 3.4 Å[26]. (This value of separation in nested tubes is also supported by other authors[27].) It is quite close to that of the "elongated" toroidal form of C_{240} proposed by us[15].

3.2.2 *Elongated tori.* The experiments, at the present time, suggest that the torus of type (D) with parallel fringes at a separation of 3.7 Å, such as C_{240}, is likely to exist. Thus, the type (C) structures having height of 3.7 Å could exist. See Fig. 6.

If we consider the $1/k$ part of the chain of the circle, the number of hexagons can be put n_1 and n_2 for the outer and inner circle of the upper (or lower) hexagonal chain (see Fig. 6), respectively. Each upper and lower hexagonal chain contains $n_1^2 + n_2^2 + 2(n_1 + n_2)$ atoms. The number of the hexagons along the height is put L, where L is a positive integer. For torus C_{240}, $n_1 = n_2 = 3$, and $L = 1$ and $k = 5$. If we elongate (by putting hexagons for allowed locations) the thickness of the tube, then $r_o - r_i$, $n_1 - n_2$ increases. On the other hand, if we elongate the height of the torus, L increases.

By inserting a cylindrical tube of hexagons, we stretch the length of the toroidal forms whose heights are larger than the radii, by putting $n_1 = n_2 = 3$, $k = 5$ and increasing L. The stretched toroidal forms we thus obtained[17], type (D), are C_{240}, C_{360}, C_{480}, C_{600}, C_{720}, C_{840} . . . (See Fig. 7). These forms are links between toroidal forms and short (nanometer-scale) length turn-over tubes. The values of the cohesive energies for tori C_{240}, C_{360}, C_{480}, C_{600}, C_{720}, and C_{840} are −7.338, −7.339, −7.409, −7.415, −7.419, and −7.420 eV/atom, respectively. Note that their cohesive energies decrease with increasing height of the tori (or L) (i.e., number of hexagons). Simulations showed that these stretched toroidal forms are thermodynamically stable.

Using the torus C_{288} of D_{6h} which is derived from the torus C_{240} of D_{5h}, shallow tori, type (D), are generated by putting $L = 1$, $k = 6$, and $n_2 = 3$, with varying n_1 (= 3,4, 5,6,7,8,9). Tori having D_{6h} symmetry are shown in Fig. 8.

In Table 1, cohesive energies for the tori (of $L = 1$, $k = 6$) for various n_1 and n_2 are given. The cohesive energy is the lowest for $n_1 - n_2 = 0$, and also has

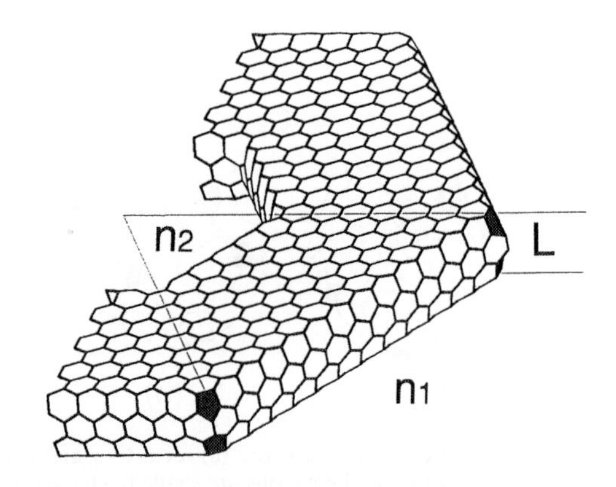

Fig. 6. Part of the elongated torus: here, n_1, n_2, and L are the number of the hexagons along the inner circle, outer circle, and height of the torus, respectively; this figure is for the case of $n_1 = 12$, $n_2 = 6$, and $L = 1$.

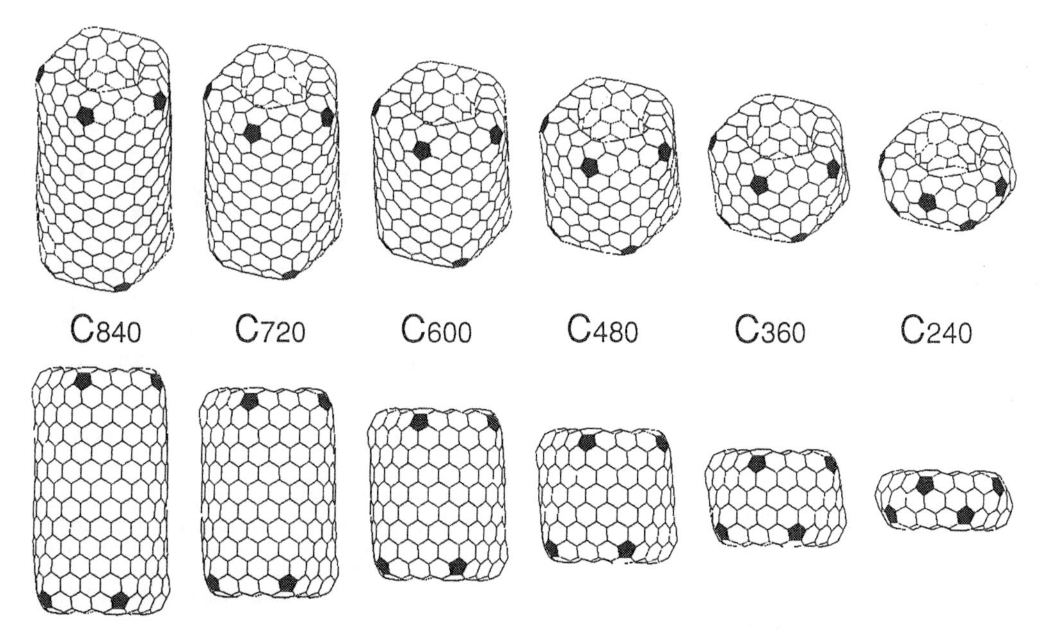

Fig. 7. Optimized structures of the tori shown by elongation to height: pentagons and heptagons are shaded; top views and side views are shown in each case.

minimal at $n_1 - n_2 = 2$ for $n_2 = 3,4$, and 6. Contrary to the five-fold rotational symmetric surface, where the upper and lower planes become convex and, hence, increase the energy of the tori with n_1, the surface remains flat for the six-fold rotational symmetric case ($k = 6$) with increasing n_1. However, the cohesive energy of the six-fold rotational symmetric torus increases with n_1 (if n_1 is larger than $n_2 + 2$) (i.e., elongating n_1). This increasing in energy with n_1 arises from the increasing stress energy of the outer edges, where the number of hexagons that have folding bonds at the edges increases linearly with n_1.

4. HELICALLY COILED FORM OF GRAPHITIC CARBON

4.1 *Expecting properties of helices*

Helically coiled forms of the carbon cage on the nanometer scale[14] concerned here have graphitic layer(s), in contrast to the micron-order amorphous carbon fiber previously created by experiments[28]. The motivation of studying the helical structure is as follows: (1) the electrical, magnetic, and elastic properties can be modulated by the tiling pattern of the pentagons, hexagons, and heptagons and/or writhing

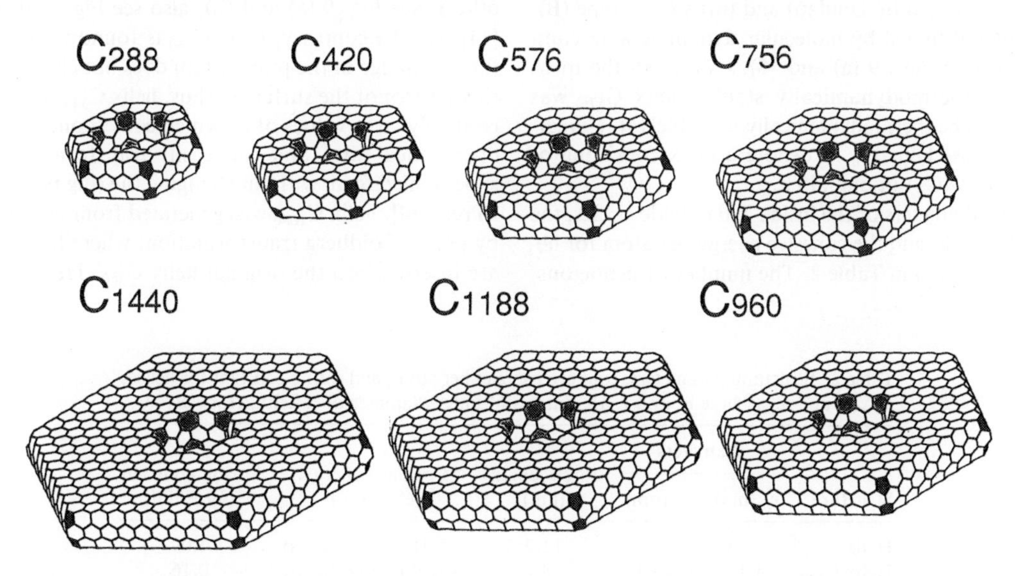

Fig. 8. Optimized shallow toroidal structures: the subscripts indicate the number of the carbon atoms in the torus; pentagons and heptagons are shaded.

Table 1. Cohesive energies of shallow tori; the parameters n_1 and n_2 are the number of hexagons along the outer and inner circle, respectively (see Fig. 6). Here N is the number of atoms in a torus

n_1	n_2	N	Energy (eV/atom)
3	3	288	−7.376
4	3	420	−7.369
5	3	576	−7.375
6	3	756	−7.376
7	3	960	−7.375
8	3	1188	−7.372
9	3	1440	−7.369
4	4	384	−7.378
5	4	540	−7.368
6	4	720	−7.374
7	4	924	−7.374
8	4	1152	−7.372
9	4	1404	−7.370
6	6	576	−7.382
7	6	780	−7.369
8	6	1008	−7.373
9	6	1260	−7.372
10	6	1536	−7.370
11	6	1836	−7.368
12	6	2160	−7.366

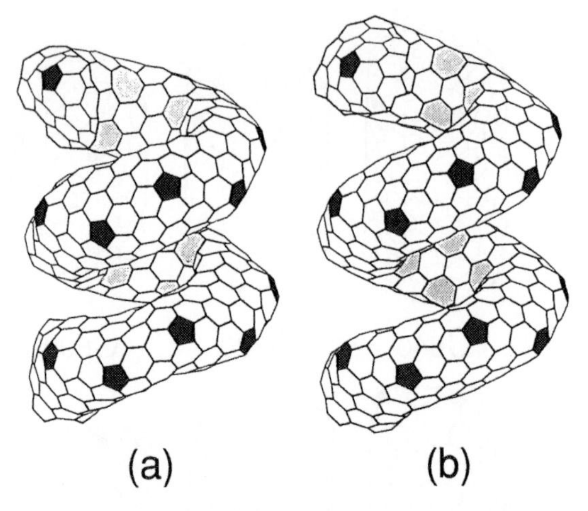

Fig. 9. Helically coiled form C_{360}: one pitch contains a torus C_{360}. (a) coil length = 12.9 Å and, (b) coil length = 13.23 Å. The tiling pattern of heptagons in the inner ridge line is changed, though the pattern of pentagons in the outer ridge line remains upon changing the coil length.

or twisting[29], in addition to changing the diameters (of the cross-sections) and the degree of helical arrangement as in straight tubes[6]; (2) a variety of applications are expected because a variety of helical structures can be formed; for instance, a helix with a curved axis can form a new helix of higher order, such as a super-coil or a super-super coil, as discussed below.

4.2 *Properties for the helices derived from normal tori*

The properties of optimized helical structures, which were derived from torus C_{540} and C_{576}, type (A), (proposed by Dunlap) and torus C_{360}, type (B), (proposed by us) by molecular dynamics were compared. (see Figs. 9 (a) and 10). (Although the torus C_{576} is thermodynamically stable, helix C_{576} was found to be thermodynamically unstable[14]. Hereafter, we use helix C_n to denote a helix consisting of one torus (C_n) in one pitch.

The diameters of the inside and outside circles, the pitch length, and the cohesive energy per atom for helices are given in Table 2. The number of pentagons, hexagons, and heptagons per 360 and 540 atoms in the helical structure are the same as in the torus C_{360} and C_{540}[13,14]. By pulling the helix coil, the coil length for helix C_{360} increases from 12.9 Å (pitch angle α = 15.17 degrees, See Fig. 9 (a)) to 13.23 Å (α = 19.73 degrees, Fig. 9 (b)).

Because the second derivative of the cohesive energy with respect to the coil length provides the spring constant, the spring constants of the helical structures per pitch were estimated numerically. As shown in Table 2, the spring constant for helix C_{360} is 25 times larger than that of helix C_{540}. We found that the helix C_{360} is so stiff that the ring pattern changes. Although the pattern of the pentagons remains the same, the heptagons along the inner ridge line move their position and their pattern changes discretely with increasing pitch angle α (from one stable pitch angle to the other). See Fig. 9 (a) and (b); also see Fig. 3 of ref. [14]. On the contrary, helix C_{540} is found to be soft (i.e., a change in the pitch length does not change the ring pattern of the surface). Thus, helix C_{540} can have relatively large values of α, which corresponds to the open-coiled form and can easily transform to the super-coiled form without changing the ring patterns. In ref. [14], helix C_{1080} was generated from helix C_{360} by use of Goldberg transformation, where hexagons are inserted into the original helix C_{360}. Helix C_{1080}

Table 2. Structural parameters, cohesive energies per atom, and spring constant for helices C_{360} and C_{540}; here r_o and r_i are outer and inner diameter of a helix, respectively

Structure	diameters r_o (nm)	diameters r_i (nm)	Pitch length (nm)	Cohesive energy (eV/atom)	Spring constant (meV/nm)
Helix C_{360}	2.26	0.78	12.9	−7.41 (−7.41 torus)	4.09
Helix C_{540}	4.14	2.94	8.5	−7.39 (−7.40 torus)	0.16

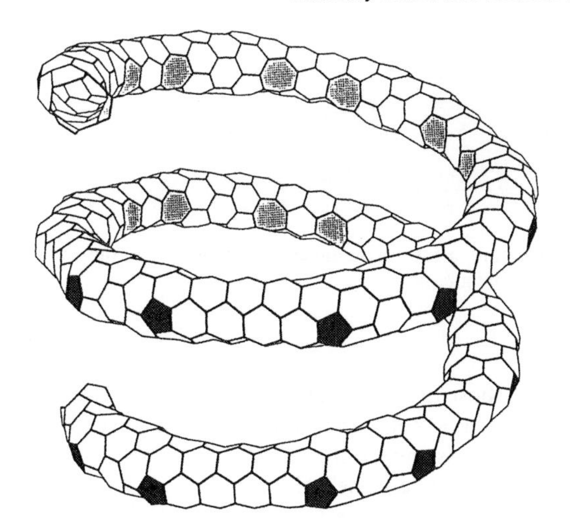

Fig. 10. Helically coiled form C_{540}: one pitch contains a torus C_{540}.

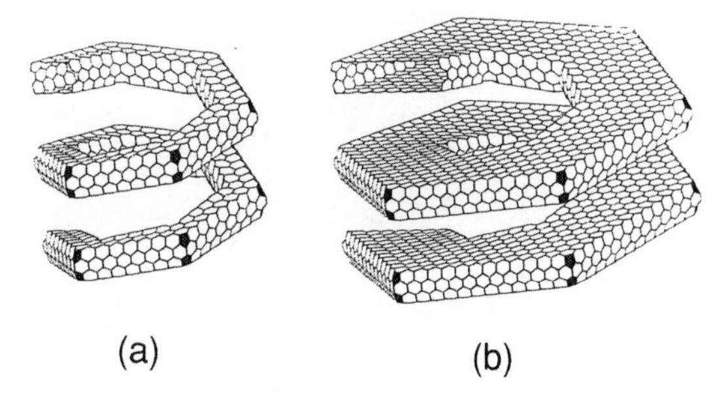

Fig. 12. Elongated helical structures (a) helix C_{756} and (b) helix C_{2160}.

was found to be stiffer than helix C_{360}[14]. The difference in elasticity of the helically coiled forms of helices may be attributed to the difference in patterns of the heptagons; these are sensitive to the geometric properties, such as the ratio of the radii of the cross-section and the curvature.

Because the tube diameter of the helix C_{540} is small compared to the helix C_{360}, atoms at the open ends must bend inwards to cover the open end. The open end is covered with six hexagons, one heptagon, one square, and one pentagon, see Fig. 11. The electronic structure of helices is strongly affected by the end pattern of the rings because the end rings of odd numbers play a scattering center, such as a disclination center as discussed by Tamura and Tsukada[30]. The edge effect may lead to adding an exotic electronic character to the helical structure which is not seen in the straight tubes.

4.3 *Helices derived from elongated tori*

From elongated tori, such as type (C), type (D), and type (E), helical structures are derived. For example, from the type (C) elongated torus of D_{6h}, mentioned in 3.2.2, helix C_{756} ($n_1 = 6$, $n_2 = 3$, $L = 1$) and helix C_{2160} ($n_1 = 12$, $n_2 = 6$, $L = 1$) can be generated (see Fig. 12). In these cases, the flat part (i.e., the part resembling the graphite layer) becomes wider and wider with increasing n_1. Thus, this type of helical structure has minimal cohesive energy at $n_1 - n_2 = 0,2$ as observed in the shallow tori.

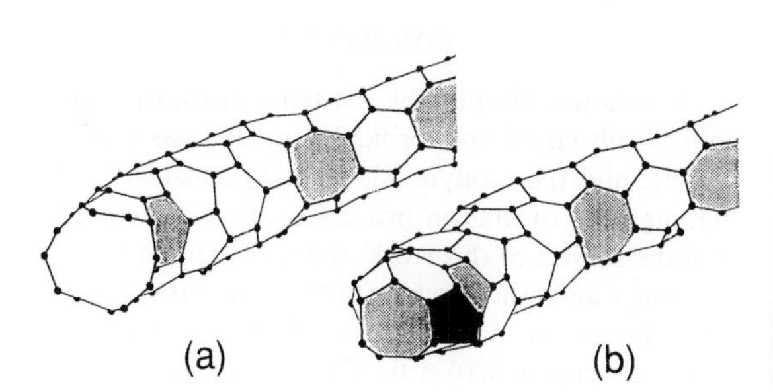

Fig. 11. Edge of the helix C_{540}: (a) initial state and, (b) reconstructed form of the edge; the edge contains a square, heptagon, pentagons and hexagons.

4.4 *Comparison with experiments*

Ivanov *et al.*[31] and Van Tendeloo *et al.*[32] reported a synthesis of helically coiled multi-layered form. They showed that: (1) cobalt on silica is the best catalyst-support combination for the production of graphite tubes, such as straight tubes and coiled ones, and (2) decreasing temperature from 973 to 873 K leads to strong decrease in the amorphous carbon production. Also, (3) helically coiled carbon tubes were obtained with inner and outer diameter of 3–7 and 15–20 nm, respectively, and up to 30 μm in length. The size of the helical structure is orders of magnitude smaller than the helix-shaped fibers composed of amorphous carbon[28]. Note their sizes are much larger than that of the theoretical one[14]. (4) Using TEM and the electron diffraction method, they suggested that the helically coiled tubes consist of a regularly polygonized structure, where the bend may be related to pairs of pentagon-hexagon carbon rings in the hexagonal network as suggested by ref. [14]. (5) As shown in Fig. 13, a helix-shaped nanotube with ra-

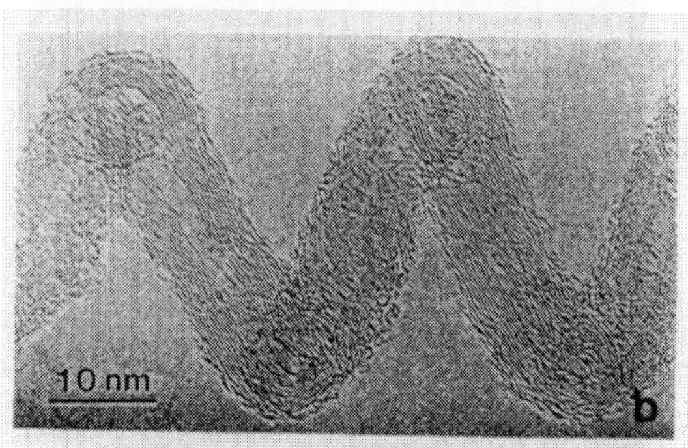

Fig. 13. TEM picture of a helix-shaped structure with radius of about 18 nm, pitch about 30 nm, containing 10 graphite tubes (after V. Ivanov *et al.*).

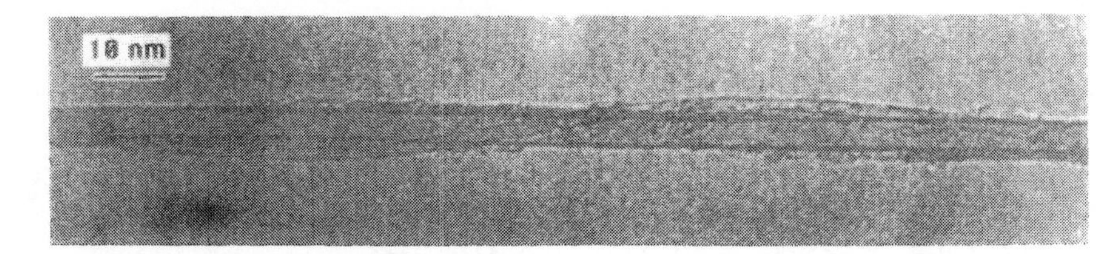

Fig. 14. TEM picture of a single layered helix-shaped structure: a 1.3-nm diameter helix coils around the 3.6 nm tube (after C.H. Kiang *et al.*).

dius of about 18 nm, pitch about 30 nm, has ten graphitic layered tubes (diameter of the innermost tube is about 2.5 nm.).

C.-H. Kiang *et al.*[33] reported that the single-layered coiled tubes were obtained by co-vaporizing cobalt with carbon in an arc fullerene generator. A single-layered helical structure with radii of curvature as small as 20 nm was seen. These helically coiled forms tend to bundle together. In the soot obtained with sulfur-containing anodes, they also found the 1.3-nm diameter tube coil around the 3.6 nm tube (see Fig. 14). This kind of structure was theoretically proposed in ref. [14].

By close analysis of diffraction pattern of catalyst-grown coiled tube, X. B. Zhang *et al.*[34] reported the larger angular bends and their number to be about 12 per helix turn (the magnitude of angular bends is about 30 degrees), and this is the essential in determining the geometry. Their smallest observed helix has a radius of about 8 nm. Thus, their sizes are much larger than those of theoretically predicted ones. However, it should be noted that in ref. [14], we have pointed out the existence of the larger helices. We provide an example of a large helical form: the helix C_{1080} can be generated from helix C_{360} using the Goldberg algorithm, as larger tori were derived from genetic one such as torus C_{120}[14].

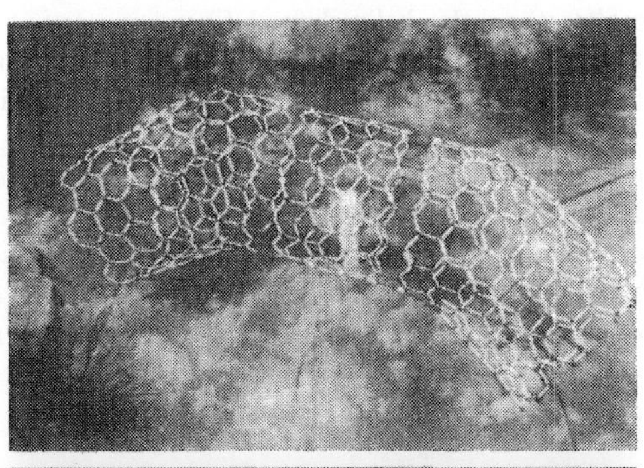

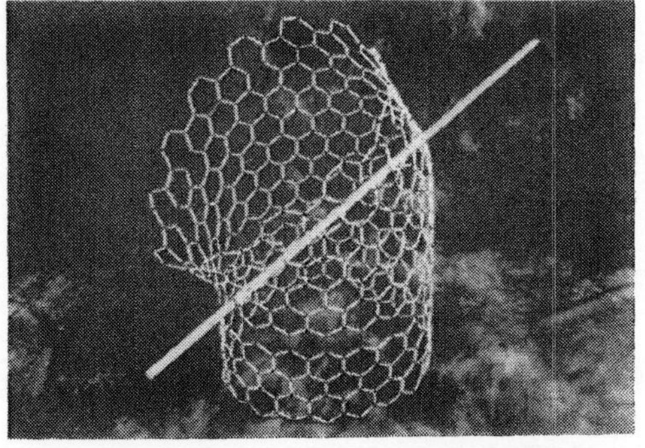

Fig. 15. Tori of Zhang *et al.*; a 30-degree connections of tubes: a (12,0) segment and a (7,7) segment (After Zhang *et al.*).

Zhang *et al.*[34] also provide a molecular model by connecting tubes at angle 30° bends by introducing the required pairs of pentagons and hexagons in the hexagonal network. Their method is quite similar to Dunlap's way of creating torus C_{540}. They combined the $(12 + 9n,0)$ tube and $(7 + 5n,7 + 5n)$ tube in the Dunlap's notation, and also combined $(9n,0)$ tube to $(5n,5n)$ tube to show the feasibility of the multi-layered helical forms. Combination of these tubes, however, leads to toroidal forms whose connection is quite similar to that of torus C_{540}. But the tori of Zhang *et al.* (see Fig. 15) can easily be turned into helices by regularly rotating the azimuth of the successive lines connecting pentagon and heptagons with small energy, as helix C_{540} as shown in ref. [14]. (Torus C_{540} can transform to helix C_{540} without rebonding the carbon atoms on the network with small energy, as we showed in section 4.2.) See also ref. [35].

Good semi-quantitative agreements are found in diffraction patterns and proposed models obtained by molecular-dynamics[14], because the results of the experiments[31–34] are consistent with the atomic models proposed by us[14]. However, in the present state of high-resolution electron microscopy, taking into account, moreover, the number of sheets and the complicated geometry of the helix, it seems unlikely to directly visualize the pentagon-hexagon pairs.

5. CONCLUSION

We showed the possible existence of various forms of helically coiled and toroidal structures based on energetic and thermodynamic stability considerations. Though the formation process of these structures is not the subject of this work, the variety of patterns in the outer and inner surface of the structures indicates that there exist many different forms of stable cage carbon structures[10–19]. The molecules in a one-dimensional chain, or a two-dimensional plane, or a three-dimensional supermolecule are possible extended structures of tori with rich applications.

Many different coiled forms of stable cage carbon

structures also exist. In the helically coiled form, coils may be able to transform into other forms. It would be interesting if the proposed structure and its variant forms—combinations of helix with toroidal forms, helical coiling around the tube, nested helical forms, coiled structure of higher order such as supercoil observed in biological systems—could be constructed in a controlled manner from the graphitic carbon cage. Because the insertion of pentagons and heptagons into hexagons changes the electronic structures[26], helical and toroidal forms will have interesting electrical and magnetic properties, which could not be seen in the cylindrical tubes by modulation of the periodicity of the appearance of the pentagon and hexagon pairs[36].

Acknowledgements—We are grateful to G. Van Tendeloo and D. S. Bethune for sending us TEM pictures of helically coiled graphitic carbon. We are also grateful for the useful discussions with Masaru Tsukada, Ryo Tamura, Kazuto Akagi, and Jun-ichi Kitakami. We are also grateful for the useful discussions with Toshiki Tajima, J. C. Greer, and Sumio Iijima.

REFERENCES

1. H. W. Kroto, J. R. Heath, S. C. O'Brien, R. F. Curl, and R. E. Smalley, *Nature* (London) **318**, 162 (1985). W. Krätschmer, L. D. Lamb, K. Fostiroropoulos, and D. R. Huffman, *Nature* **347**, 354 (1990).
2. F. Diederich, R. L. Whetten, C. Thilgen, R. Ettl, I. Chao, and M. M. Alverez, *Science* **254**, 1768 (1991). K. Kikuchi, N. Nakahara, T. Wakabayashi, M. Honda, H. Matsumiya, T. Moriwaki, S. Suzuki, H. Shiromaru, K. Saito, K. Yamauchi, I. Ikemoto, and Y. Achiba, *Chem. Phys. Lett.* **188**, 177 (1992). D. Ugarte, *Nature* (London) **359**, 707 (1992). S. Iijima, *J. Phys. Chem.* **91**, 3466 (1987).
3. R. Buckminster Fuller, US patent No. 2682235 (1954).
4. Koji Miyazaki, *Puraton to Gojyuunotou* (Plato and Five-Storied Pagoda) (in Japanese) pp. 224. Jinbun-shoin, Kyoto, (1987). Koji Miyazaki, *Fivefold Symmetry* (Edited by I. Hargittai) p. 361. World Sci. Pub, Singapore (1992).
5. S. Iijima, *Nature* (London) **354**, 56 (1991). T. W. Ebbesen and P. M. Ajayan, *Nature* **358**, 220 (1992). S. Iijima and T. Ichihashi, *Nature* (London) **363**, 603 (1993). D. S. Bethune, C. H. Kiang, M. S. de Vries, G. Gorman, R. Savoy, J. Vazquez, and R. Beyers, *Nature* **363**, 605 (1993).
6. J. W. Mintmire, B. I. Dunlap, and C. T. White, *Phys. Rev. Lett.* **68**, 631 (1992). N. Hamada, S. Sawada, and A. Oshiyama, *Phys. Rev. Lett.* **68**, 1579 (1990). D. H. Robertson, D. W. Brenner, and J. W. Mintmire, *Phys. Rev.* **B45**, 12592 (1992). M. Fujita, M. Saito, G. Dresselhaus, and M. S. Dresselhaus, *Phys. Rev.* **B45**, 13834 (1992).
7. A. L. Mackay and H. Terrones, *Nature* (London) 352, 762 (1991). T. Lenosky, X. Gonze, M. P. Teter, and V. Elser, *Nature* **355**, 333 (1992). D. Vanderbilt and J. Tersoff, *Phys. Rev. Lett.* **68**, 511 (1992). S. J. Townsend, T. J. Lenosky, D. A. Muller, C. S. Nichols, and V. Elser, *Phys. Rev. Lett.* **69**, 921 (1992). R. Phillips, D. A. Drabold, T. Lenosky, G. B. Adams, and O. F. Sankey, *Phys. Ref.* **B46**, 1941 (1992). W. Y. Ching, Ming-Zhu Huang, and Young-nian-Xu, *Phys. Rev.* **B46**, 9910 (1992). Ming-Zhu Huang, W. Y. Ching, and T. Lenosky, *Phys. Rev.* **B47**, 1593 (1992).
8. E. Heilbonner, *Helv. Chim. Acta* **37**, 921 (1954).
9. K. Miyazaki, *Polyhedra and Architecture* (in Japanese), pp. 270. Shokoku-sha, Tokyo (1979).
10. B. I. Dunlap, *Phys. Rev.* **B46**, 1933 (1992).
11. L. A. Chernozatonskii, *Phys. Lett.* **A170**, 37 (1992).
12. S. Itoh, S. Ihara, and J. Kitakami, *Phys. Rev.* **B47**, 1703 (1993).
13. S. Ihara, S. Itoh, and J. Kitakami, *Phys. Rev.* **B47**, 12908 (1993).
14. S. Ihara, S. Itoh, and J. Kitakami, *Phys. Rev.* **B48**, 5643 (1993).
15. S. Itoh and S. Ihara, *Phys. Rev.* **B48**, 8323 (1993).
16. S. Itoh and S. Ihara, *Phys. Rev.* **B49**, 13970 (1994).
17. S. Ihara and S. Itoh, *Proceedings of 22nd International Conference on Physics of Semiconductors* (Edited by D. J. Lockwood), p. 2085. World Sci. Pub., Singapore (1995).
18. M. Fujita, M. Yoshida, and E. Osawa, *Fullerene Sci. Tech.* (in press).
19. E. G. Gal'pern, I. V. Stankevich, A. L. Chistyakov, and L. A. Chernozatonskii, *Fullerene Sci. Technol.* **2**, 1 (1994).
20. F. H. Stillinger and T. A. Weber, *Phys. Rev.* **B31**, 5262 (1985). F. H. Stillinger and T. A. Weber, *Phys. Rev.* **B33**, 1451 (1986).
21. F. F. Abraham and I. P. Batra, *Surf. Sci.* **209**, L125 (1989).
22. J. C. Greer, S. Itoh, and S. Ihara, *Chem. Phys. Lett.* **222**, 621 (1994).
23. M. R. Pederson, J. K. Johnson, and J. Q. Broughton, *Bull. Am. Phys. Soc.* **39**, 898 (1994).
24. J. K. Johnson, *Proceedings Materials Research Society Symposium*, Fall, 1993 (to be published).
25. M. Endo (private communication).
26. S. Iijima, P. M. Ajayan, and T. Ichihashi, *Phys. Rev. Lett.* **69**, 3100 (1992). S. Iijima, T. Ichihashi, and Y. Ando, *Nature* (London) **356**, 776 (1992).
27. Y. Saito, T. Yoshikawa, S. Bandow, M. Tomita, and T. Hayashi, *Phys. Rev.* **B48**, 1907 (1993).
28. H. Iwanaga, M. Kawaguchi, and S. Motojima, *Jpn. J. Appl. Phys.* **32**, 105 (1993). S. Motojima, M. Kawaguchi, K. Nozaki, and H. Iwanaga, *Carbon* **29**, 379 (1991). M. S. Dresselhaus and G. Dresselhaus, *Adv. Phys.* **30**, 139 (1981).
29. W. R. Bauer, F. H. C. Crick, and J. H. White, *Sci. Am.* **243**, 100 (1980). F. H. C. Crick, *Proc. Natl. Acad. Sci. U.S.A.* **73**, 2639 (1976).
30. R. Tamura and M. Tsukada, *Phys. Rev.* **B49**, 7697 (1994).
31. V. Ivanov, J. B. Nagy, Ph. Lambin, X. B. Zhang, X. F. Zhang, D. Bernaerts, G. Van Tendeloo, S. Amelinckx, and J. Van Landuyt, *Chem. Phys. Lett.* **223**, 329 (1994).
32. G. Van Tendeloo, J. Van Landuyt, and S. Amelincx, *The Electrochemical Society 185th Meeting*, San Francisco, California, May (1994).
33. C. H. Kiang, W. A. Goddard III, R. Beyers, J. R. Salem, and D. S. Bethune, *J. Phys. Chem.* **98**, 6618 (1994).
34. X. B. Zhang, X. F. Zhang, D. Bernaerts, G. Van Tendeloo, S. Amelinckx, J. Van Landuyt, V. Ivanov, J. B. Nagy, Ph. Lambin, and A. A. Lucas, *Europhys. Lett.* **27**, 141 (1994).
35. B. I. Dunlap, *Phys. Rev.* **B50**, 8134 (1994).
36. K. Akagi, R. Tamura, M. Tsukada, S. Itoh, and S. Ihara, *Phys. Rev. Lett.* **74**, 2703 (1995).

MODEL STRUCTURE OF PERFECTLY GRAPHITIZABLE COILED CARBON NANOTUBES

A. FONSECA, K. HERNADI, J. B.NAGY, PH. LAMBIN and A. A. LUCAS
Institute for Studies in Interface Sciences, Facultés Universitaires Notre-Dame de la Paix, Rue de Bruxelles 61, B-5000 Namur, Belgium

(*Received* 20 *April* 1995; *accepted in revised form* 3 *August* 1995)

Abstract—The connection of two straight chiral or achiral cylindrical carbon nanotube sections of approximately the same diameters connecting at a "knee" angle of $\pi/5$ is described. Such knees are based on the insertion in the plane of the knee of diametrically opposed pentagonal and heptagonal rings in the hexagonal network. Relationships are also established between the nanotubes and their concentric graphitic layers. A growth mechanism leading to perfect carbon tubules and tubule connections on a catalyst particle at a molecular level is described. The mechanism suggested explains the formation of curved nanotubes, tori or coils involving the heptagon–pentagon construction of Dunlap.

Key Words—Carbon fibers, nanofibers, nanotubes, nanotube knees, fullerenes, tubules.

1. INTRODUCTION

During the last years, several authors have reported the production of carbon nanotubes by the catalytic decomposition of hydrocarbons in the presence of metals[1–5]. More recently, carbon nanotubes were also found as by-products of arc-discharge[6] and hydrocarbon flame[7] production of fullerenes.

The appearance of a large amount of curved and coiled nanotubes among the tubes produced by the catalytic method stimulated several studies on the theoretical aspect of the coiling mechanism[8–11]. Based on observations from high resolution electron microscopy and electron diffraction, it was proposed in these studies that curving and coiling could be accomplished by the occurrence of "knees" connecting two straight cylindrical tube sections of the same diameter. Such knees can be obtained by the insertion in the plane of the knee of diametrically opposed pentagonal and heptagonal carbon rings in the hexagonal network. The heptagon with its negative curvature is on the inner side of the knee, and the pentagon is on the outer side. The possibility of such construction was suggested by Dunlap[12,13]. Theoretical models of curved nanotubes forming tori of irregular diameters have also been described by Itoh *et al.* [14].

In this paper we elaborate models of perfect tubule connections leading to curved nanotubes, tori or coils using the heptagon–pentagon construction of Dunlap[12,13]. In order to understand the mechanisms of formation of perfectly graphitized multilayered nanotubes, models of concentric tubules at distances close to the characteristic graphite distance and with various types of knee were built. (Hereafter, for the sake of clarity, "tubules" will be reserved to the individual concentric layers in a multilayered nanotube.)

2. KNEE STRUCTURES

2.1 *Labeling tubules*

Following a standard notation[12,13], a cylindrical tubule can be described by the (L,M) couple of integers, as represented in Fig. 1. When the plane graphene sheet (Fig. 1) is rolled into a cylinder so that the equivalent points 0 and M of the graphene sheet are superimposed, a tubule labeled (L,M) is formed. L and M are the numbers of six membered rings separating 0 from L and L from M, respectively. Without loss of generality, it can be assumed that $L \geq M$.

Among all the different tubules, and for the sake of simplicity, mostly $(L,0)$ and (L',L') nonchiral tubules will be considered in this paper. Such tubules can be described in terms of multiples of the distances l and d, respectively (Fig. 2).

The perimeter of the $(L,0)$ tubule is composed of L "parallel" hexagon building blocks bonded side by side, with the bonded side parallel to the tubule axis. Its length is equal to Ll.

The perimeter of the (L',L') tubule is composed of L' "perpendicular" hexagon building blocks bonded head to tail by a bond perpendicular to the tubule axis. Its length is equal to $L'd$.

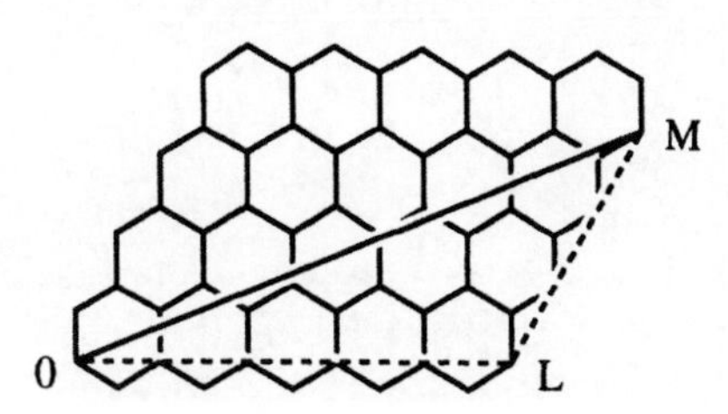

Fig. 1. Unrolled representation of the tubule (5,3). The $0M$ distance is equal to the perimeter of the tubule. $0M = a\sqrt{3(L^2+M^2+LM)}$, where a is the C–C bond length.

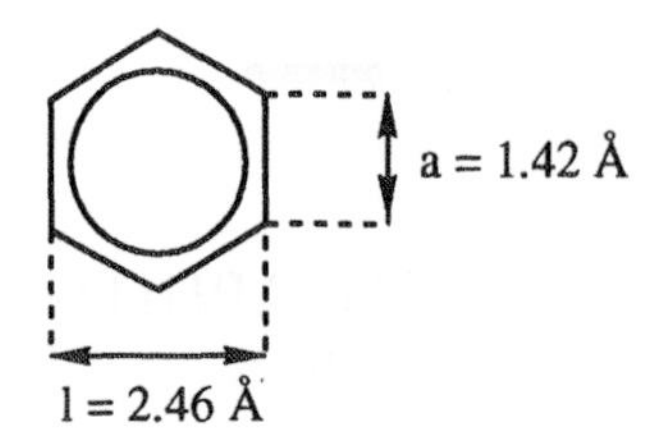

parallel hexagon unit cell
(**l** for the (L,0) tubules)

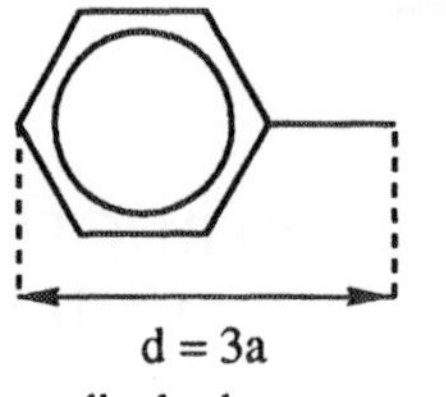

perpendicular hexagon unit cell
(**d** for the (L',L') tubules)

a = side of the hexagon in graphite
l = width of the hexagon in graphite = $a\sqrt{3}$

Fig. 2. Building blocks for the construction of (L,0) and (L',L') tubules.

2.2 *Connecting a* (L ,0) *to a* (L',L') *tubule by means of a knee*

Dunlap describes the connection between (L,0) and (L',L') tubules by means of knees. A knee is formed by the presence of a pentagon on the convex and of a heptagon on the concave side of the knee. An example is illustrated in Fig. 3(a). The (12,0)–(7,7) knee is chosen for illustration because it connects two tubules whose diameters differ by only 1%. The bent tubule obtained by that connection was called ideal by Dunlap[12,13].

If one attempts to build a second coaxial knee around the ideal (12,0)–(7,7) knee at an interlayer distance of 3.46 Å, the second layer requires a (21,0)–(12,12) knee. In this case, the axis going through the centers of the heptagon and of the pentagon of each tubule are not aligned. Moreover, the difference of diameter between the two connected segments of each knee is not the same for the two knees [Fig. 3(b)].

As distinct from the ideal connection of Dunlap, we now describe the series of nanotubule knees ($9n$,0)–($5n$,$5n$), with n an integer. We call this series the *perfectly graphitizable carbon nanotubules* because the difference of diameter between the two connected segments of each knee is constant for all knees of the series (Fig. 4). The two straight tubules connected to form the $n=1$ knee of that series are directly related to C_{60}, the most perfect fullerene[15], as shown by the fact that the (9,0) tubule can be closed by 1/2 C_{60} cut at the equatorial plane perpendicular to its threefold rotation symmetry axis, while the (5,5) tubule can be closed by 1/2 C_{60} cut at the equatorial plane perpendicular to its fivefold rotation symmetry axis [Fig. 5(a)].

As a general rule, any knee of the series ($9n$,0)–($5n$,$5n$) can be closed by 1/2 of the fullerene $C_{(60.n^2)}$. Note that, for this multilayer series, there is a single axis going through the middle of the heptagons and pentagons of any arbitrary number of

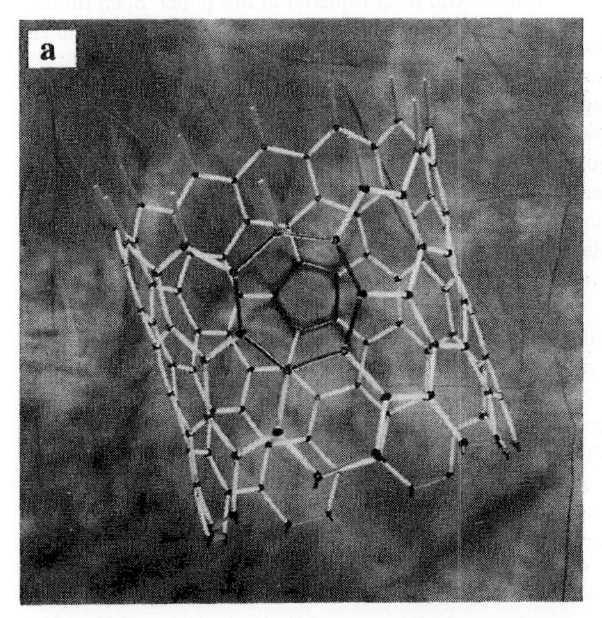

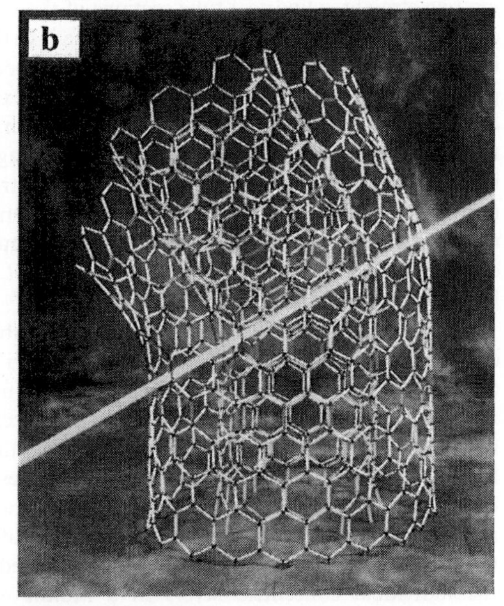

Fig. 3. (a) Model structure of the (12,0)–(7,7) knee, shown along the 5–7 ring diameter, (b) model structure of the (21,0)–(12,12) knee ($\Delta_{Diam.} = +1\%$), separated from the (12,0)–(7,7) knee ($\Delta_{Diam.} = -1\%$) by the graphite interplanar distance (3.46 Å).

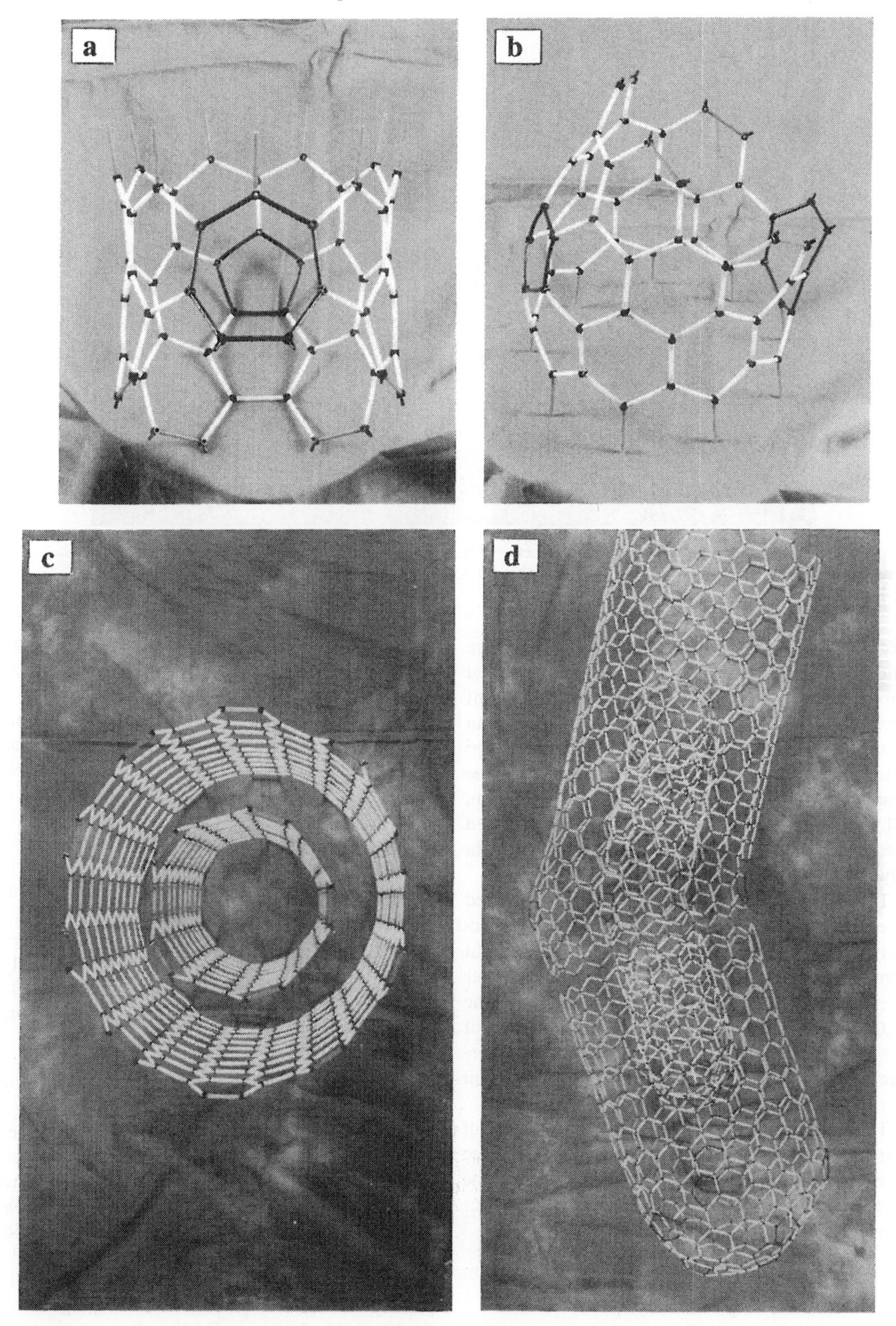

Fig. 4. (a) and (b) Model structures of the (9,0)–(5,5) knee, (c) model structure of the (5,5) tubule inside of the (10,10) tubule, (d) model structure of the (18,0)–(10,10) knee and the (9,0)–(5,5) knee with a graphite interplanar distance (3.46 Å) between them.

layers. Hence, this series can be described as realizing the best epitaxy both in the 5–7 knee region and away from it in the straight sections.

Dunlap's plane construction gives an angle of 30° for the $(L,0)$–(L',L') connections while we observed 36° from our ball and stick molecular models constructed with rigid sp^2 triangular bonds (Fig. 5). In fact, 30° is the angle formed by the "pressed" tubule

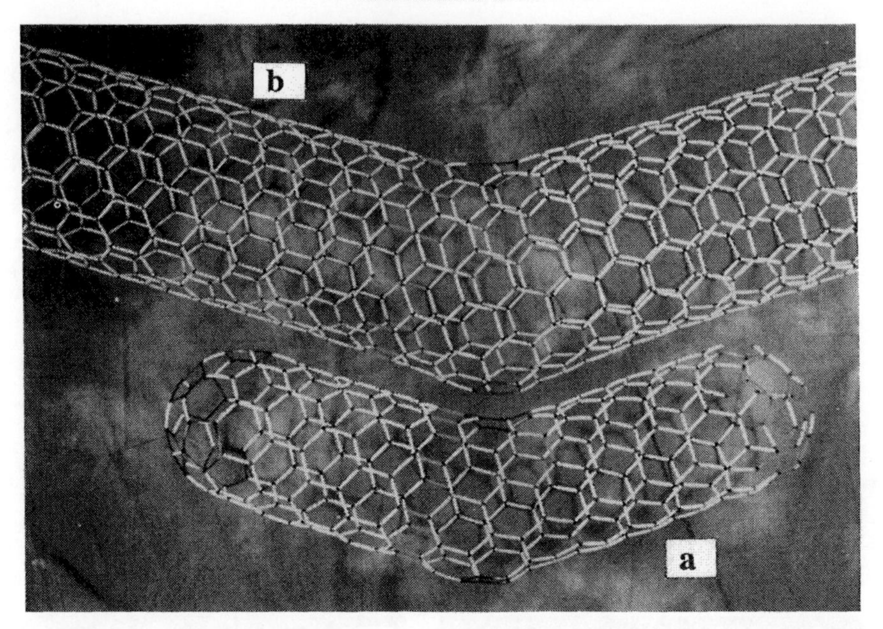

Fig. 5. Model structure of the (9,0)–(5,5) curved nanotubule ended by two half C_{60} caps (a) and of the (12,0)–(7,7) curved nanotubule (b). A knee angle of 36° is observed in both models.

knee if the five and seven membered rings are removed. By folding the knee to its three-dimensional shape, the real angle expands to 36°. At the present time, it is not known whether this discrepancy is an artefact of the ball and stick model based on sp^2 bonds or whether strain relaxation around the knee demands the 6° angle increase. Electron diffraction and imaging data[8–10] have not so far allowed assessment of the true value of the knee angle in polygonized nanotubes.

The diameters (D_n) of the perfectly graphitized series are $D_{n\perp}=15na/\pi$ and $D_{n//}=9na\sqrt{3}/\pi$, respectively, for the "perpendicular" and "parallel" straight segments. $D_{n//}$ is 3.8% larger than $D_{n\perp}$. This diameter difference is larger than the 1% characterising the (12,0)–(7,7) ideal connection of Dunlap[12,13], but it is independent of n. A few percent diameter difference can easily be accommodated by bond relaxation over some distance away from the knee.

Table 1 gives the characteristics of classes of bent tubules built on $(9n+x,\ 0)$–$(5n+y,\ 5n+y)$ knee connections. In these classes, $y=(x/2)\pm 1$, and the integers x and y are selected to give small relative diameter differences, so that

$$\Delta_{\text{Diam.}}=(D_{n//}-D_{n\perp})/D_{n//},\ D_{n\perp}=(5n+y)3a/\pi$$
$$\text{and}\quad D_{n//}=(9n+x)a\sqrt{3}/\pi$$

The first layers of the connections leading to the minimum diameter difference are also described in Table 1. For the general case, contrary to the $(9n,0)$–$(5n,5n)$ knees, the largest side of the knee is not always the parallel side $(9n+x,\ 0)$. As seen in Table 1, the connections (7,0)–(4,4) and (14,0)–(8,8) are, from the diameter difference point of view, as ideal as the (12,0)–(7,7) described by Dunlap[12,13].

For the perfectly graphitizable $(9n,0)$–$(5n,5n)$ nanotubule series, the diameter of the graphite layer n is equal to $6.9n$ Å, within 2%, so that the interlayer distance is 3.46 Å. For this interlayer distance, the smallest possible knee is (5,0)–(3,3), with a diameter of 3.99 Å (Table 1), because the two smaller ones could not give layers at the graphitic distance. Note that all the $n=0$ tubules are probably unstable due to their excessive strain energy[16].

The nanotube connections whose diameter differences are different from the 3.8% value characteristic to the $(9n,0)$–$(5n,5n)$ series, will tend to that value with increasing the graphite layer order n.

The inner (outer) diameter of the observed curved or coiled nanotubules produced by the catalytic method[8] varies from 20 to 100 Å (150 to 200 Å), which corresponds to the graphite layer order $3\leq n\leq 15$ (Table 1).

2.3 *Description of a perfectly graphitizable chiral tubule knee series*

Among all the chiral nanotubules connectable by a knee, the series $(8n,n)$–$(6n,4n)$, with n an integer, is perfectly graphitizable. For that series, the diameter of the graphite layer of order n is equal to $6.75n$ Å, within 1%, so that the interlayer distance is 3.38 Å. Moreover, the two chiral tubules are connected by a pentagon–heptagon knee, with the equatorial plane passing through the pentagon and heptagon as for the $(9n,0)$–$(5n,5n)$ series. On the plane graphene construction, the two chiral tubules are connected at an angle of 30°. As for the $(9n,0)$–$(5n,5n)$ series, 36° is observed from our ball and stick molecular model constructed with rigid sp^2 triangular bonds (Fig. 6).

2.4 *Constructing a torus with (9n,0)–(5n,5n) knees*

Building up a torus using the (9,0)–(5,5) knee [Fig. 7(a)] is compatible with the 36° knee angle.

Table 1. Characteristics of some knees with minimal diameter difference

n	x	y	$(9n+x,0)$	$(5n+y,5n+y)$	$\Delta_{Diam.}$ (%)	$(D_{n//}+D_{n\perp})/2$ (Å)
0	3	2	(3,0)	(2,2)	−15.5	2.53
0	4	2	(4,0)	(2,2)	13.4	2.92
0	5	3	(5,0)	(3,3)	−3.9	3.99
0	6	3	(6,0)	(3,3)	13.4	4.38
0	7	4	(7,0)	(4,4)	1.0	5.45
0	8	5	(8,0)	(5,5)	−8.2	6.52
1	0	0	(9,0)	(5,5)	3.8	6.914
1	1	1	(10,0)	(6,6)	−3.9	7.98
1	2	1	(11,0)	(6,6)	5.5	8.38
1	3	2	(12,0)	(7,7)	−1.0	9.44
1	4	3	(13,0)	(8,8)	−6.6	10.51
1	5	3	(14,0)	(8,8)	1.0	10.91
1	6	4	(15,0)	(9,9)	−3.9	11.98
1	7	4	(16,0)	(9,9)	2.6	12.37
1	8	5	(17,0)	(10,10)	−1.9	13.44
2	0	0	(18,0)	(10,10)	3.8	13.83

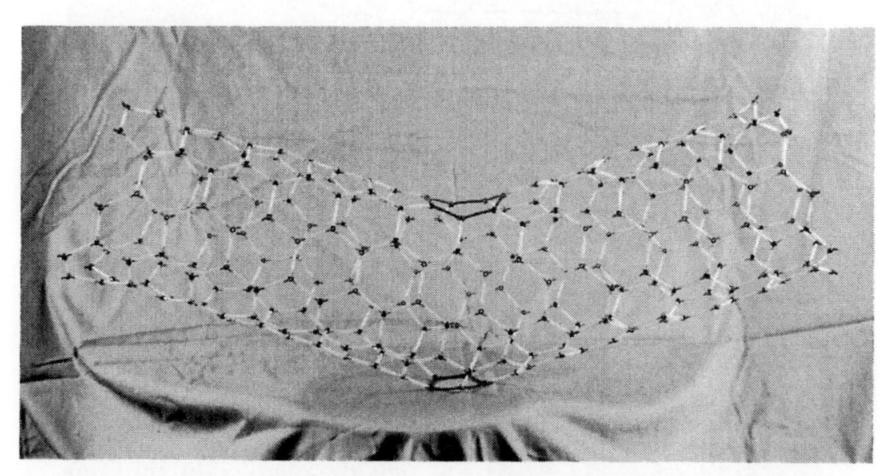

Fig. 6. Model structure of the (8,1)–(6,4) knee extended by two straight chiral tubule segments.

This torus contains 520 carbon atoms and 10 knees with the heptagons on the inner side forming abutting pairs. It has a fivefold rotation symmetry axis and if it is disconnected at an arbitrary cross section, all the carbons remain at their position because there is no strain. This is also the case for the C_{900} torus presented in Fig. 7(b), where each segment is elongated by two circumferential rings. At each knee, the orientation of the hexagons changes from parallel to perpendicular and *vice versa*.

The number of atoms forming the smallest $n=1$ knee of the $(9n,0)$–$(5n,5n)$ series is 57 and it leads to the C_{520} torus having adjacent heptagons. The number of atoms of any knee of that series is given by:

$$N_n=24n^2+33n \quad (1)$$

As the N_n knee can have $n-1$ inner concentric knees, all of them separated by approximately the graphite interplanar distance, n is called the "graphite layer order". In fact, the number of atoms of the torus n is given by $10(24n^2+33n-5n)$ because $10n$ atoms are common for adjacent knees at the $(5n,5n)$–$(5n,5n)$ connection (see below, Section 2.5).

The (9,0)–(5,5) torus represented in Fig. 7(b) contains 10 straight segments of 38 atoms each joining at 10 knees. The sides of each knee have been elongated by the addition of hexagonal rings. (The picture of that torus and of the derived helices are given in the literature[11].) The corresponding general formula giving the number of atoms in such elongated knees is:

$$N_{n,c}=N_n+cn \quad (2)$$

The constant c [equal to 38 for the torus of Fig. 7(b)] gives the length of the straight segment desired, with $c=20(Hex_\perp)+18(Hex_{//})$ where, $Hex_\perp$ and $Hex_{//}$ are the numbers of hexagonal rings extending the knee in the appropriate direction. The smallest N_n knee [Fig. 8(a) for $n=1$], and the way of constructing prolonged $N_{n,c}$ knees [Fig. 8(b) for $n=1$ and $c=38$] are represented in Fig. 8. In that plane or "pressed tubule" figure, only one half of the knee is shown. For symmetry reasons, the median plane of the $N_{n,c}$ knees crosses the knee at the same position as the limits of the corresponding knee does on a planar representation (Fig. 8). The dotted bonds are the borders between the knee and the next nanotubes or

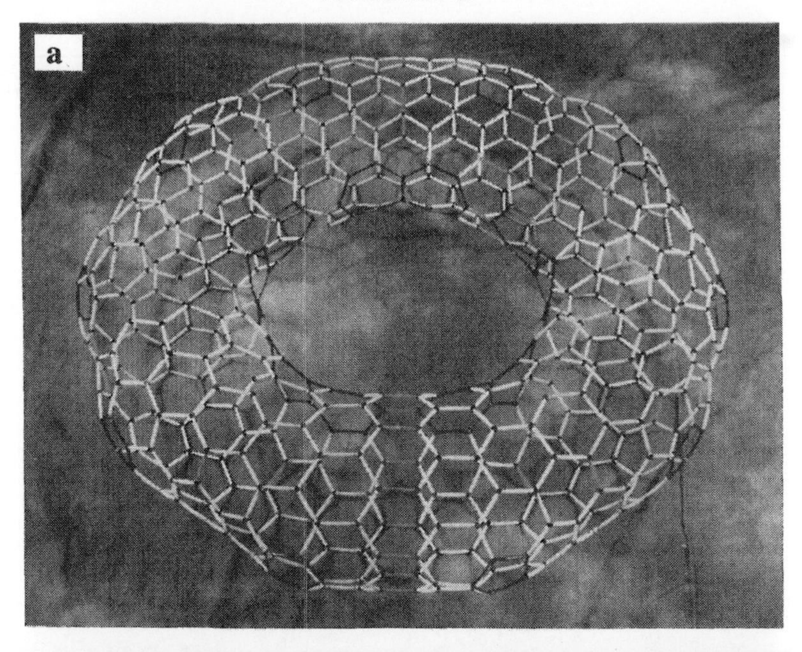

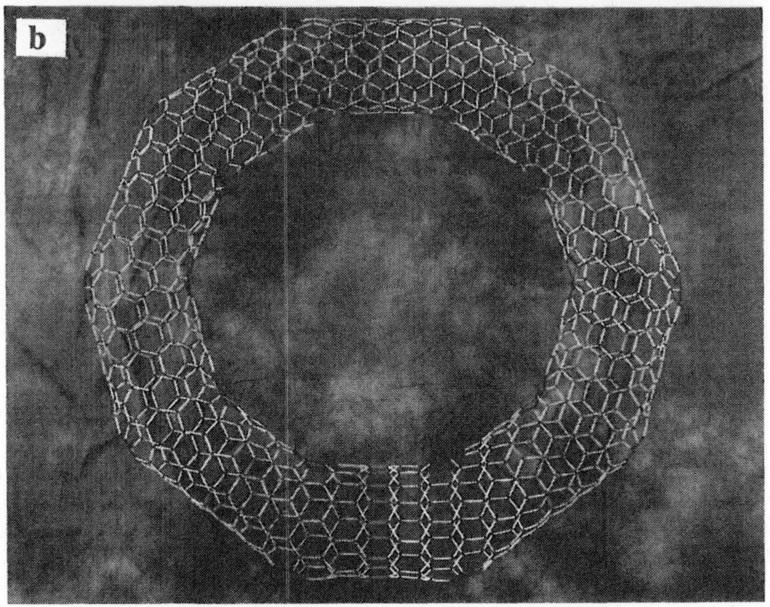

Fig. 7. (a) Torus C_{520} formed by 10 (9,0)–(5,5) knees, (b) torus C_{900} formed by 10 (9,0)–(5,5) knees extended by adding one circumferential ring on both sides.

knees. On the inner equatorial circle of the torus made from the N_n knees, each pair of abutting heptagons sharing a common bond is separated from the pair of heptagons on either side by one bond. This can also be observed on Fig. 7(a).

2.5 *From torus to helix*

As seen in the previous section, if two identical knees of the $(L,0)$–(L',L') family are connected together symmetrically with respect to a connecting plane, and if this connecting process is continued while maintaining the knees in a common plane, the structure obtained will close to a torus which will be completed after 10 fractional turns (Figs 7 and 9).

However, if a rotational bond shift is introduced at the connecting meridian between two successive fractions of a torus, then its equator will no longer be a plane. In Fig. 10, single bond shifts in two (9,0)–(5,5) knees are represented. The repeated introduction of such bond shifts will lead to a helix. Several bond shifts can also be introduced at the same knee connection. If there is a long straight tubule segment joining two knees, several rotation bond shifts can also be introduced at different places of that segment. All of these bond shifts can be present in the same helix. The helix will be regular or irregular depending on the periodic or random occurrence of such bond shifts in the straight sections. The pitch and diameter of a regular coiled tubule will be determined by the length of the straight

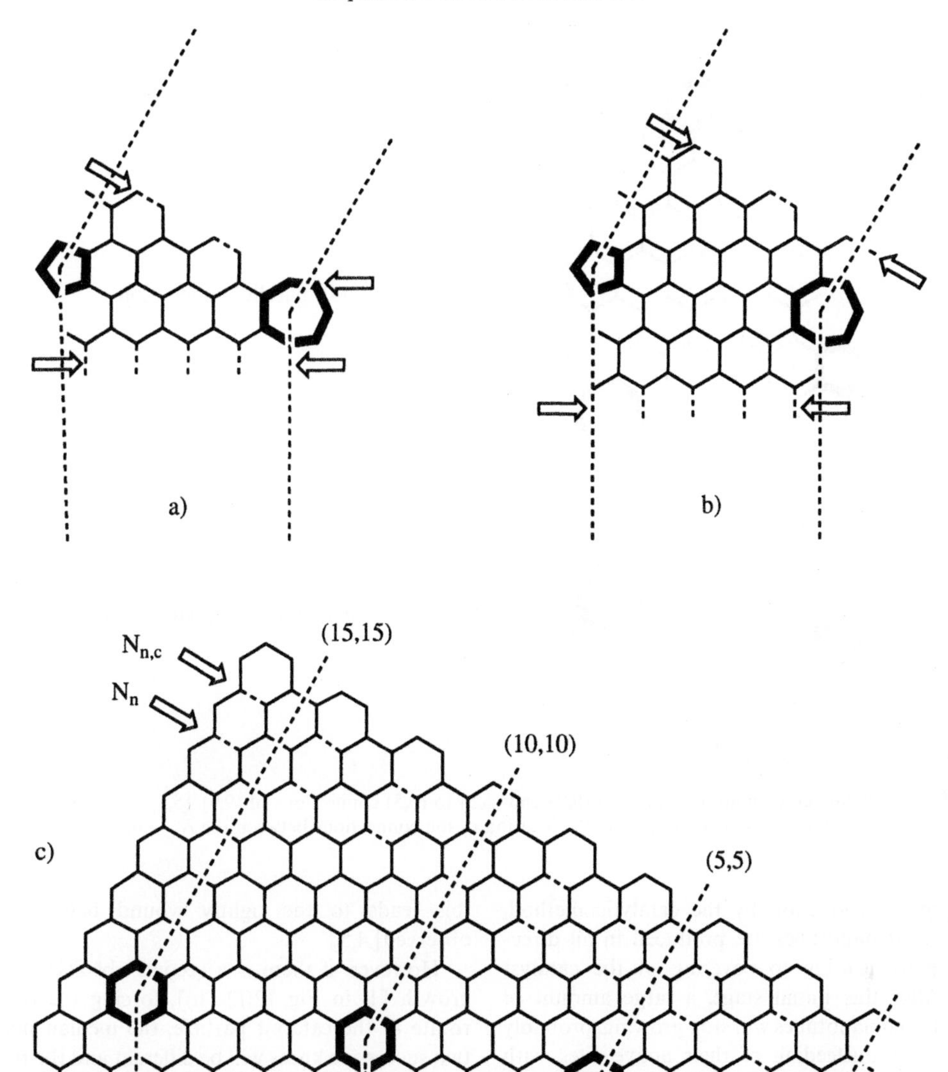

Fig. 8. Planar representation of the $(9n,0)$–$(5n,5n)$ knees, having a 36° bend angle produced by a heptagon–pentagon pair on the equatorial plane. The arrows show the dotted line of bonds where the knee N_n or $N_{n,c}$ is connected to the corresponding straight tubules: (a) knee N_n for $n=1$; (b) stretched knee $N_{n,c}$ for $n=1$ and $c=38$; (c) general knees N_n and $N_{n,c}$.

sections and by the distribution of bond shifts. The diameter and pitch of observed coiled tubules vary greatly[7–12].

3. A POSSIBLE MECHANISM FOR THE GROWTH OF NANOTUBES ON A CATALYST PARTICLE

The formation of a helix or torus — the regularity of which could be controlled by the production of heptagon–pentagon pairs in a concerted manner at the catalyst surface — is first explained from the macroscopic and then from the chemical bond points of view.

3.1 *Macroscopic point of view*

Some insight can be gained from the observation of tubule growth on the catalyst surface by the decomposition of acetylene (Fig. 11). At the beginning

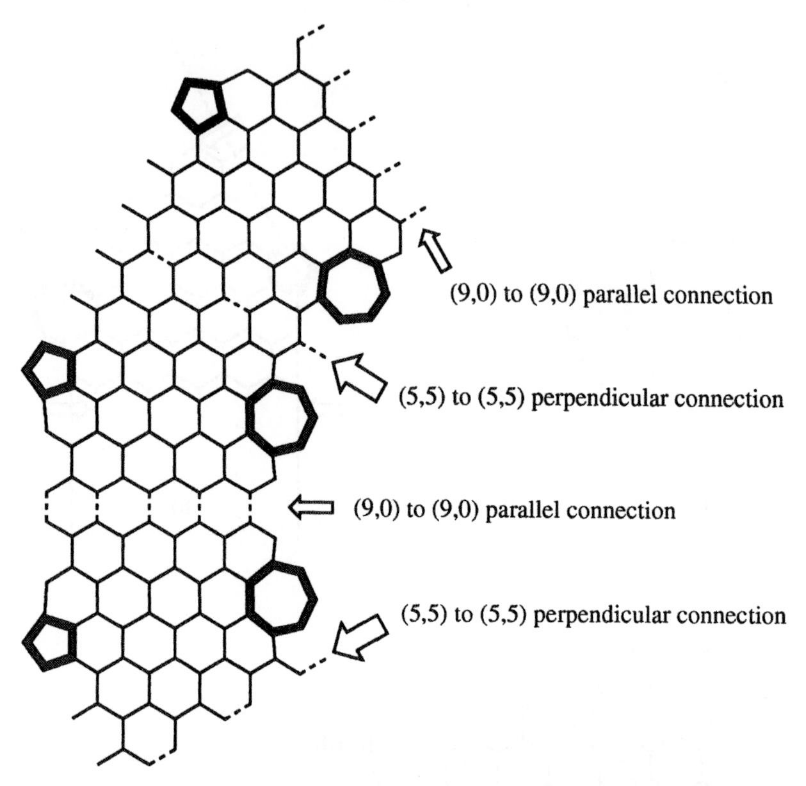

Fig. 9. Planar representation of (9,0) to (9,0) and (5,5) to (5,5) connections of (9,0)–(5,5) knees leading to a torus. The arrows indicate the location of the connections between the $N_{n,c}$ knees.

of the tubule production by the catalytic method, many straight nanotubes are produced in all directions, rapidly leading to covering of the catalyst surface. After this initial stage, a large amount of already started nanotubes will stop growing, probably owing to the misfeeding of their active sites with acetylene or to steric hindrance. The latter reason is in agreement with the mechanism already suggested by Amelinckx *et al.*[10], whereby the tubule grows by extrusion out of the immobilized catalyst particle. It is also interesting to point out that in Fig. 11 there is no difference between the diameters of young [Fig. 11(a)] and old [Fig. 11(c)] nanotubes.

Since regular helices with the inner layer matching the catalyst particle size have been observed[4,5], we propose a steric hindrance model to explain the possible formation of regular and tightly wound helices.

If a growing straight tubule is blocked at its extremity, one way for growth to continue is by forming a knee at the surface of the catalyst, as sketched in Fig. 12. Starting from the growing tubule represented in Fig. 12(a), after blockage by obstacle A [Fig. 12(b)], elastic bending can first occur [Fig. 12(c)]. Beyond a certain limit, a knee will appear close to the catalyst particle, relaxing the strain and freeing the tubule for further growth [Fig. 12(d)]. If there is a single obstacle to tubule growth (A in Fig. 12), the tubule will continue turning at regular intervals [Fig. 12(e) and (f)] but as it is impossible to complete a torus because of the catalyst particle, this leads to the tightly wound helices already observed[4,5].

However, if there is a second obstacle to tubule growth [B in Fig. 12(f)–(h)], forcing the tubule to rotate at the catalyst particle, the median planes of two successive knees will be different and the resulting tubule will be a regular helix. Note that the catalyst particle itself could act as the second obstacle B. The obstacles A and B of Fig. 12 are hence considered as the bending driving forces in Fig. 11, with A regulating the length of the straight segments ($9n$,0) and ($5n$,$5n$) and B controlling the rotation angle or number of rotational bond shifts (Fig. 10).

From the observation of the early stage of nanotube production by the catalytic decomposition of acetylene, it is concluded that steric hindrance arising from the surrounding nanotubes, graphite, amorphous carbon, catalyst support and catalyst particle itself could force bending of the growing tubules.

3.2 *Chemical bond point of view*

To form straight cylindrical carbon nanotubes, one possibility is for the carbon hexagons to be "bonded" to the catalyst surface during the growth process. In that "normal" case, one of the edges of the growing hexagons remains parallel to the catalyst surface during growth (Fig. 13). This requires that for every tubule — single or multilayered nanotube — with one or more ($5n$,$5n$)–($9n$,0) knees, the catalyst should offer successive active perimeters differing by

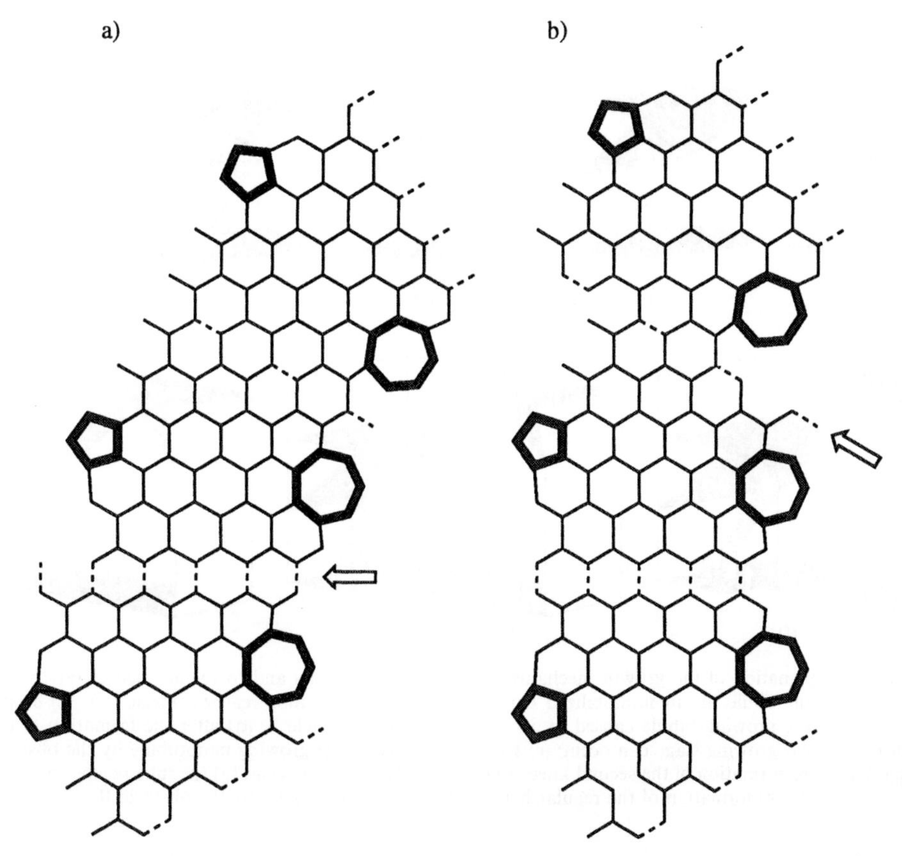

Fig. 10. (a) Planar representation of a single rotational bond shift at the (9,0) to (9,0) connection of two (9,0)–(5,5) knees. This leads to a $2\pi/9$ rotation out of the upper knee plane; (b) single bond shift at the (5,5) to (5,5) connection of two (9,0)–(5,5) knees. This leads to a $2\pi/5$ rotation out of the lower knee plane. The arrow indicates the location of the bond shift.

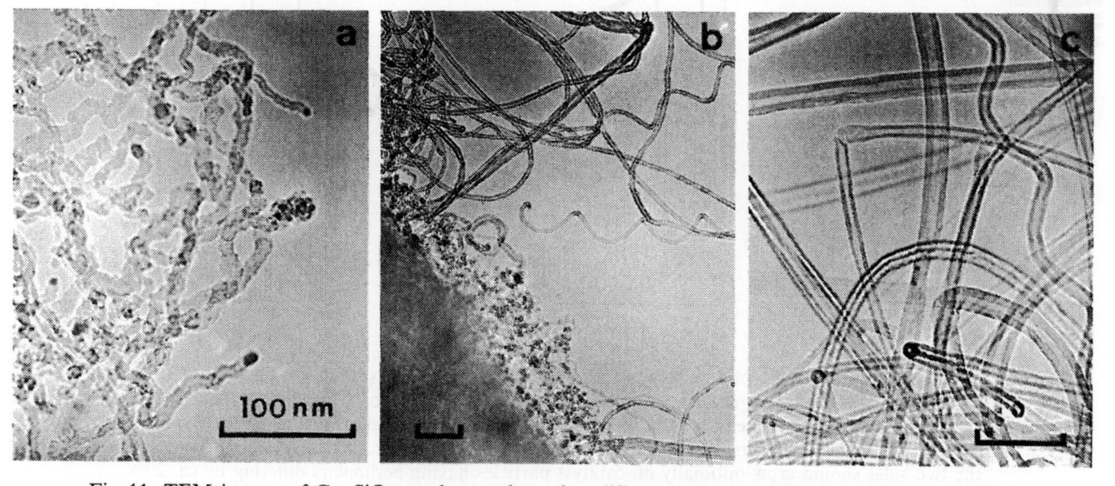

Fig. 11. TEM images of Co–SiO_2 catalyst surface after different exposure times to acetylene at 700°C: (a) 1 minute; (b) 5 minutes; and (c) 20 minutes.

about 20% (Fig. 13). The number of carbons bonded to the catalyst surface also changes by ca. 20%.

A model involving that variation of the catalyst active perimeter across the knee will first be considered. Afterwards, a model involving the variation of the number of "active" coordination sites at a constant catalyst surface will be suggested.

3.2.1 *Model based on the variation of the active catalyst perimeter.* To form the (5,5)–(9,0) knee represented in Fig. 13(c) on a single catalyst particle, the catalyst should start producing the (5,5) nanotubule of Fig. 13(a), form the knee, and afterwards the (9,0) nanotubule of Fig. 13(b), or *vice versa.* It is possible to establish relationships between

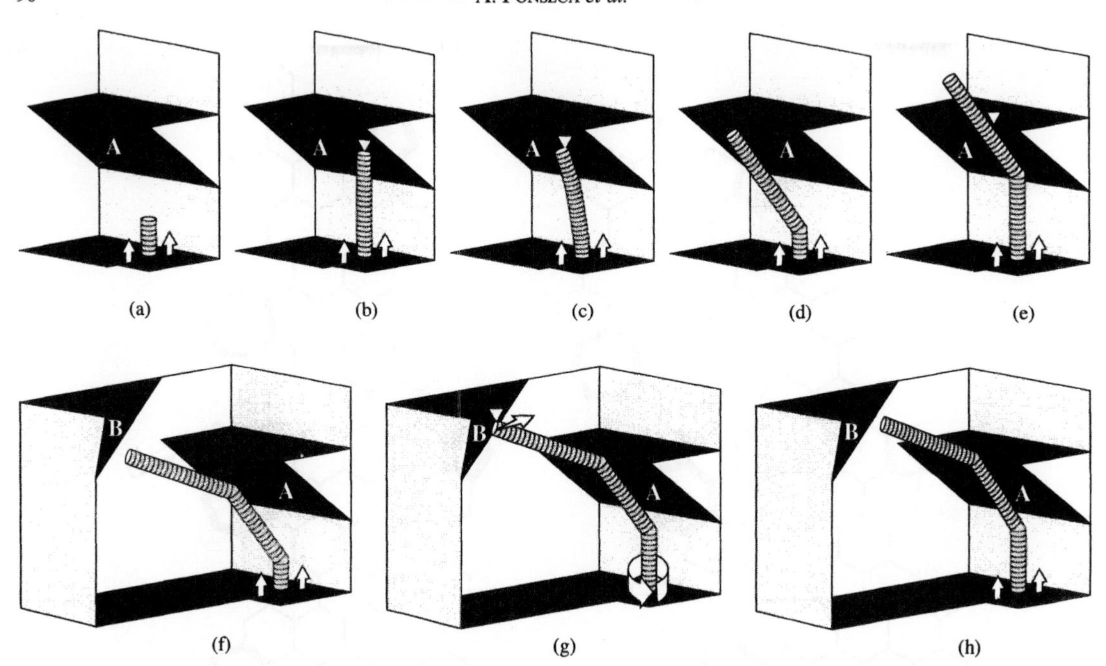

Fig. 12. Explanation of the growth mechanism leading to tori (a)–(e) and to regular helices (a)–(h). (a) Growing nanotubule on an immobilized catalyst particle; (b) the tube reaches obstacle A; (c) elastic bending of the growing tubule caused by its blockage at the obstacle A; (d) after the formation of the knee, a second growing stage can occur; (e) second blockage of the growing nanotubule by the obstacle A; (f) after the formation of the second knee, a new growing stage can occur; (g) the tube reaches obstacle B; (h) formation of the regular helicity in the growing tubules by the obstacle B.

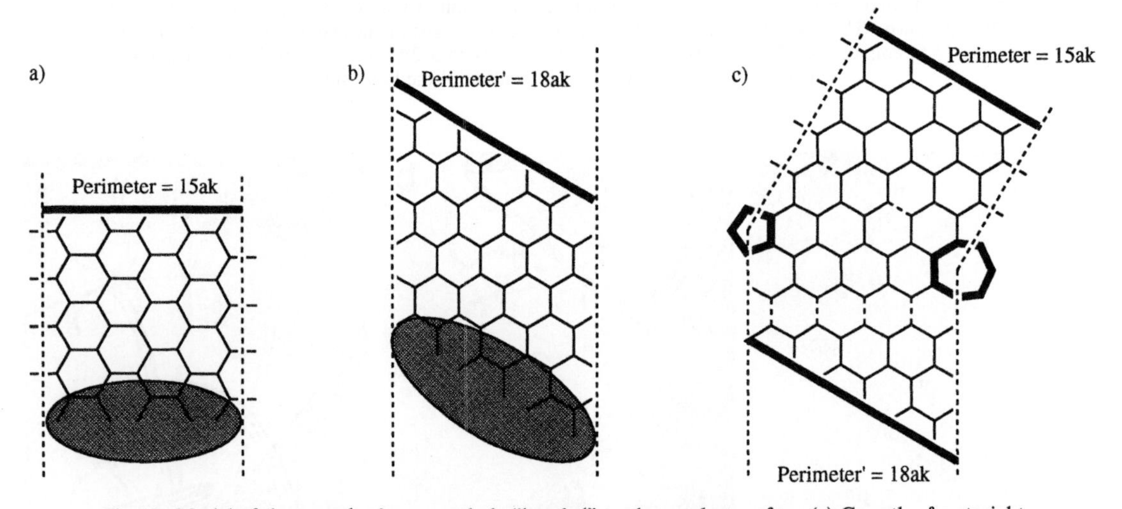

Fig. 13. Model of the growth of a nanotubule "bonded" to the catalyst surface. (a) Growth of a straight (5,5) nanotubule on a catalyst particle, with perimeter $15ak$; (b) growth of a straight (9,0) nanotubule on a catalyst particle whose perimeter is $18ak$ (k is a constant and the grey ellipsoids of (a) and (b) represent catalyst particles, the perimeters of which are equal to $15ak$ and $18ak$, respectively); (c) (5,5)–(9,0) knee, the two sides should grow optimally on catalyst particles having perimeters differing by ca. 20%.

the catalyst particle perimeter and the nature of the tubules produced:

- When the active perimeter of the catalyst particle matches perfectly the values $15nak$ or $18nak$ (where n is the layer order, a is the side of the hexagon in graphite and k is a constant), the corresponding straight nanotubules ($5n$,$5n$) or ($9n$,0) will be produced, respectively [Fig. 13(a) and (b)].
- If the active perimeter of the catalyst particle has a dimension between $15nak$ and $18nak$, — i.e. $15nak <$ active perimeter $< 18nak$ — the two different tubules can still be produced, but under stress.
- The production of heptagon–pentagon pairs among these hexagonal tubular structures leads to the formation of regular or tightly wound helices. Each knee provides a switch between the tubule formation of Fig. 13(a) and (b) [Fig. 13(c)].

According to Amelinckx *et al.*[9], this switch could be a consequence of the rotation of an ovoid catalyst particle. However, from this model, during the production of the parallel hexagons, a complete "catalyst–tubule bonds rearrangement" must occur after each hexagonal layer is produced. Otherwise, as seen from the translation of the catalyst particle in a direction perpendicular to its median plane, the catalyst would get completely out of the growing tubule. Since a mechanism involving this "catalyst–tubule bonds rearrangement" is not very likely, we shall now try to explain the growth of a coiled tubule using a model based on the variation of the number of active coordination sites at a constant catalyst surface by a model which does not involve "catalyst–tubule bonds rearrangement".

3.2.2 *Model based on the variation of the number of "active" coordination sites at the catalyst surface.* The growth of tubules during the decomposition of acetylene can be explained in three steps, which are the decomposition of acetylene, the initiation reaction and the propagation reaction. This is illustrated in Fig. 14 by the model of a (5,5) tubule growing on a catalyst particle:

- First, dehydrogenative bonding of acetylene to the catalyst surface will free hydrogen and produce C_2 moieties bonded to the catalyst coordination sites. These C_2 units are assumed to be the building blocks for the tubules.
- Secondly, at an initial stage, the first layer of C_2 units diffusing out of the catalyst remains at a Van der Waals distance from the C_2 layer coordinated to the catalyst surface. Then, if the C_2 units of that outer layer bind to one another, this will lead to a half fullerene. Depending on whether the central axis of that half fullerene is a threefold or a fivefold rotation axis, a (9*n*,0) or a (5*n*,5*n*) tubule will start growing, respectively. The half fullerene can also grow to completion instead of starting a nanotubule[17]. This assumption is reinforced by the fact that we have detected, by HPLC and mass spectrometry, the presence of fullerenes C_{60}, C_{70}, ... C_{196} in the toluene extract of the crude nanotubules produced by the catalytic decomposition of acetylene.
- Third, the C_2 units are inserted between the catalyst coordination sites and the growing nanotubule (Fig. 14). The last C_2 unit introduced will still be bonded to the catalyst coordination sites. From the catalyst surface, a new C_2 unit will again displace the previous one, which becomes part of the growing tubule, and so on.

We shall now attempt to explain, from the chemical bond point of view, the propagation reaction at the basis of tubule growth. A growth mechanism for the (5*n*,5*n*) tubule, the (9*n*,0) tubule and the (9*n*,0)–(5*n*,5*n*) knee, which are the three fundamental tubule building blocks, is also suggested.

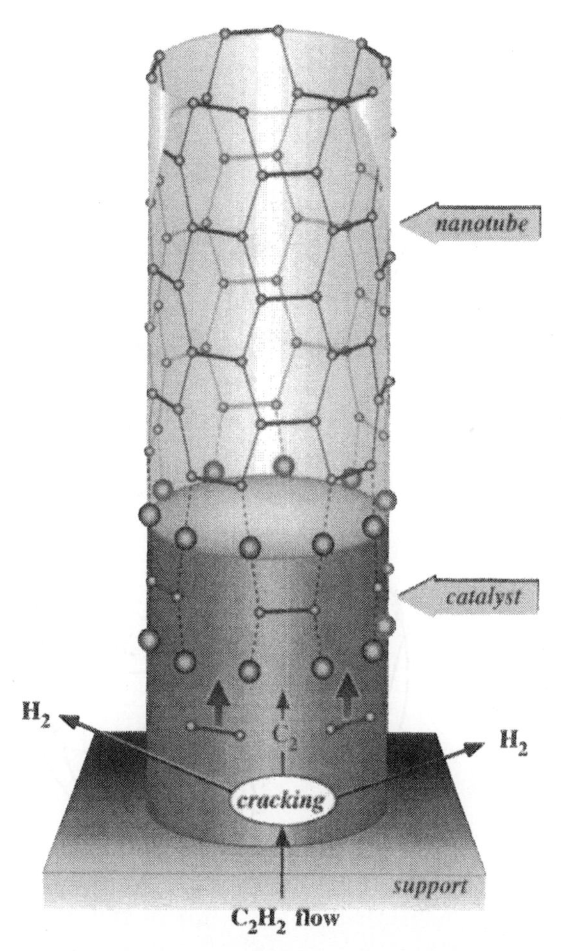

Fig. 14. Schematic representation of a (5,5) tubule growing on the corresponding catalyst particle. The decomposition of acetylene on the same catalyst particle is also represented. The catalyst contains many active sites but only those symbolized by grey circles are directly involved in the (5,5) tubule growth.

3.2.2.1 *Growth mechanism of a* (5n,5n) *tubule, over 20*n *coordination sites of the catalyst.* The growth of a general (5*n*,5*n*) tubule on the catalyst surface is illustrated by that of the (5,5) tubule in Figs 14 and 15. The external circles of the Schlegel diagrams in Fig. 15(a)–(c) represent half C_{60} cut at the equatorial plane perpendicular to its fivefold rotational symmetry axis or the end of a (5,5) tubule. The equatorial carbons bearing a vacant bond are bonded to the catalyst coordinatively [Fig. 15(a) and (a′)].

For the sake of clarity, ten coordination sites are drawn a little further away from the surface of the particle in Fig. 15(a)–(c). These sites are real surface sites and the formal link is shown by a solid line. In this way the different C_2 units are easily distinguished in the figure and the formation of six-membered rings is obvious. The planar tubule representations of Fig. 15(a′)–(c′) are equivalent to those in Fig. 15(a)–(c), respectively. The former figures allow a better understanding of tubule growth. Arriving C_2 units are first coordinated to the catalyst coordination

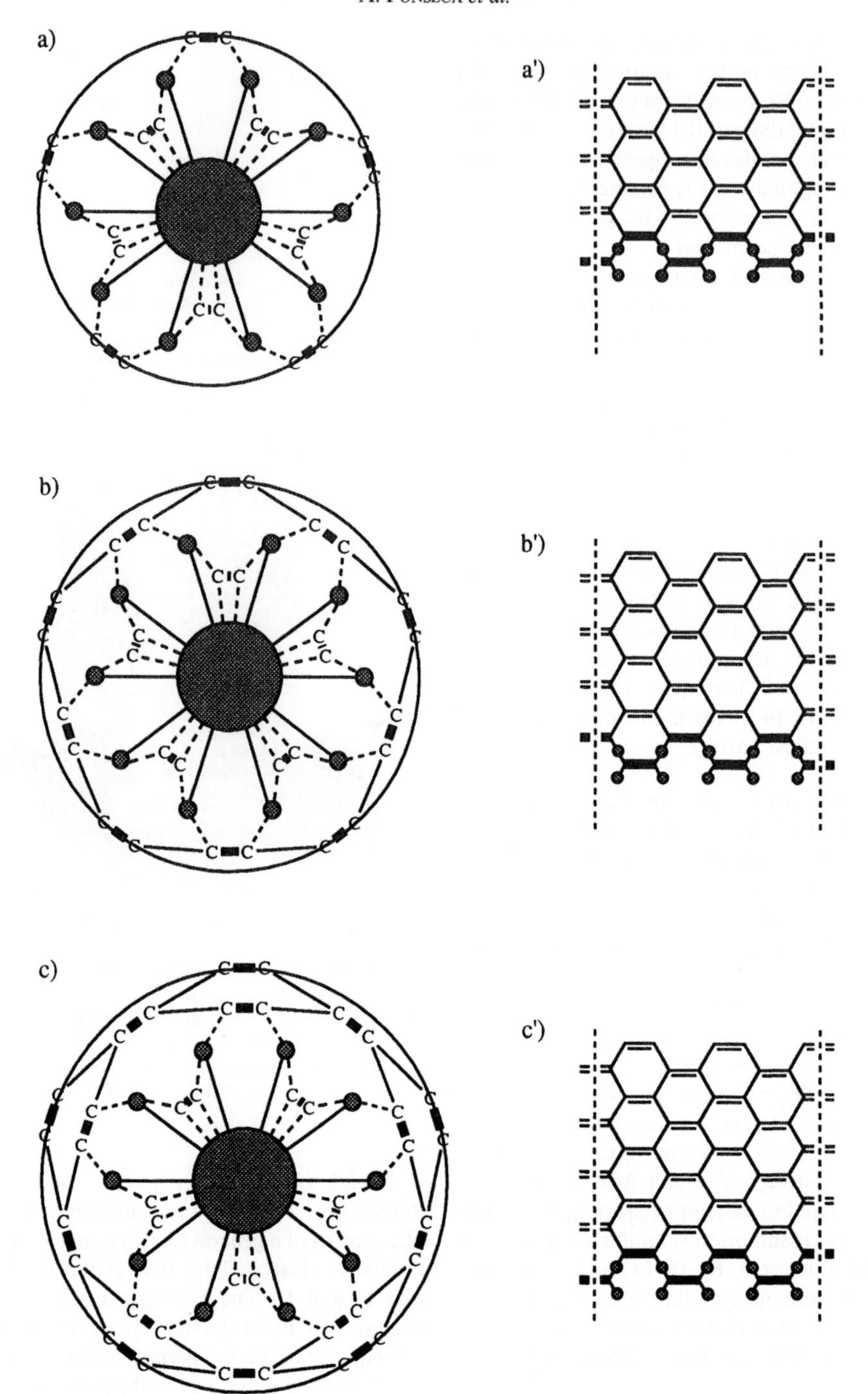

Fig. 15. Growth of a $(5n,5n)$ tubule on the catalyst surface, illustrated by that of the (5,5) tubule. The central grey circle represents the catalyst particle with 10 coordination sites, and the small grey circles represent the other 10 catalyst coordination sites. The normal and bold lines represent single and double bonds, respectively, while coordinative bonds are represented by dotted lines [(a), (b) and (c)]; (a′), (b′) and (c′) are the corresponding planar representations.

sites already coordinated to the growing tubule and, finally, they are inserted between the tubule and the catalyst coordination sites Fig. 15(b) and (b′)]. This leads to the formation of the first additional crown of perpendicular six membered rings to the growing tubule. In this growing structure, the C_2 units arrive and are inserted perpendicularly to the tubule axis. Further arriving C_2 units will behave as the latter one did, and so on [Fig. 15(c) and (c′)].

From Fig. 15, it can be seen that 20 coordination sites of the catalyst are involved in the growth of the (5,5) tubule, and in general, $20n$ coordination sites will be involved in the growth of the $(5n,5n)$ tubule. It is obvious that the related catalyst particle must have more than the 20 coordination sites used, at least for the decomposition of acetylene.

3.2.2.2 *Growth mechanism of a (9*n*,0) tubule, over 24*n* coordination sites of the catalyst.* The growth of a general (9*n*,0) tubule on the catalyst surface is illustrated by that of the (9,0) tubule in Fig. 16 which shows the "unsaturated" end of a (9,0) tubule in a planar representation. At that end, the carbons bearing a vacant bond are coordinatively bonded to the catalyst (grey circles) or to a growing *cis*-polyacetylene chain (oblique bold lines in Fig. 16). The vacant bonds of the six *cis*-polyacetylene chains involved are taken to be coordinatively bonded to the catalyst [Fig. 16(b)]. These polyacetylene chains are continuously extruded from the catalyst particle where they are formed by polymerization of C_2 units assisted by the catalyst coordination sites. Note that in order to reduce the number of representations of important steps, Fig. 16(b) includes nine new C_2 units with respect to Fig. 16(a).

The 12 catalyst coordination sites — drawn further away from the surface of the particle (closer to the tubule) — are acting in pairs, each pair being always coordinatively bonded to one carbon of an inserted (1°) or of a to-be-inserted (2°) C_2 unit and to two other carbons which are members of two neighbouring *cis*-polyacetylene chains (3°). It should be emphasized that, as against the (5*n*,5*n*) tubule growth, the C_2 units extruded from the catalyst particle are positioned in this case parallel to the tubule axis before their insertion.

Two coordinative bonds of two neighbouring *cis*-polyacetylene chains of the growing tubule are specifically displaced from a pair of catalyst coordination sites by the insertion of one carbon of a C_2 unit [2° in Fig. 16(b)]. The second carbon of that C_2 unit is still bonded to the pair of coordination sites. It will later be displaced from that pair of coordination sites by the arrival of the next two C_2 segments of the two growing *cis*-polyacetylene chains considered [3° in Fig. 16(c)]. The C_2 unit and *cis*-polyacetylene C_2 segments are hence inserted into the growing tubule. The further arriving C_2 units or *cis*-polyacetylene C_2 segments will act as the latter one did, and so on [Fig. 16(c)].

As can be seen from Fig. 16, 24 coordination sites of the catalyst are involved in the growth of the (9,0) tubule and, in general, 24*n* coordination sites will be involved in the growth of the (9*n*,0) tubule.

3.2.2.3 *Growth mechanism of a (9*n*,0)–(5*n*,5*n*) knee, involving from 24*n* to 20*n* coordination sites of the catalyst.* Across the knee, the growth of the tubule starts as explained for the (9*n*,0) tubule (Fig. 16) and must end as explained for the (5*n*,5*n*) tubule (Fig. 15) or *vice versa*.

As a starting point, a growing (9,0) tubule on the catalyst surface is illustrated in Fig. 17(a). In that growing structure, the C_2 units arrive and are inserted parallel to the tubule axis. If there is enough space — probably thanks to the distortion caused by the elastic folding of the growing tubule [Fig. 12(b)] — two carbons (1°) can be inserted instead of one between a pair of coordination sites [Fig. 17(b)]. The knee is started by the arrival of a C_2 unit perpendicular to the tubule axis, each carbon of which has a vacant bond for coordination. These two carbons (instead of the carbon with two vacant bonds usually involved) displace one carbon from a pair of coordination sites. This perpendicular C_2 unit constitutes the basis of the future seven-membered ring. Note again that nine new C_2 units are shown in Fig. 17(b) as compared with Fig. 17(a).

Among the three C_2 units arriving together, two are parallel to the tubule axis (2°) and one is orthogonal to the tubule axis [1° in Fig. 17(b)]. First, only the parallel units are inserted into the tubule structure [Fig. 17(c)]. These two inserted C_2 units (2°) were displaced by the arrival of two orthogonal C_2 units (3°) and all further arriving C_2 units will be orthogonal to the tubule axis.

Meanwhile, the coordinative bonding of the two C_2 units within the pairs of coordinative sites forces the occupied sites to spread. As a result, in the direction opposite to the place where the first orthogonal C_2 unit arrived, there is only enough room for a single site (*) instead of for a pair of sites [Fig. 17(c)]. That site now coordinates the two *cis*-polyacetylene C_2 segments, but without the possibility of inserting a new C_2 unit. The displacement of that site from the growing nanotubule leads to the closure

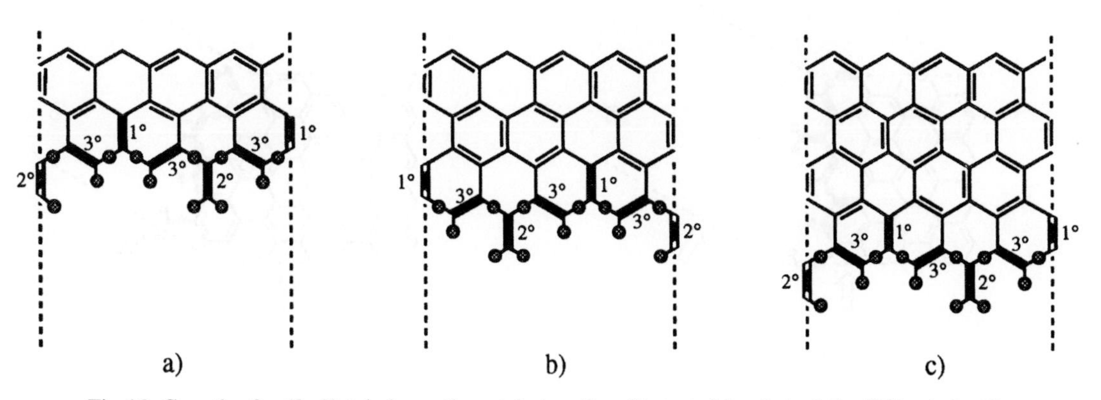

Fig. 16. Growth of a (9*n*,0) tubule on the catalyst surface illustrated by that of the (9,0) tubule. The normal and bold lines represent single and double bonds, respectively.

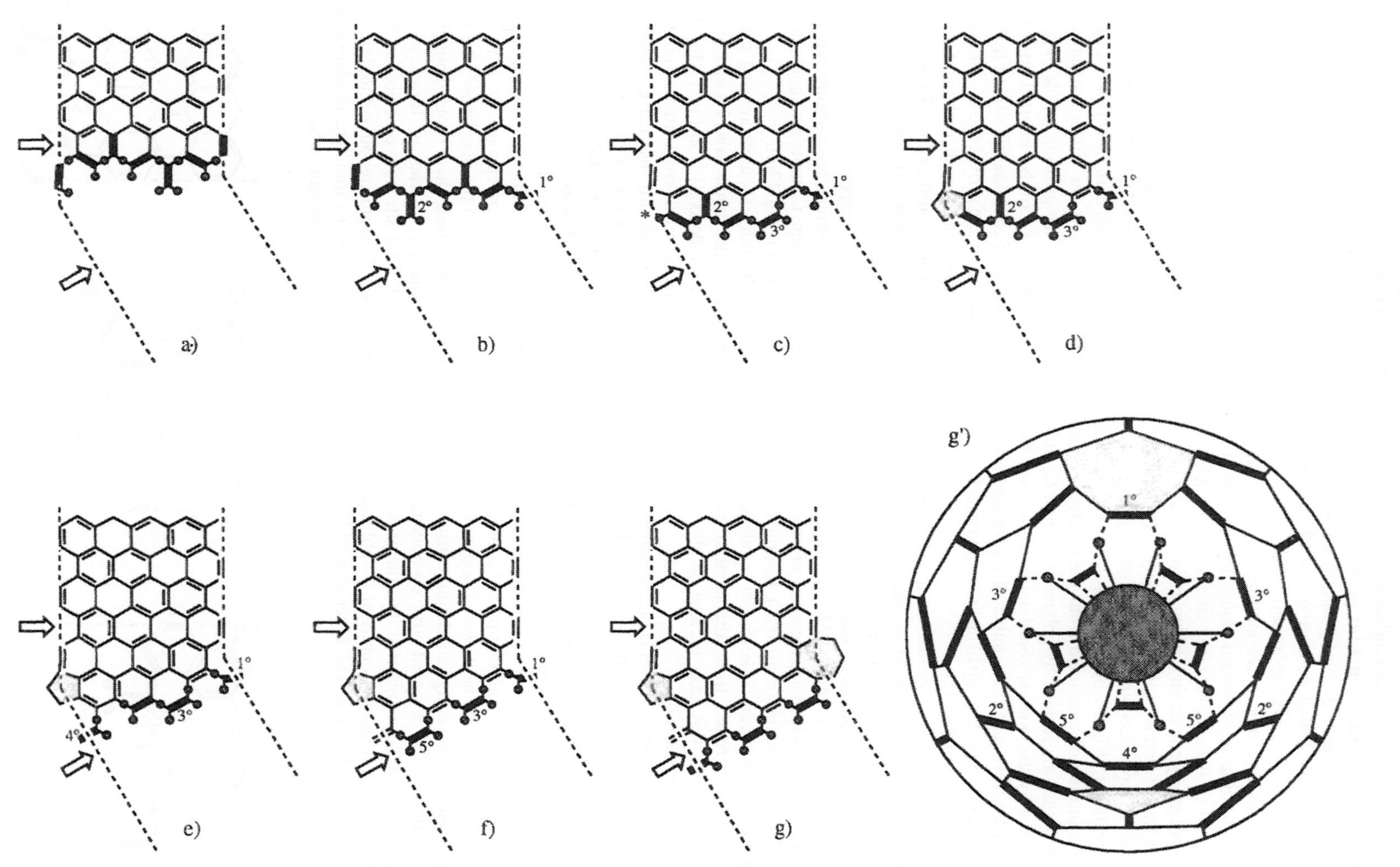

Fig. 17. Growth mechanism of a $(9n,0)$–$(5n,5n)$ knee involving from 24 to 20 coordination sites of the catalyst. (a)–(g) Planar representation of the successive tubule growing steps; (g′) Schlegel diagram representation of the whole knee with the C_2 numbering corresponding to that of the individual steps (a)–(g).

of the five-membered ring [Fig. 17(d)]. Note also that only six new C_2 units are shown in Fig. 17(c) with respect to Fig. 17(b).

Once the pair of sites is disconnected from the growing tubule [Fig. 17(d)], an orthogonal C_2 unit is inserted below the five membered ring [4° in Fig. 17(e)]. The latter inserted C_2 unit and the remaining two *cis*-polyacetylene C_2 segments are finally displaced by the arrival two orthogonal C_2 units [5° in Fig. 18(f)].

The carbon atoms linked to the remaining twenty coordination sites have now finished to rearrange to a fivefold symmetry [Fig. 17(f)]. Consecutive insertion of the five orthogonal C_2 units [1°, 3° and 5°) displaced from the catalyst by the arrival of five new orthogonal C_2 units closes the seven-membered ring and completes the knee [Fig. 17(g) and (g′)]. Further growth will yield a (5*n*,5*n*) tubule. It will proceed as already explained in Fig. 15. Figure 17(g′) is a Schlegel diagram explanation of the (9,0)–(5,5) knee, equivalent to Fig. 17(g). In that diagram it is possible to identify the C_2 units introduced at the different steps [Fig. 17(a)–(g)] by their numbering.

3.2.2.4 *Growth mechanism of a (5*n*,5*n*)–(9*n*,0) knee, involving from 20*n* to 24*n* coordination sites of the catalyst.* The explanation given here to pass from a (9*n*,0) tubule to a (5*n*,5*n*) tubule can also be used backwards to pass from a (5*n*,5*n*) tubule to a (9*n*,0) tubule. The progressive steps are illustrated in Fig. 18. The starting point of the knee is again the blockage of tubule growth at the seven-membered ring [1° in Fig. 18(a)]. The later formation of the five-membered ring is only a consequence of tubule growth blockage at the seven-membered ring. Secondly, probably because of the large space created on the other side of the tubule by the elastic bending [Fig. 12(b)] after tubule blockage, a *cis*-butadiene [2° in Fig. 18(b)] can be inserted instead of the usual orthogonal C_2 units. This *cis*-butadiene arriving with four new coordination sites [black points on Fig. 18(b)] will be the head of two *cis*-polyacetylene chains [3° in Fig. 18(c)]. (The freshly arrived C_2 segments of the *cis*-polyacetylene chain, not yet inserted in the tubule, are represented by dotted lines for the sake of clarity.) These two chains are started at the five-membered ring [Fig. 18(e)]. The insertion of that *cis*-butadiene also disturbs the other coordination sites, so that four other *cis*-polyacetylene chains [4° and 5° in Fig. 18(c)] are also inserted into the growing tubule. The other logical growing steps of the (5,5)–(9,0) knee are very close to the mechanism explained for the (9*n*,0) tubule. The insertion of the first *cis*-polyacetylene units [3° in Figs 18(c) and (d)] and the coordination of the first parallel C_2 units [6° in Fig. 18(d)], followed by the insertion of the *cis*-polyacetylene units [4° in Figs 18(d) and (e)] and the coordination of parallel C_2 units [7° in Fig. 18(e)] leads to positioning of the five membered ring. The closure of that ring, followed by the insertion of the *cis*-polyacetylene units [5° and 8° in Figs 18(e) and (f)] and coordination of three parallel C_2 units [9° in Fig. 18(f)] leads to positioning of the seven-membered ring. The closure of that ring, during a normal (9*n*,0) tubule growing step [Fig. 19(g) and (g′)], completes the knee and the following steps will be the growth of the (9*n*,0) tubule proceeding as explained in Fig. 16. Figure 18(g′) is a Schlegel diagram of the (5,5)–(9,0) knee, equivalent to Fig. 18(g). In that diagram it is possible to identify the C_2 units introduced at the different growing steps [Fig. 18(a)–(g)] by their numbering.

An important conclusion obtained from the model based on the variation of the number of coordination sites at the catalyst surface is that all double bonds can be localized on the (9*n*,0)–(5*n*,5*n*) knee and connected nanotubes. In fact, in that model, only single C–C bonds are formed and there is no double bond formation during the nanotubule growth. All of the double bonds are already localized on the C_2 units or on the inserted *cis*-polyacetylene chains. As seen from Figs 17 and 18, there is no double bond on the sides of the five membered ring and there is only one double bond on the seven membered ring. It should also be pointed out that localizing the double bonds on the (9*n*,0)–(5*n*,5*n*) knee and connected tubules was a very difficult task before the establishment of this model. Once the double bonds are localized, after replacing the vacant bonds of the (9*n*,0)–(5*n*,5*n*) knee of Fig. 17(g′) by hydrogens, it is possible to have a three dimensional view of the knee (Fig. 19). The introduction of the parameters of Fig. 19 into a more sophisticated program can also be used in order to minimize the energy and simulate the real knee angle[18,19].

Concerning the multi-shell tubules, the graphitic layers of the growing nanotubule described are supposed to grow on the same catalyst particle, at the same time as the inner layer does. This is in agreement with the observations of Fig. 11, where it is possible to see that the diameter range of the young tubes (1 minute) is about the same as that of the old ones (20 minutes). Moreover, it has also been observed that during a long exposure time (5 hours) of the tubules to the reaction conditions, only amorphous carbon is deposited on the outer layer[4,5], and no tubes with larger diameters were observed.

As the diameter of the catalyst particle is supposed to be close to that of the single-shell tubule[20], or to that of the inner tubule[8], the number of graphitic layers might depend on the flow rate of acetylene at the catalyst particle. The graphitic layers are supposed to be formed by the C_2 units formed on the catalyst particle, exceeding those needed for the growth of the multi-shell tubule inner layer. This generalisation to multi-layer tubules is just a hypothesis, since we do not have any experimental proof yet.

4. CONCLUSIONS

The building of knees, tori and helices is described by a simple formalism.

Relationships are established between the tubules

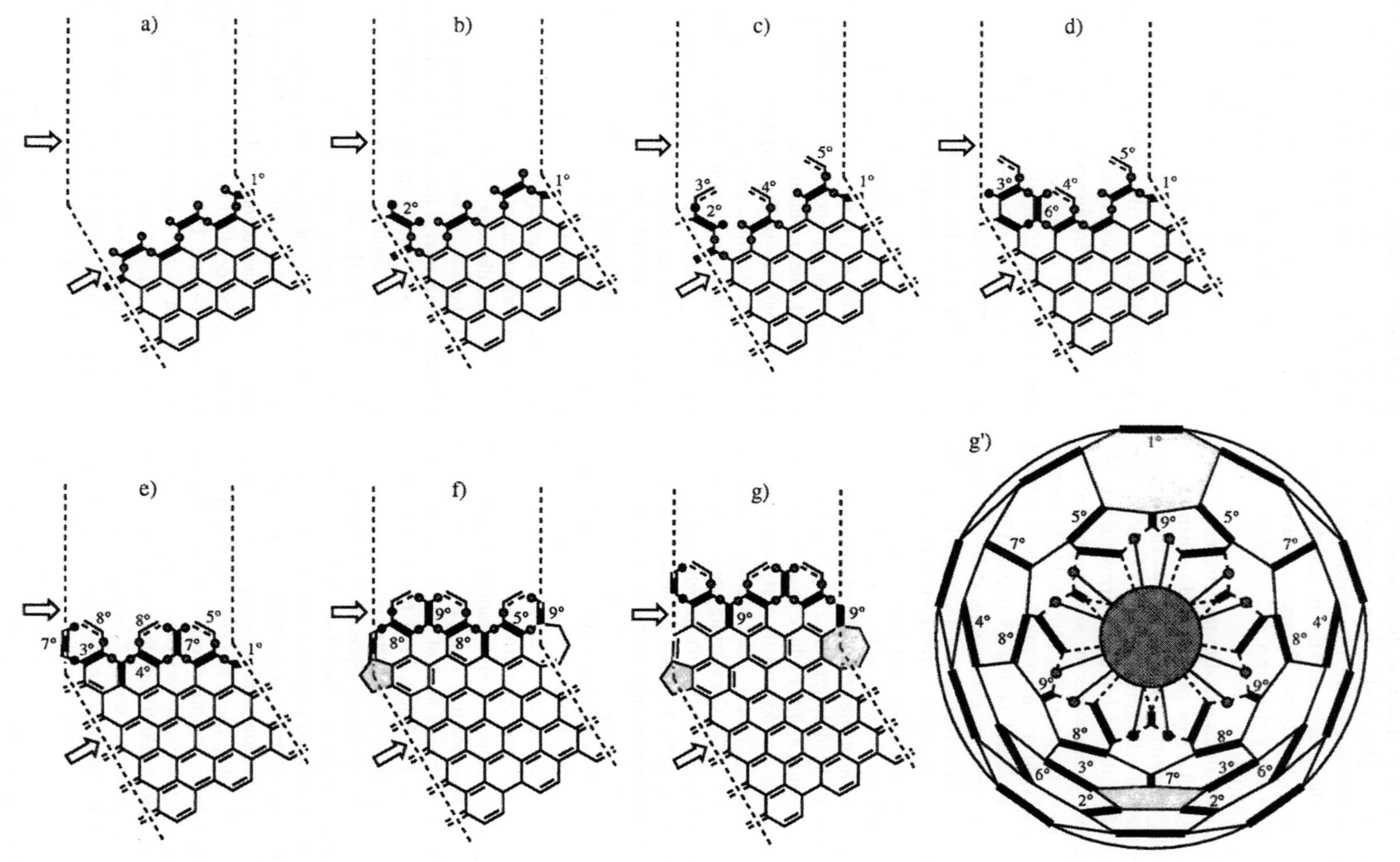

Fig. 18. Growth mechanism of a $(5n,5n)$–$(9n,0)$ knee involving from 20 to 24 coordination sites of the catalyst. (a)–(g) Planar representation of the successive tubule growing steps; (g') Schlegel diagram representation of the whole knee with the C_2 numbering corresponding to that of the individual steps (a)–(g).

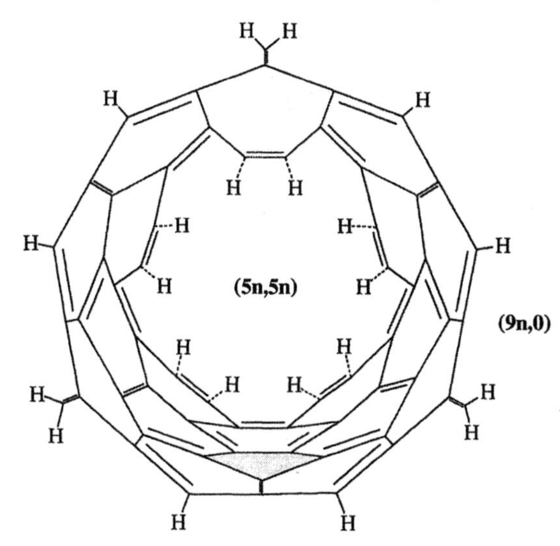

Fig. 19. Schlegel diagram of the (9,0)–(5,5) knee after replacing the vacant bonds by hydrogen. All of the double bonds are localized according to the growth process represented in Figs 17(g′) and 18(g′).

and their graphitic layers. Hence, after measurement of a nanotube diameter by HREM, it is possible to establish the order of its graphitic layer.

All the double bonds can be localized on the $(9n,0)$–$(5n,5n)$ knee and connected tubule. There is no double bond on the sides of the five-membered ring and only one double bond on the seven-membered ring.

Acknowledgements—The authors acknowledge the Region of Wallonia and the Belgian Prime Minister Office (PAI-projects) for financial support. Thanks are also addressed to D. Bernaerts for the TEM images, to D. Van Acker for the photographs and to F. Vallette for the drawings.

REFERENCES

1. R. T. K. Baker, *Carbon* **27**, 315 (1989).
2. N. M. Rodriguez, *J. Mater. Res.* **8**, 3233 (1993).
3. A. Oberlin, M. Endo and T. Koyama, *J. Cryst. Growth* **32**, 335 (1976).
4. V. Ivanov, J. B. Nagy, P. Lambin, A. Lucas, X. B. Zhang, X. F. Zhang, D. Bernaerts, G. Van Tendeloo, S. Amelinckx and J. Van Landuyt, *Chem. Phys. Lett.* **223**, 329 (1994).
5. V. Ivanov, A. Fonseca, J. B.Nagy, A. Lucas, P. Lambin, D. Bernaerts and X. B. Zhang, *Carbon.* (in press).
6. S. Iijima, *Nature* **354**, 56 (1991).
7. J. B. Howard, K. Das Chowdhury and J. B. Vander Sande, *Nature* **370**, 603 (1994).
8. D. Bernaerts, X. B. Zhang, X. F. Zhang, G. Van Tendeloo, S. Amelinckx, J. Van Landuyt, V. Ivanov and J. B.Nagy, *Phil. Mag.* **71**, 605 (1995).
9. S. Amelinckx, X. B. Zhang, D. Bernaerts, X. F. Zhang, V. Ivanov and J. B.Nagy, *Science* **265**, 635 (1994).
10. X. B. Zhang, X. F. Zhang, D. Bernaerts, G. Van Tendeloo, S. Amelinckx, J. Van Landuyt, V. Ivanov, J. B.Nagy, Ph. Lambin and A. A. Lucas, *Europhys. Lett.* **27**, 141 (1994).
11. J. H. Weaver, *Science* **265**, 511 (1994).
12. B. I. Dunlap, *Phys. Rev. B* **46**, 1933 (1992).
13. B. I. Dunlap, *Phys. Rev. B* **49**, 5643 (1994).
14. S. Itoh and S. Ihara, *Phys. Rev. B* **48**, 8323 (1993).
15. R. A. Jishi and M. S. Dresselhaus, *Phys. Rev. B.* **45**, 305 (1992).
16. A. A. Lucas, Ph. Lambin and R. E. Smalley, *J. Phys. Chem. Solids* **54**, 587 (1993).
17. T. Guo, P. Nikolaev, A. G. Rinzler, D. Tomanek, D. T. Colbert and R. E. Smalley, *Nature* (submitted).
18. Ph. Lambin, A. Fonseca, J.-P. Vigneron, J. B.Nagy and A. A. Lucas, *Chem. Phys. Lett.*. (in press).
19. Ph. Lambin, J.-P. Vigneron, A. Fonseca, J. B.Nagy and A. A. Lucas, *Synth. Metals* (in press).
20. P. M. Ajayan, J. M. Lambert, P. Bernier, L. Barbedette, C. Colliex and J. M. Planeix, *Chem. Phys. Lett.* **215**, 509 (1993).

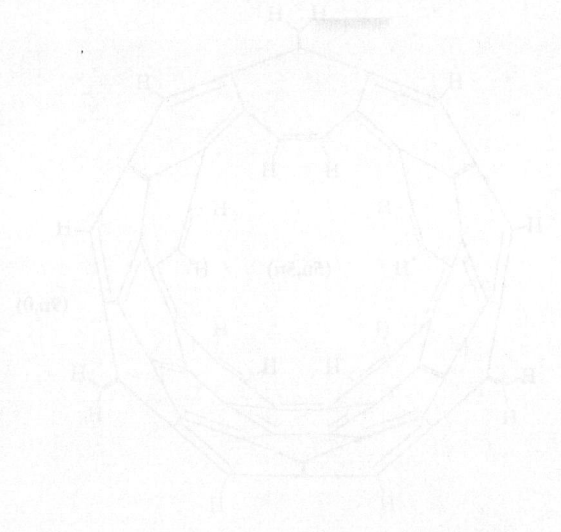

HEMI-TOROIDAL NETWORKS IN PYROLYTIC CARBON NANOTUBES

A. SARKAR, H. W. KROTO
School of Chemistry and Molecular Science, University of Sussex, Brighton, BN1 9QJ U.K.

and

M. ENDO
Department of Engineering, Shinshu University, Nagano, Japan

(*Received* 26 *May* 1994; *accepted in revised form* 16 *August* 1994)

Abstract—Evidence for the formation of an archetypal hemi-toroidal link structure between adjacent concentric walls in pyrolytic carbon nanotubes is presented. The observed and simulated TEM images for such structures are in excellent agreement. This study suggests that double-walled carbon nanotubes, in which the inner and outer tubes are linked by such hemi-toroidal seals, may be one viable way of overcoming the reactivity at the graphene edges of open-ended tubes to engineer stable and useful graphene nanostructures.

Key Words—Nanotubes, pyrolytic carbon nanotubes, hemi-toroidal nanostructures.

1. INTRODUCTION

The discovery by Iijima[1] that carbon nanotubes form in a Krätschmer-Huffman fullerene generator[2], and the subsequent development of techniques by Ebbeson and Ajayan[3] and others to produce them in viable quantities, promise to have important implications for future nanotechnology. The tubes are essentially elongated fullerenes[4] and, thus, description (in particular structural characterisation) in terms of fullerene chemistry/physics principles is appropriate and useful. A growth mechanism has been proposed for the primary tube growth phase[4] which involves ingestion of carbon fragments (e.g., atoms and chain molecules) at the reactive hemi-fullerene closed end-caps of the tubes. An open-ended growth mechanism has also been proposed[5]. A closed-end growth mechanism is consistent with the observations of Ulmer *et al.*[6] and McElvany *et al.*[7], that fullerenes can grow by ingestion of carbon fragments into their completely closed networks. Both growth schemes also are consistent with Iijima's observation that, in general, the tubes appear to have their hexagons helically disposed in the tube walls[1].

The discovery that carbon generates closed-cage fullerenes spontaneously[2,8] as well as nanotubes[1,3] has stimulated the question of whether other topologically accessible networks are feasible. Predictions on the possibility of both small fullerenes[9] and giant fullerenes[10,11] appear to have been vindicated[12,13]. As well as general closed-cage structures, extended repeat-pattern networks involving seven- and other-membered rings have been considered by Terrones and MacKay[14,15] and Lenosky *et al.*[16]. Other structures have been considered and, in particular, Chernozatonskii[17] and Ihara *et al.*[18,19] have discussed toroidal structure.

During the course of a detailed study of the structure and growth of pyrolytic carbon nanotubes (PCNTs) we have found that complex graphitisation processes occur during a *second* stage of heat treatment of these materials[20,21]. In particular, we here draw attention to evidence that *hemi*-toroidal graphite surfaces occur regularly during nanotube production. In this communication we present evidence for the formation of *single*-layered hemi-toroid structures, within bulk multi-walled material during PCNT graphitisation. The structures occur in material prepared by pyrolysis of hydrocarbons at ca. 1000°C and subjected to secondary heat treatment at 2800°C. The structures have been studied experimentally by high resolution transmission electron microscopy (HRTEM)[20] and the resulting images compared with simulations. The simulated structures were generated from hypothetical fullerene related hemi-toroidal networks, which were constructed on the basis of feasible graphene topologies relaxed by molecular mechanics. During the course of the present study of PCNTs, Iijima *et al.*[5] have obtained TEM images showing related structures in arc-formed carbon nanotubes (ACNTs).

2. OBSERVATIONS

The PCNTs obtained by decomposition of benzene at ca. 1000–1070°C on a ceramic substrate were gathered using a toothpick and heat treated at a temperature of 2800°C for 15 minutes[21]. The heat-treated PCNTs were then mounted on a porous amorphous carbon electron microscope grid (so-called carbon ultra-microgrid). HRTEM observations were made using a LaB_6 filament operating at an accelerating voltage of 300 kV. The special aberration coefficient $C_s = 2.0$ mm and lambda = 0.0197 A. The under-focus value was chosen delta f = 65 nm, where the $\sin(x)$ transfer function has a large plateau at $\sin(x)$ ca. -1 against the interlayer spacing of carbon in the range 5.0–2.8 A. These conditions were optimum for observing high-contrast 002 lattice fringes form the PCNTs.

The simulations were carried out on a Silicon Graphics Iris Indigo workstation using the CERIUS molecular modeling and the associated HRTEM module. The multislice simulation technique was applied using the following parameters: electron energy 400 kV (lambda = 0.016 A); C_s (aberration coefficient) = 2.7 mm; focus value delta f = 66 nm; beam spread = 0.30 mrad.

In Fig. 1 is shown a HRTEM image of part of the end of a PCNT. The initial material consisted of carbon nanotubes upon which bi-conical spindle-like secondary growth had deposited[21], apparently by inhomogeneous deposition of aromatic carbonaceous, presumably disordered, layers on the primary substrate nanotube. Prior to further heat treatment, the secondary growth showed no regular structure in the HRTEM images (Fig. 1), suggesting that it might consist largely of some form of "amorphous" carbon. After the second stage of heat treatment, at 2800°C, the amorphous sheaths graphitise to a very large degree, producing multi-walled graphite nanotubes which tend to be sealed off at points where the spindle-like formations are thinnest. The sealed-off end region of one such PCNT is shown in Fig. 1.

Fig. 1. The sealed-off tip of an annealed PCNT.

The detailed analysis of the way in which the overall and internal structure of PCNTs apparently arise is discussed elsewhere[20]. Here, we draw attention to some particularly interesting and unusual structures which occur in the body of the nanotubes. An expansion of the section of the central core which lies ca. $\frac{1}{3}$ below the tip of the nanotube in Fig. 1 is shown in Fig. 2. Loop structures occur at points a–d and a′–d′ in the walls in directly opposing pairs. This parallel behaviour must, on the basis of statistical arguments, be related and we interpret the patterns as evidence for a hemi-toroidal connection between the inner and outer adjacent concentric graphene tubes (i.e., turnovers similar to a rolled-over sock). That the loops, seen in the HRTEM, are evidence for very narrow single-walled closed-ended tubes trapped within the walls can be discounted, also on statistical grounds.

Perhaps the most important features in the images are the dark line images that interconnect adjacent loops. We particularly draw attention to the structure b–b′ in the expanded image in Fig. 2. Here, the adjacent loops are fairly distinct and well matched on ei-

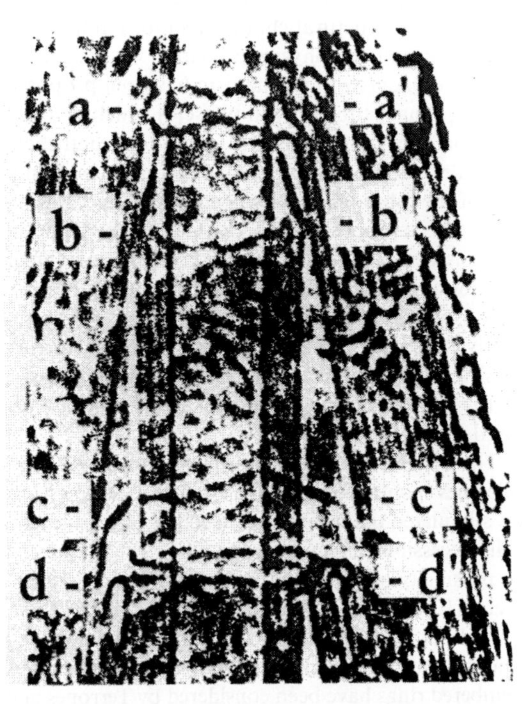

Fig. 2. Expanded image of part of the annealed PCNT of Fig. 1 showing several regions (a/a′–d/d′), in detail, where complex looped images appear to be associated with graphite turnover structures. The clearest eliptical image can be seen between markers b and b′.

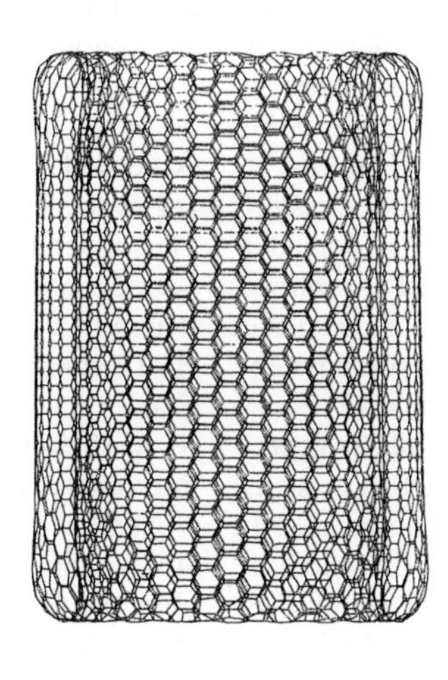

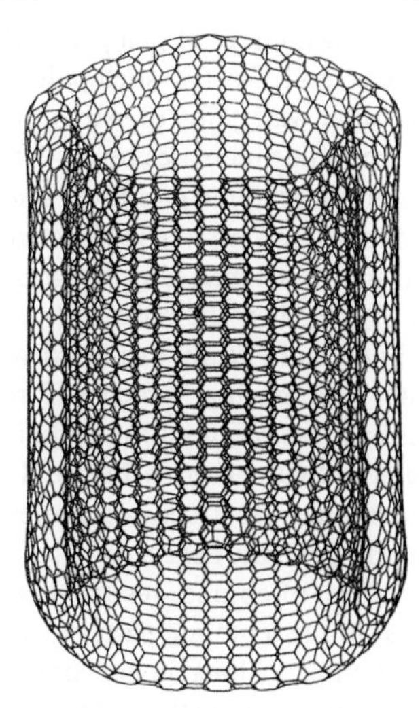

Fig. 3. Molecular graphics images of an archetypal flattened toroidal model of a nanotube with $n = 5$ and $m = 4$ at three different orientations in a plane perpendicular to the paper. Note the points on the rim where the cusps are located.

ther side. The image is, as discussed below, quite consistent with a single twin-walled tube connected at one end.

In Fig. 3, a set of molecular graphics images of molecular models of hemi-toroidal structures, which are the basis of the archetypal double-walled nanotube, is depicted. The Schlegel-type diagram for the *rim* of the smallest likely double-walled tube is depicted in Fig. 4, where the fact that it consists of a hexagonal network in which 5 sets of pentagon/heptagon (p/h) pairs are interspersed in a regular manner is seen. This diagram results from the connection of five p/h pair templates of the kind depicted in Fig. 5a. The inside and outside walls of the hemi-toroid are linked by the matching p/h pairs in such a way that the curvature allows optimal graphene sheet distortion and inter-sheet interaction. For every p/h pair, the outer cylinder gains an extra two carbon atoms over the inner

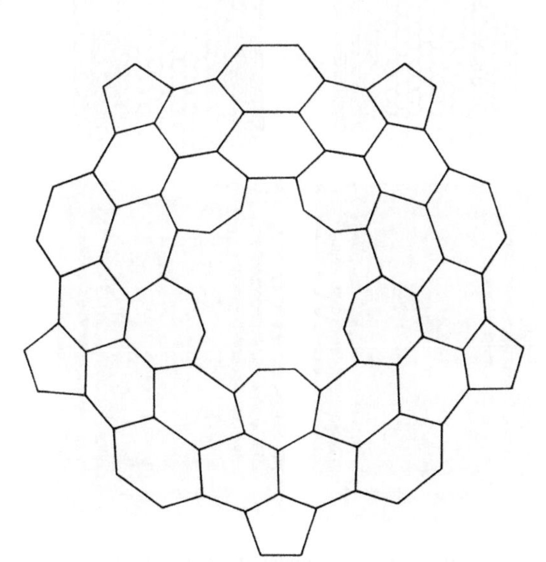

Fig. 4. Schlegel-type diagram of the hemi-toroidal rim of a double walled nanotube, in which the inner and outer walls are connected by a set of $n = 5$ pentagon/heptagon pairs.

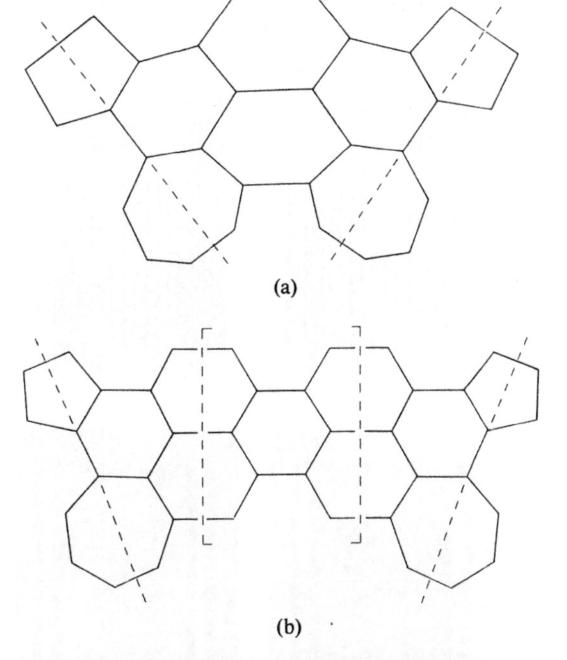

Fig. 5. General templates for inner/outer wall rim connections: a) simplest feasible template (section between the dashed lines) in which *one* bond separates the heptagons on the inside rim, *one* bond separates the heptagons and pentagons across the rim, and two bonds and a hexagon separate pengatons on the outside rim; b) The overall circumference of the hemitoroidal structure can readily be increased without increasing the interwall separation by inserting m further hexagon + two half-length bond units of the kind shown in the square dashed brackets.

one. The number of p/h pairs determines the wall separation and, for five such pairs, the circumference of the outer wall gains 10 extra atoms and the difference in the radius increases by ca. 3.5 A—very close to the optimum graphite interlayer spacing. Furthermore, if a particular template is bisected and further hexagonal network inserted at the dashed line, as indicated in Fig. 5, then the radii of the connected concentric cylinders are easily increased *without* increasing the interwall spacing, which will remain approximately 0.35 nm, as required for graphite. A structure results such as that shown in Fig. 3. These templates generate non-helical structures; however, modified versions readily generate helical forms of the kind observed[24]. There is, of course, yet another degree of freedom in this simple template pattern, which involves splitting the template along an arc connecting the midpoints of the bonds between the h/p pairs. This procedure results in larger inter-wall spacings[24]; the resulting structures are not considered here.

We note that in the HRTEM images (Figs. 1 and 2) the points b and b′ lie at the apices of the long axis of an elliptical line-like image. Our study suggests that this pattern is the HRTEM fingerprint of a hemi-toroidal link structure joining two concentric graphite cylinders, whose radii differ by the graphite interlayers separation (ca. 0.35 nm). If the number of p/h pairs, $n = 5$ and the number of insertions $m = 4$ (according to Fig. 4), then the structure shown in Fig. 3 is generated. The basic topologically generated structure was relaxed interactively using a Silicon Graphics workstation operating the Cerius Molecular Mechanics Programme. The results, at a series of angles (0, 5, 10, and 15°) to the normal in a plane perpendicular to the paper, are shown diagrammatically in Fig. 6a. The associated simulated TEM images for the analogous angles to the electron beam are depicted in Fig. 6b. The model has inner/outer radii of ca. 1.75/2.10 nm (i.e., 5× and 6× the diameter of waist of C_{70}). The structure depicted in the experimentally observed HRTEM image is ca. 3.50/3.85 (i.e., ca. 10× and 11× C_{70}).

As the tube orientation changes, we note that the interference pattern associated with the rim changes from a line to an ellipse and the loop structures at the apices remain quite distinct. The oval patterns in the observed (Fig. 2) and simulated (Fig. 6) HRTEM images are perfectly consistent with one another. For this preliminary investigation, a symmetric wall configuration was used for simplicity. Hemi-toroid connection of inner and outer tubes with helical structured walls requires somewhat more complicated dispositions of 5/6/7 rings in the lip region[24]. The general validity of the conclusions drawn here is, however, not affected. Initial studies of this problem[24] indicate that linking between the inner and outer walls is also not hindered in general.

The toroids show an interesting change in overall morphology as they become larger, at least at the lip. The hypothetical small toroid shown in Fig. 3a is actually quite smooth and essentially a fairly rounded structure. As the structures become larger, the strain

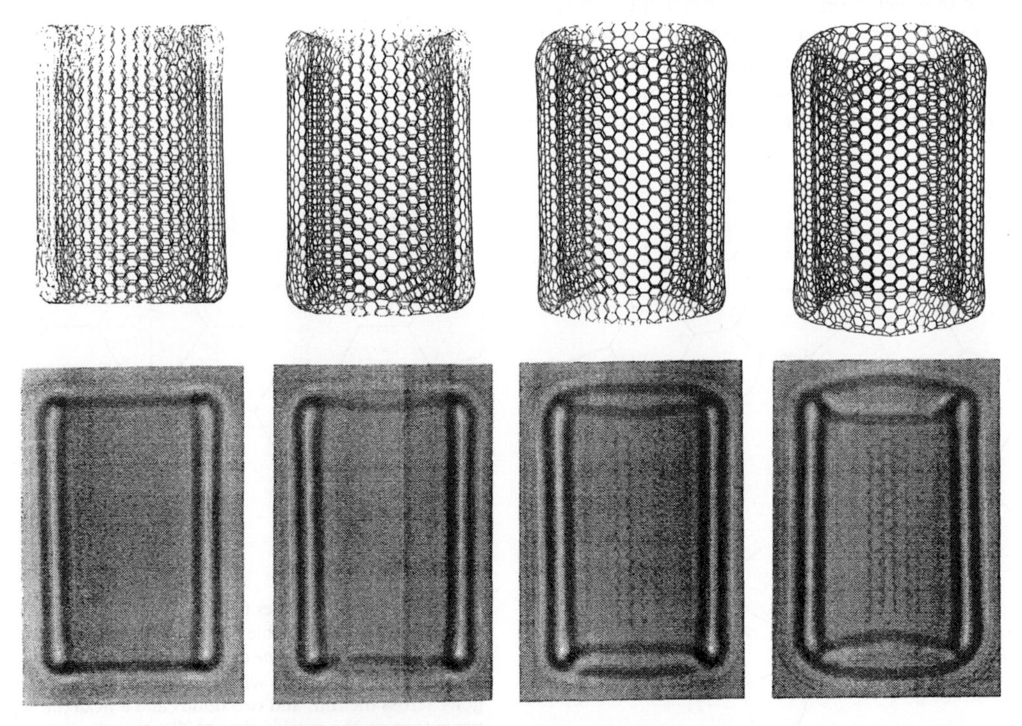

Fig. 6. above) Molecular graphics images of an archetypal flattened toroidal model of a nanotube with $n = 5$ and $m = 4$ at three different orientations (0, 5, 10, 15°) in a plane perpendicular to the paper. below) Resulting simulated TEM images of the nanotube at the above orientations to the electron beam; note that even the spring onion-like bulges at the ends are reproduced.

becomes focused in the regions where the pentagons and heptagons lie, producing localised cusps and saddle points that can be seen relatively clearly in Fig. 3. Fascinating toroidal structures with D_{nh} and D_{nd} symmetry are produced, depending on whether the p/h pairs at opposite ends of the tube are directly aligned or are offset by $2\pi/2n$. In Fig. 3 the symmetry is D_{5h}. Chiral structures are produced by off-setting the pentagon from the heptagons[24]. In a D_{5d} structure, the walls are fluted between heptagons at opposite ends of the inner tube and the pentagons of the outer walls. It is interesting to note that, in the computer images, the cusps give rise to variations in the smoothness of the more-or-less elliptical image generated by the rim when viewed at an angle. The observed image (b–b′) exhibits variations consistent with the localised cusping that our model predicts will occur.

3. DISCUSSION

In this study, we note that epitaxial graphitisation is achieved by heat treatment of the apparently mainly amorphous material which surrounds a nanotube[20]. As well as bulk graphitisation, localised hemi-toroidal structures that connect adjacent walls appear to form readily. This type of structure may be important as it suggests that double-walled, closed structures may be produced by heat treatment, so forming stable nanoscale graphite tubes in which dangling bonds have been eliminated. It will be most interesting to probe the relative reactivity of these structures for the future of nanoscale devices, such as quantum wire supports. Although the curvatures of the rims appear to be quite tight, it is clear from the loop images that the occurrence of turnovers between concentric cylinders with a gap spacing of the standard graphite interlayer spacing is relatively common. Interestingly, the edges of the rims appear to be readily visible and this has allowed us to confirm the relationship between opposing loops. Bulges in the loops of the kind observed at d′ in Fig. 2 are simulated theoretically as indicated in Fig. 6.

During the course of the study on the simplest archetypal twin-walled structures in PCNT material described here, the observation of Iijima *et al.*[5] on related multi-walled hemi-toroidal structures in ACNTs appeared. It will be most interesting to probe the differences in the formation process involved.

The question might be addressed as to whether the hemispheroidal structures observed here may provide an indication that complete toroids may be formed from graphite. The structures appear to form as a result of optimal graphitisation of two adjacent concentric graphene tubes, when further extension growth is no longer feasible (i.e., as in the case shown in Fig. 2 at b–b′ where a four-walled tube section must reduce to a two-walled one.) This suggests that perfect toroids are unlikely to form very often under the conditions discussed here. However, the present results do suggest that, in the future, nanoscale engineering techniques may be developed in which the graphene edges of open-ended double-walled tubes, produced by controlled oxidation or otherwise, may be cauterised by rim-seals which link the inner and outer tubes; this may be one viable way of overcoming any dangling bond reactivity that would otherwise preclude the use of these structures in nanoscale devices. In this way, a long (or high) fully toroidal structure might be formed.

Acknowledgement—We thank Raz Abeysinghe, Lawrence Dunne, Thomas Ebbeson, Mauricio Terrones, and David Walton for their help. We are also grateful to the SERC and the Royal Society for financial support.

REFERENCES

1. S. Iijima, *Nature* **354**, 56 (1991).
2. W. Krätschmer, L. Lamb, K. Fostiropoulos, and D. Huffman, *Nature* **347**, 354 (1990).
3. T. W. Ebbeson and P. M. Ajayan, *Nature* **358**, 220 (1992).
4. M. Endo and H. W. Kroto, *J. Phys. Chem.* **96**, 6941 (1992).
5. S. Iijima, P. M. Ajayan, and T. Ichihashi, *Phys. Rev. Letts.* **69**, 3100 (1992).
6. G. Ulmer, E. E. B. Cambell, R. Kuhnle, H-G. Busmann, and I. V. Hertel, *Chem. Phys. Letts.* **182**, 114 (1991).
7. S. W. McElvany, M. M. Ross, N. S. Goroff, and F. Diederich, *Science* **259**, 1594 (1993).
8. H. W. Kroto, J. R. Heath, S. C. O'Brien, R. F. Curl, and R. E. Smalley, *Nature* **354**, 359 (1985).
9. H. W. Kroto, *Nature* **329**, 529 (1977).
10. H. W. Kroto and K. G. McKay, *Nature* **331**, 328 (1988).
11. H. W. Kroto, *Chem. Brit.* **26**, 40 (1990).
12. T. Guo, M. D. Diener, Y. Chai, M. J. Alford, R. E. Haufler, S. M. McClure, T. Ohno, J. H. Weaver, G. E. Scuseria, and R. E. Smalley, *Science* **257**, 1661 (1992).
13. L. D. Lamb, D. R. Huffman, R. K. Workman, S. Howells, T. Chen, D. Sarid, and R. F. Ziolo, *Science* **255**, 1413 (1991).
14. H. Terrones and A. L. Mackay, In *The Fullerenes* (Edited by H. W. Kroto, J. E. Fischer, and D. E. Cox), pp. 113–122. Pergamon, Oxford (1993).
15. A. L. Mackay and H. Terrones, In *The Fullerenes* (Edited by H. W. Kroto and D. R. M. Walton), Cambridge University Press (1993).
16. T. Lenosky, X. Gonze, M. P. Teter, and V. Elser, *Nature* **355**, 333 (1992).
17. L. A. Chernozatonskii, *Phys. Letts. A.* **170**, 37 (1992).
18. S. Ihara, S. Itoh, and Y. Kitakami, *Phys. Rev. B* **47**, 13908 (1993).
19. S. Itoh, S. Ihara, and Y. Kitakami, *Phys. Rev. B* **47**, 1703 (1993).
20. A. Sarkar, H. W. Kroto, and M. Endo, in prep.
21. M. Endo, K. Takeuchi, S. Igarishi, M. Shiraishi, and H. W. Kroto, *J. Phys. Chem. Solids* **54**, 1841 (1993).
22. S. Iijima, *Phys. Rev. Letts.* **21**, 3100 (1992).
23. S. Iijima, *Mat. Sci. and Eng.* **B19**, 172 (1993).
24. U. Heinen, H. W. Kroto, A. Sarkar, and M. Terrones, in prep.

PROPERTIES OF BUCKYTUBES AND DERIVATIVES

X. K. Wang, X. W. Lin, S. N. Song, V. P. Dravid, J. B. Ketterson, and R. P. H. Chang*
Materials Research Center, Northwestern University, Evanston, IL 60208, U.S.A.

(*Received* 25 *July* 1994; *accepted* 10 *February* 1995)

Abstract—The structural, magnetic, and transport properties of bundles of buckytubes (buckybundles) have been studied. High-resolution electron microscopy (HREM) images have revealed the detailed structural properties and the growth pattern of buckytubes and their derivatives. The magnetic susceptibility of a bulk sample of buckybundles is -10.75×10^{-6} emu/g for the magnetic field parallel to the bundle axes, which is approximately 1.1 times the perpendicular value and 30 times larger than that of C_{60}. The magnetoresistance (MR) and Hall coefficient measurements on the buckybundles show a negative MR at low temperature and a positive MR at a temperature above 60 K and a nearly linear increase in conductivity with temperature. The results show that a buckybundle may best be described as a semimetal. Using a stable glow discharge, buckybundles with remarkably large diameters (up to 200 μm) have been synthesized. These bundles are evenly spaced, parallel, and occupy the entire central region of a deposited rod. HREM images revealed higher yield and improved quality buckytubes produced by this technique compared to those produced by a conventional arc discharge.

Key Words—Buckytubes, buckybundle, glow discharge, magnetic properties, transport properties.

1. INTRODUCTION

Since the initial discovery[1,2] and subsequent development of large-scale synthesis of buckytubes[3], various methods for their synthesis, characterization, and potential applications have been pursued[4–12]. Parallel to these experimental efforts, theoreticians have predicted that buckytubes may exhibit a variation in their electronic structure ranging from metallic to semiconducting, depending on the diameter of the tubes and the degree of helical arrangement[13–16]. Thus, careful characterization of buckytubes and their derivatives is essential for understanding the electronic properties of buckytubes.

In this article, we describe and summarize our studies on the structural, magnetic,and transport properties of buckytubes. In addition, we describe how a conventional arc discharge can be modified into a stable glow discharge for the efficient synthesis of well-aligned buckytubes.

2. EXPERIMENTAL

2.1 *Synthesis of buckytubes*

Bundles of buckytubes were grown, based on an arc method similar to that of Ebbesen and Ajayan[3]. The arc was generated by a direct current (50–300 A, 10–30 V) in a He atmosphere at a pressure of 500 Torr. Two graphite electrode rods with different diameters were employed. The feed rod (anode) was nominally 12.7 mm in diameter and 305 mm long; the cathode rod was 25.4 mm in diameter and 100 mm long (it remained largely uneroded as the feed rod was consumed). Typical rod temperature near the arc was in the range of 3000 K to 4000 K.[11] After arcing for an hour, a deposited carbon rod 165 mm in length and 16 mm in diameter was built up on the end of the cathode. The deposition rate was 46 μm/sec.

2.2 *Structural measurements by electron microscopy*

Most of the HREM observations were made by scraping the transition region between the "black ring" material and the outer shell, and then dispersing the powder onto a holey carbon TEM grid. Additional experiments were conducted by preparing cross-sections of the rod, such that the rod was electron-transparent and roughly parallel to the electron beam. HREM observations were performed using an HF-2000 TEM, equipped with a cold field emission gun (c FEG) operated at 200 keV, an Oxford Pentafet X-ray detector, and a Gatan 666 parallel EELS spectrometer.

2.3 *Magnetic susceptibility measurements*

Magnetic susceptibility measurements were performed using a magnetic property measurement system (Quantum Designs Model MPMS). This system has a differential sensitivity of 10^{-8} emu in magnetic fields ranging from -5.5 T to $+5.5$ T over a temperature range of 1.9 K–400 K. The materials studied included: three buckybundle samples of 0.0712 g, 0.0437 g, and 0.0346 g; 0.0490 g of C_{60} powder, 0.1100 g of gray-shell material, 0.1413 g of polycrystalline graphite anode, and a 0.0416-g graphite single crystal. Measurements were performed at temperatures from 2 K to 300 K and in magnetic fields ranging from 0.005 T to 4 T. The susceptibility of buckytubes was measured with the magnetic field (H) either parallel to ($\chi_B^{\parallel}$) or perpendicular to ($\chi_B^{\perp}$) the buckybundle axis. All samples used in this work were enclosed in gelatin capsules, and the background of the container

*Author to whom correspondence should be addressed.

was subtracted from the data. The absolute accuracy of the mass susceptibility relative to the "standard" value of the graphite crystal was about 1%.

2.4 *Transport property measurements*

The transport properties were measured using standard dc (for the Hall effect) and ac (for MR) four-terminal techniques. The instrument used in this study was the same one we employed for the measurements of magnetic properties. The contact configuration is shown schematically by the inset in Fig. 6 (a). The I-V characteristic was measured to ensure Ohmic behavior so that no hot electron effects were present. The magnetic field was applied perpendicular to the tube axis. The measured MR and apparent Hall coefficient showed essentially the same temperature and field dependence, regardless of the samples used and the distance between the potential contacts, implying that the samples were homogeneous. For example, the residual resistivity ratio, R(300 K)/R(5 K), measured on different single buckybundles agreed with each other within 1%. In what follows we present the data taken on a single buckybundle having a diameter of 60 μm, the distance between the two potential contacts being 350 μm.

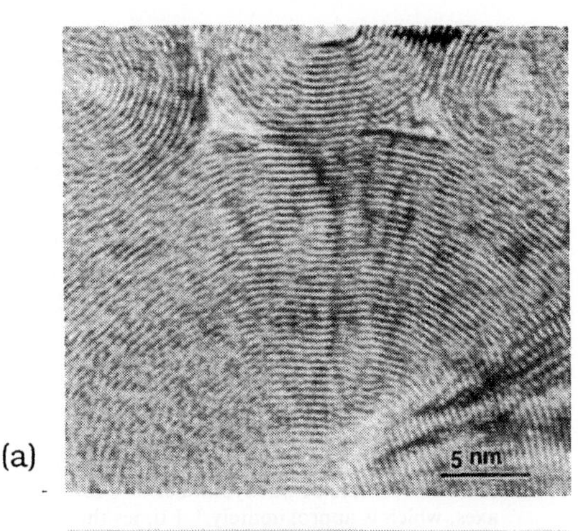

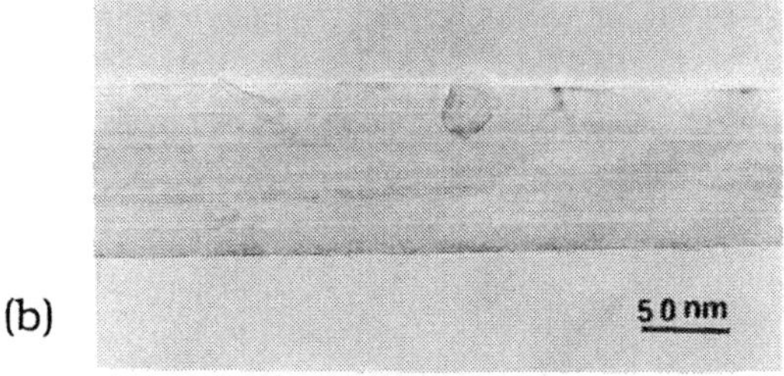

Fig. 1. (a) A cross-sectional TEM image of a bundle of buckytubes; (b) an HREM image of a single bundle of buckytubes with their axes parallel to the bundle axis.

3. RESULTS AND DISCUSSIONS

3.1 *Structural properties*

Buckytubes were observed for the first time by HREM[1,2] and their structural properties were subsequently characterized. In this section, we will briefly describe observations of the structure of a bundle of buckytubes, evidence for a helical growth of buckytubes and their derivatives, and the single-shell structures.

3.1.1 *The structure of buckybundles.* Both cross-sectional and high-resolution electron microscopy images of a single bundle are shown in Fig. 1 (a) (end-on-view) and 1 (b) (side view of a single bundle). The end-on view shows that the tubes are composed of concentric graphitic sheets. The spacing between the adjacent graphitic sheets is about 0.34 nm. The thinnest tube in this specimen, consisting of 8 carbon-hexagon sheets, has an outside diameter of 8 nm. The largest one, consisting of 48 sheets, has an outside diameter of about 30 nm. It is worth noting that although the tubes have a wide range of diameters, they tend to be packed closely together. The side view of the single bundle directly reveals that the bundle consists of closely packed buckytubes running parallel to one another, these images clearly demonstrate that the bundle is actually a bundle of buckytubes. Since the valence requirements of all atoms in a buckytube (with two sealed ends) are satisfied, the interaction among buckytubes should be Van der Waals in nature. Therefore, it is energetically favorable for buckytubes packed closely together to form a "buckybundle."

3.1.2 *The helicacy of buckytubes.* The helicacy of buckytubes is an interesting phenomenon. It has been suggested[17] that the growth pattern, as well as many properties of buckytubes, are intimately related to their helicacy. Here, we present the visible observations of frozen growth stage of buckytubes and derivatives suggesting a helical growth mechanism.

Figure 2 shows two HREM images of buckytubes seen end-on (i.e., the axis of the tube being parallel to the electron beam). The hollow center region is apparent, indicating the obvious tubular nature of the tubes. Strain contrast was evident in all these images, which is reminiscent of disclination type defects[18,19]. If we follow the individual inner-shell graphitic sheets around, shown in Fig. 2 (a), we observe that the termination is incomplete; that is, one extra graphitic sheet is associated with one portion of the inner shell compared with its opposite side. Figure 2 (b) shows that six graphitic sheets are seen to wrap around a thicker buckytube. In other words, the tubes are more of a rolled carpet geometry rather than the Russian Doll-type structure in our sample.

We also observed that the rounded particulates in the transition region between the "black ring" and the outer shell of the deposited rod are a collection of completely closed graphitic sheets with a helical pattern of inner shells. Figure 3 shows a larger buckyfootball containing smaller inner footballs that seem to grow inside the larger one. The inner footballs clearly display extra unterminated graphitic sheets, indicative of helical growth. These observations strongly suggest that Fig. 3 represents buckyfootballs that form through a helical growth of the sheets analogous to that of

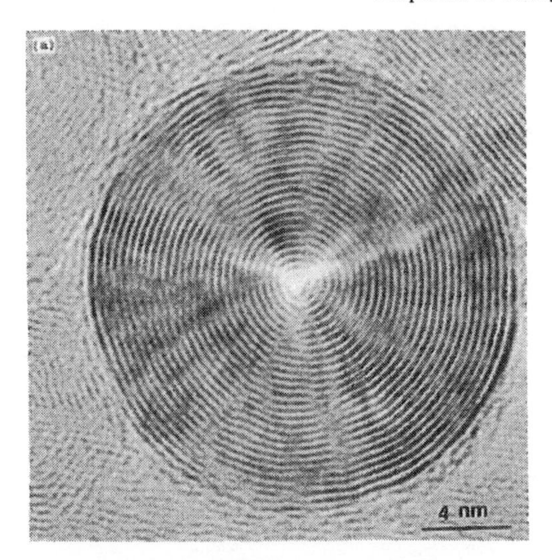

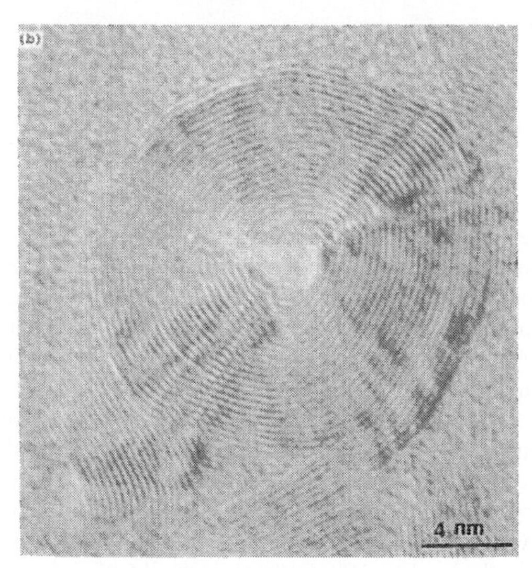

Fig. 2. Two HREM images of frozen growth of buckytubes seen end-on.

single-crystal growth through a screw dislocation mechanism[2].

3.1.3 *The structure of single-shell buckytubes.* Single-shell buckytubes were first synthesized by Iijima and Ichihashi[20] and Bethune *et al.*[21]. Unlike their reports, where the single-shell buckytubes were found on the chamber wall, we found a large number of monolayer buckytubes in a deposited rod which built up on the cathode of the dc arc[22]. The synthesis of the single-shell buckytubes used in this study was similar to that described in the experimental section. However, instead of using a monolithic graphite anode, we utilized a composite anode consisting of copper rods inserted inside a graphite anode in a variety of ways. The single-shell buckytubes assumed a large variety of shapes (shown in Fig. 4), such as buckytents, buckydomes, and giant fullerenes, and much shorter tubes than in previous reports. Protrusions, indentations, and bending were quite common, unlike the faceted or smooth-faceted morphologies of multilayer buckytubes. Because the insertion of pentagons and heptagons produced positive and negative curvature of the hexagonal network, respectively, the complicated morphologies of single-shell buckytubes were probably a result of the insertion of pentagons, heptagons, or various combinations into the single-shell buckytube. We believe that it is much easier to incorporate pentagonal and heptagonal defects in these single-shell structures than in multilayer buckytubes.

3.2 *Magnetic properties*

In the first paper reporting the observation of C_{60}, Kroto *et al.* suggested that the molecule was aromatic-like with its inner and outer surfaces covered with a "sea" of π-electrons[18]. Accordingly, C_{60} should exhibit a large diamagnetic susceptibility associated with a π-electron ring current. This suggestion appeared to be supported by NMR chemical shift measurements [19,23]. Theoretical calculations of the magnetic susceptibility of C_{60} performed by various groups were not consistent; some authors predicted a vanishingly small diamagnetism[24–26] and others predicted strong diamagnetism (as in an aromatic system)[27,28]. Susceptibility measurements on C_{60} and C_{70} showed that the value for C_{70} was about twice that of C_{60}, but that both were very small[25,29]. Another theoretical calculation by Ajiki & Ando[30] showed that the calculated magnetic susceptibility for the magnetic field perpendicular to the carbon nanotube axis was about three orders of magnitude larger than the parallel value. It is, then, of interest to extend the measurements to buckytubes.

With the availability of bulk samples of carbon nanotubes (bundles of buckytubes), a systematic study of the magnetic properties became possible. A cylindrical bulk deposit, with a diameter of 10 mm, consisting of an inner core and an outer cladding, was formed on the graphite cathode during the arc process. The outer cladding, a gray shell, was composed mainly of amorphous carbon and buckydoughnuts[31]. The inner core, with a diameter of 8 mm, consisted of an array of rather evenly spaced, parallel, and closely packed bundles approximately 50 micrometers in diameter. HREM revealed that the buckybundles in our best sample were comprised of buckytubes running parallel to one another[32]. A buckytube in each bundle could be pictured as a rolled-up graphitic sheet with a diameter ranging from 8 Å to 300 Å and a length of a few microns. The tube was capped by surfaces involving 6 pentagons (per layer) on each end[1,2]. The purity of the bulk sample was examined by energy dispersive X-ray analysis. No elements heavier than carbon were observed. The C_{60} powder was extracted from the soot formed in the same chamber employed for production of the buckytubes. The purity of the C_{60} powder used in this study was better than 99% as examined by high-performance liquid chromatography.

The measured mass susceptibility values for buckybundle (both $\chi_B^{\parallel}$ and $\chi_B^{\perp}$), C_{60}, the gray-shell material, the polycrystalline graphite anode, and the

Fig. 3. Three small buckyfootballs appear to grow inside a large buckyfootball.

graphite crystal are shown in Table 1. The measured mass susceptibility of -0.35×10^{-6} emu/g for C_{60} is consistent with the literature value[29], and is 30 times smaller than that of buckytubes. The C_{60} powder shows the strongest magnetic field dependence of the susceptibility and does not exhibit diamagnetism until H is greater than 0.5 T; saturation is observed for fields greater than 3 T. The C_{60} results, involving a very small diamagnetic susceptibility and a strong magnetic field dependence, appear to support the Elser-Haddon result where a cancellation occurs between the diamagnetic and paramagnetic contributions[26]. The small measured susceptibility of C_{60} suggests that if it (or possibly other fullerenes) is present as a contaminant in the buckybundle matrix (which is likely at some level) its contribution will be small.

From Table 1, we see that the measured susceptibility of the polycrystalline graphite anode (used to produce the fullerenes measured here) is 6.50×10^{-6} emu/g, which is near the literature value (implying the behaviors of the remaining materials do not involve impurities arising from the source material). The measured susceptibility of the gray-shell material, which consists of amorphous carbon mixed with fragments of buckytubes and buckydoughnuts, is close to (but larger than) that of the source rod.

The measured magnetic susceptibility of multilayer buckytubes for $\chi_B^\perp$ is approximately half $\chi_G^\parallel$ of graphite. This can be interpreted as follows. We recall that crystalline graphite is a semimetal with a small band overlap and a low density of carriers (10^{-18}/cm^3)[33, 34]; the in-plane effective mass is small ($m^* \sim \frac{1}{20} m_0$). The bending of the graphite planes necessary to form a buckytube changes the band parameters. The relevant dimensionless parameter is the ratio a/R, where a (≈ 3.4 Å) is the lattice constant and R is the buckytube radius. For $R \approx 20$ Å, the shift is expected to alter the nature of the conductivity[13–16]. In our buckybundle samples, most of material involves buckytubes with $R > 100$ Å confirmed by statistical analysis of TEM data, and we assume that the elec-

Fig. 4. Various morphologies of single-shell buckytubes.

Table 1. Measured room-temperature susceptibilities

Material	Symbol	Susceptibility $\times 10^{-6}$ emu/g
Buckybundle: axis parallel to H	$\chi_B^\parallel$	−10.75
Buckybundle: axis perpendicular to H	$\chi_B^\perp$	−9.60
C_{60} powder	$\chi_{C_{60}}$	−0.35
Polycrystalline graphite anode	—	−6.50
Gray-shell material	—	−7.60
Graphite: c-axis parallel to H	$\chi_G^\parallel$	−21.10
Graphite: c-axis perpendicular to H	$\chi_G^\perp$	−0.50

tronic properties are close to those of a graphite plane. The study of electron energy loss spectra (EELS) supports this model[19]. We can identify three relevant energy scales. First, the quantizing effect of the cylindrical geometry involves an energy $\Delta E = \hbar^2/2mR^2 \approx 0.7 \times 10^{-2}$ eV. Second, there is the Fermi energy E_f, which is 1.2×10^{-2} eV. Third, there is the thermal energy, which at room temperature (the highest temperature studied) is about 2.5×10^{-2} eV. We see that all three of these energy scales are of the same order. If we consider a higher temperature, where the carriers are Boltzmann particles (with a small inelastic mean free path l_i), and the magnetic field (~tesla) is a small perturbation on the particle motion, the magnetic susceptibility would be due to small quantum corrections to the energy of the system. For this quasi-classical case, the quantizing action of the geometry is not important; the response of the system to the perturbation may be considered as a sum over small plaquettes of the size l_i. (This additivity is hidden by the effect of a non-gauge-invariant formalism[35]; nevertheless, it is a general physical property, and the inelastic mean free path is the correlation radius of a local magnetization.) Therefore, at high temperatures a buckytube with $R > l_i$ may be considered as a rolled-up graphitic sheet (or concentric tubes). We use this model to calculate the susceptibility, $\chi_B^\perp$, for the field perpendicular to the buckybundle axis. We write the susceptibility tensor of single crystal graphite as

$$\overleftrightarrow{\chi} = \begin{pmatrix} \chi_G^\perp & 0 & 0 \\ 0 & \chi_G^\perp & 0 \\ 0 & 0 & \chi_G^\parallel \end{pmatrix} \tag{1}$$

To obtain the magnetic susceptibility of a buckytube for the magnetic field perpendicular to the buckybundle axis, $\chi_B^\perp$, we have to average the magnetic energy $E = 0.5\, \chi_{ij} H_i H_j$ over the cylindrical geometry of the buckytube (over a plane containing the a and c axes):

$$dE = 0.5[\chi_G^\parallel H^2 \cos^2\alpha + \xi_G^\perp H^2 \sin^2\alpha]\,d\alpha \tag{2}$$

or

$$E = 0.5H^2[\chi_G^\parallel \cos^2\alpha + \chi_G^\perp \sin^2\alpha]\,d\alpha$$

$$= 0.5H^2(0.5\chi_G^\parallel + 0.5\chi_G^\perp); \tag{3}$$

Thus

$$\chi_B^\perp = 0.5\chi_G^\parallel + 0.5\chi_G^\perp \tag{4}$$

Because $\chi_G^\parallel \approx 41\chi_G^\perp$, this argument predicts

$$\chi_B^\perp \approx 0.5\chi_G^\parallel \tag{5}$$

This is consistent with our experimental results, which strongly suggest the existence of delocalized electrons. Discussion of $\xi_B^\parallel$ can be found in reference[36].

The above discussion ignores all paramagnetic effects including band paramagnetism. Evidence for a Curie-like contribution is seen at low temperatures in some of the curves displayed in Fig. 5 and could arise, in part, from paramagnetic impurities (see below).

The anisotropic susceptibility of buckytubes is governed by various geometrical and structural factors (such as the aspect ratio and degree of perfection of its structure). In the direction perpendicular to the buckybundle axis, we do not expect $\chi_B^\perp$ to be larger than $0.5\,\chi_G^\parallel$. However, in the direction parallel to the buckybundle axis $\chi_B^\parallel$ might be much larger for a high-quality sample of buckytubes, as discussed above. The measured anisotropy factor in this work (approximately 1.1 at room temperature and increasing with falling temperature) likely represents a value smaller than that achievable with a highly ordered structure. The small value may be caused by imperfectly aligned buckytubes in the buckybundle.

The magnetic susceptibility data for buckytubes, amorphous graphite, crystalline graphite, the gray-shell material, and C_{60} as a function of temperatures are shown in Fig. 5. Paramagnetic upturns were observed (for all of the curves) at temperatures lower than 10 K. Amorphous graphite and C_{60} show no observable temperature dependence at temperatures ranging from 10 K to 300 K, whereas the buckytube sample exhibits a large increase in diamagnetic susceptibility with falling temperature.

A plot of χ^{-1} vs T for C_{60} was used to estimate a Curie constant of 8.6×10^{-9}/mole, which corresponds to 1.7×10^{-8} electron spins per carbon atom in C_{60}. It is possible that the paramagnetic upturn is caused by a small amount of O_2 within the sample. To examine this point, we sealed a C_{60} sample under a vacuum of 2×10^{-5} Torr in a chamber located at

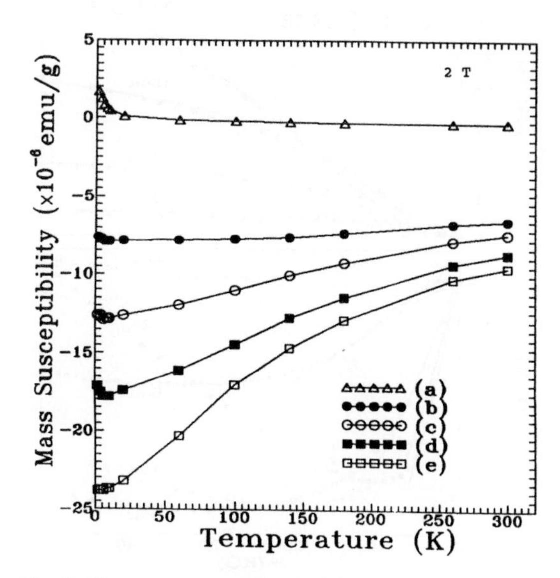

Fig. 5. Temperature dependence of the magnetic susceptibilities measured in a magnetic field of 2 T: (a) C_{60} powder, (b) polycrystalline graphite anode, (c) gray-shell material, (d) buckybundle: axis perpendicular to H, and (e) buckybundle: axis parallel to H.

the center of a quartz tube (holder). No susceptibility difference between the C_{60} sealed in vacuum and unsealed samples was observed. C_{60} is a large bandgap (1.4 eV) semiconductor[29,37], and the paramagnetic upturn at very low temperature is probably due to a very small concentration of foreign paramagnetic impurities.

We conclude this section stating that buckytubes in a bundle have a large diamagnetic susceptibility for H both parallel to and perpendicular to the buckybundle axis. We attribute the large susceptibility of the buckytubes to delocalized electrons in the graphite sheet[38]. The increase in the diamagnetism at low temperature is attributed to an increasing mean free path. C_{70}, which is formed by 12 pentagons and 25 hexagons, exhibits a larger diamagnetic susceptibility than that of C_{60}, which consists of 12 pentagons and 20 hexagons. This suggests that the diamagnetic susceptibility of fullerenes may increase with an increasing fraction of hexagons. The susceptibility of the buckytubes is likely the largest in this family.

3.3 *Transport properties*

Theory predicts that buckytubes can either be metals, semimetals, or semiconductors, depending on diameter and degree of helicacy. The purpose of our study is to give a preliminary answer to this question. A detailed analysis has been published elsewhere[39]. In this section, we present mainly experimental results.

The transverse magnetoresistance data, ρ/ρ_0 ($\Delta\rho = \rho(B) - \rho_0$), measured at different temperatures, are shown in Fig. 6. It is seen that, at low temperatures, the MR is negative at low fields followed by an upturn at another characteristic field that depends on temperature. Our low-temperature MR data has two striking features. First, at low temperatures and low fields $\Delta\rho/\rho_0$ depends logarithmically on temperature and Bright's model predicts a $1/T$ dependence at low fields. Clearly, the data cannot be described by 1D WL theory. Second, from Fig. 6 (b), we see that the characteristic magnetic field at which the MR exhibits an upturn is smaller for lower temperatures. On the contrary, the transverse MR at different temperatures for the pyrocarbon exhibits quite a different behavior[40]: the upturn field decreases with temperature. With increasing temperature, the $\Delta\rho/\rho_0$ vs B curves shift upward regularly and there is no crossover between the curves measured at different temperatures. All these facts indicate that the buckytubes show 2D WL behavior at low temperatures.

It is also seen that above 60 K the MR is positive; it increases with temperature and tends to saturate at a characteristic magnetic field that is smaller at lower temperatures. Based on a simple two-band model, this means that unequal numbers of electrons and holes are present and that the difference in electron and hole concentrations decreases with increasing temperature[41]. The temperature dependence of the conductance (shown by the right scale in Fig. 7) cannot be described by thermal excitation (over an energy gap) or variable range hopping. Instead, above 60 K conductance, $\sigma(T)$, increases approximately linearly with temperature. The absence of an exponential or a variable range-hopping-type temperature dependence in the conductivity indicates that the system is semimetallic and that the hopping between the tubes within the bundle is not the dominant transport mechanism.

From our transport measurements, we can conclude that at low temperatures, the conductivity of the bundle of buckytubes shows two-dimensional weak localization behavior and the MR is negative; above 60 K the MR is positive and increases approximately

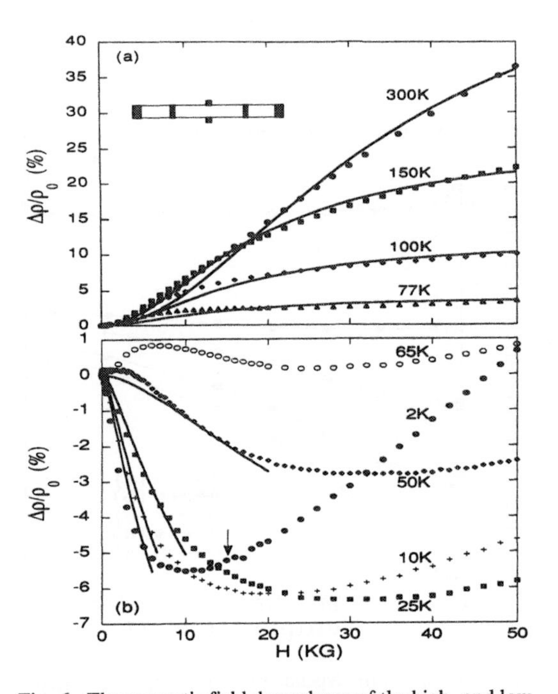

Fig. 6. The magnetic field dependence of the high- and low-temperature MR, respectively; the solid lines are calculated. The inset shows a schematic of the contact configuration for the transport measurements.

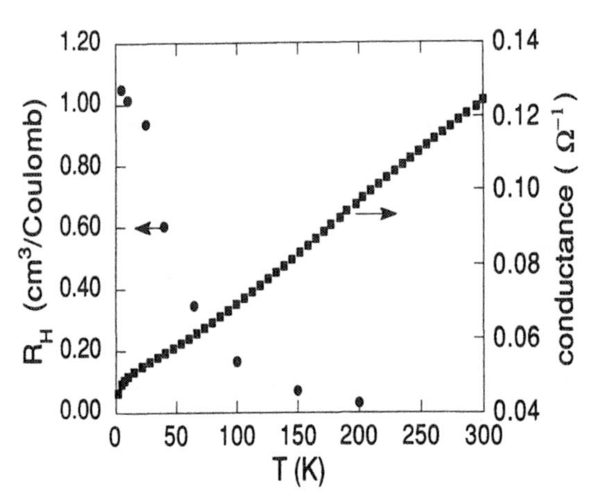

Fig. 7. The Hall coefficient (left scale) and conductance (right scale) vs temperature. R_H was determined using the measured sample dimensions without any correction.

linearly with temperate, which is mainly due to an increase in the carrier concentration. The results show that the bundle of buckytubes may best be described as a semimetal.

3.4 *The use of a stable glow discharge for the synthesis of carbon nanotubes*

Although the generation of carbon nanotubes by vapor-phase growth[42], catalytic growth[43,44], and corona discharge[45] has been reported, to our knowledge carbon nanotubes in macroscopic quantities are produced only by a carbon-arc discharge in a helium gas atmosphere. Unfortunately, carbon nanotubes synthesized by the conventional arc discharge always coexist with carbonaceous nanoparticles, possess undefined morphologies, and have a variety of defects[46]. One of the serious problems associated with this technique is that the conventional arc discharge is a discontinuous, inhomogeneous, and unstable process. An inhomogeneity of the electric field distribution or a discontinuity of the current flow may correlate with the incorporation of pentagons and other defects during growth[2]. A natural question is whether the plasma and current flow can be stabilized to grow high-quality buckytubes.

Here, we introduce the use of a stable glow discharge for synthesis of carbon nanotubes. It greatly overcomes many of the problems mentioned above and allows the synthesis of high-quality carbon nanotubes. The fullerene generator used in this study is basically the same chamber we employed for the production of buckyballs and buckytubes described earlier. However, a modification was made with incorporating a high-voltage feedthrough and a tungsten wire. The wire acted as an extension of the Tesla coil, with its free end pointing toward the arc region. Two graphite rods, with the same diameter of $\frac{3}{8}$ inches, were employed as electrodes. The two electrodes with well-polished ends were positioned very close to each other initially. The glow discharge was stimulated by a corona discharge triggered by the Tesla coil rather than by striking the anode against the cathode to generate a conventional arc discharge. After a plasma was generated gently over the smooth ends of two electrodes, the spacing between the electrodes was slowly increased. Under an appropriate dc current at a He gas pressure of 500 Torr, the glow discharge was self-sustained without the necessity of feeding in the graphite anode. In practice, the rate at which the anode was consumed was equal to the growth rate of the deposited rod, thus keeping the electrode spacing and the discharge characteristics constant.

The time dependence of the current across the gap for both the arc discharge and the glow discharge were measured during the deposition of buckytubes. The current of the discharge as a function of time was recorded using a Hewlett-Packard 7090A Measurement Plotting System. The resultant spectrum for the glow discharge, shown in Fig. 8 (a), shows no observable current fluctuation, which demonstrates that the glow discharge is a continuous process. However, the resultant plot for the conventional arc discharge, shown in Fig. 8 (b), indicates that the arc current fluctuates with time. An average arc-jump frequency of about 8 Hz was observed. These results unambiguously demonstrate that the conventional arc discharge is a transient process.

Photographs of the cross-section of two deposited rods produced by consuming graphite rods with the same diameter ($\frac{3}{8}$ inches) at the same dc current (100 A, 20 V) in a He atmosphere at the same pressure (500 Torr) are shown in Fig. 9. Photos (a) and (b) were taken from deposited rods synthesized in the glow mode and the arc mode, respectively. The two photos clearly indicate that the deposited rod produced in the glow mode has a more homogeneous black core, consisting of an array of bundles of buckytubes[32], and only a thin cladding. The glow mode produced a higher yield of buckytubes than the conventional arc mode.

Figure 10 shows scanning electron microscopy (SEM) micrographs of a cross-section of the deposited rod synthesized in the glow mode. In Fig. 10 the upper left corner of the micrographs corresponds to the center of the cross-section of the deposited rod. The micrograph shows an evenly distributed array of parallel bundles of buckytubes from the black region of the deposited rod. The thickest bundles with diameters up to 200 μm were observed for the first time at the central region of the deposited rod. It is interesting to

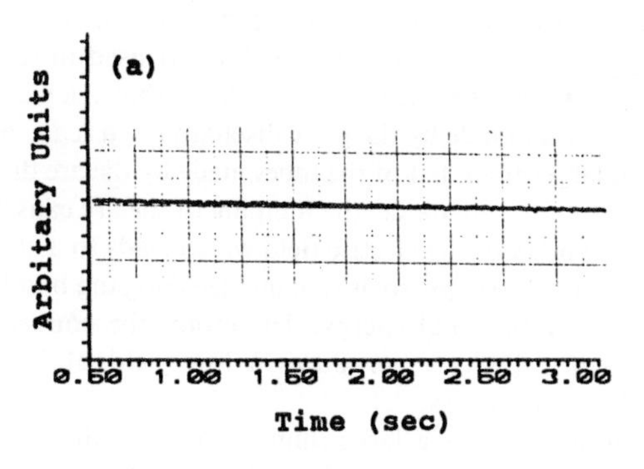

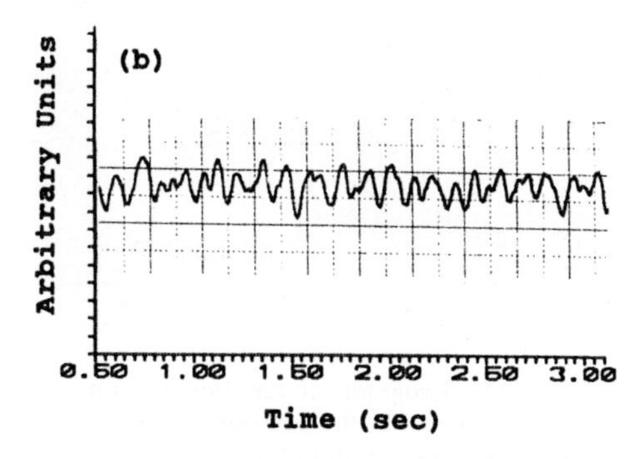

Fig. 8. (a) The current of the glow discharge as a function of time; (b) the current of the arc discharge as a function of time.

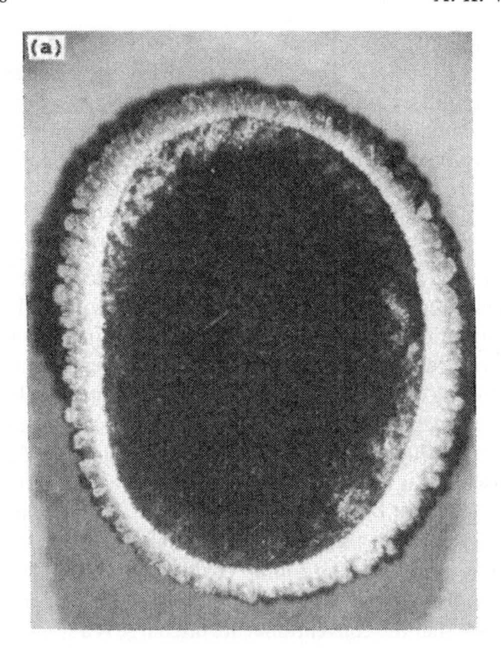

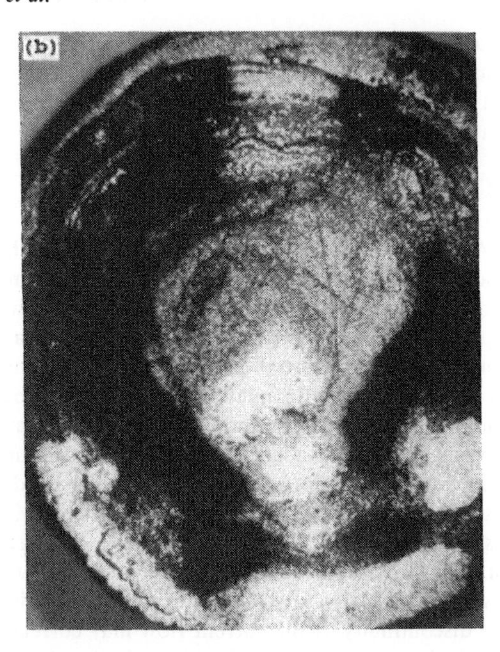

Fig. 9. Photographs of the cross-section of the deposited rods; photos (a) and (b) were taken from deposited rods produced by the glow discharge and by the conventional arc discharge, respectively.

note that the average diameter of the bundles decreases towards the perimeter of the deposited rod. The cladding shell around the rod is composed of fused graphitic flakes. The geometry and the distribution of the bundles in the deposited rod may be interpreted as follows. In the glow mode, the highest temperature occurs at the center of the deposited rod and the temperature decreases towards the perimeter because the heat is transported by both radiation and conduction via the He gas. In the cladding shell area, the temperature may not be high enough to form the bundles but is sufficient to form fused graphitic materials. Although several groups have speculated on the effect of the electric field in the formation of buckytubes[13–16,47], the influence of the temperature on the yield and distribution of the bundles has not been discussed. From our results, we may conclude that the temperature is one of the key factors in the formation of buckybundles.

0.86mm

Fig. 10. A SEM micrograph of the cross-section of the deposited rod synthesized by the glow discharge shows an image of 1/4 of the cross-section of the deposited rod; upper left corner of the image corresponds to the center of the end of the deposited rod.

Figure 11 shows two typical TEM images for the deposited rods synthesized in the glow mode and the conventional arc mode, respectively. Two samples for TEM observation were prepared in an identical way. The figures show the dramatic improvement in the yield and quality of buckytubes synthesized in the glow mode. The buckytubes and their bundles shown in Fig. 11 (a) are thicker and longer than that shown in Fig. 11 (b). The length of the tubes shown here is larger than the field of view of HREM used in this study. Our systematic studies indicated that the yield of the tubes made by the glow discharge is at least 20 times larger than that of the tubes made by the arc discharge. In the arc mode, the jumping of the arc causes an instability of the electric field which leads to a closure of the tube tips, so as to minimize dangling bonds and lower the total energy. Therefore, the conventional arc discharge produces a low yield of low-quality buckytubes.

Note also that a large number of the tube tips, shown in Fig. 11 (b), are closed by caps that are polygonal or cone-shaped. Iijima has found that the

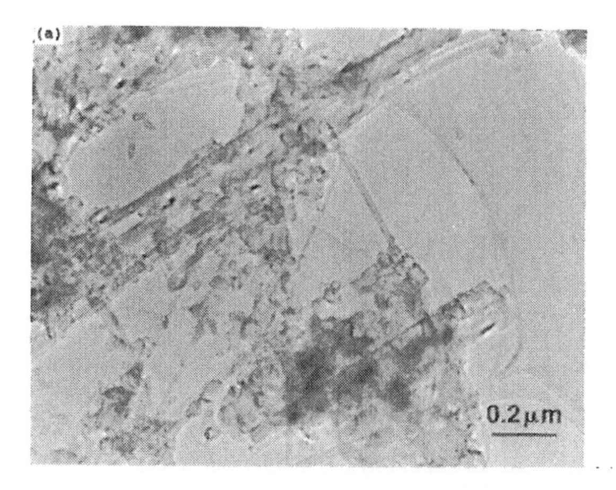

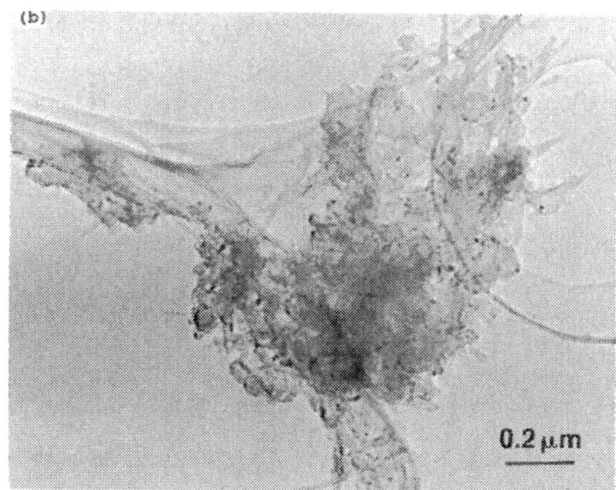

Fig. 11. (a) and (b) are HREM images of the deposited rods produced by the glow discharge and by the conventional arc discharge, respectively.

polygonal and cone-shaped caps are formed by incorporating pentagons into the hexagonal network. He speculated that the formation of the pentagons may result from a depletion of carbon in plasma near the end of cathode[2]. Our experimental results offer evidence for the above speculation. Fluctuations of carbon species caused by a discontinuous arc discharge may be responsible for the formation of short tubes with the caps, consisting of pentagons and other defects.

Based on these experimental results, one can speculate on the influence of the arc mode on the yield and distribution of the bundles. For the glow discharge, the plasma is continuous, homogeneous, and stable. In other words, the temperature distribution, the electric field which keeps growing tube tips open[47], and the availability of carbon species (atoms, ions, and radicals) are continuous, homogeneous, and stable over the entire central region of the cathode. Accordingly, a high yield and better quality buckytubes should occur over the entire central region of the cathode. These are consistent with what we observed in Fig. 9 (a), Fig. 10, and Fig. 11 (a). For the conventional arc discharge, we can speculate that the arc starts at a sharp edge near the point of closest approach, and after vaporizing this region it jumps to what then becomes the next point of closest approach (usually within about a radius of the arc area), and so on. The arc wanders around on the surface of the end of the anode, leading, on the average, to a discontinuous evaporation process and an instability of the electric field. This kind of violent, randomly jumping arc discharge is responsible for the low yield and the low quality of the deposited buckytubes. This is, again, consistent with what we showed in Fig. 9 (b) and Fig. 11 (b). Note also from Fig. 11 that carbon nanotubes and nanoparticles coexist in both samples. The coexistence of these two carbonaceous products may suggest that some formation conditions, such as the temperature and the density of the various carbon species, are almost the same for the nanotubes and the nanoparticles. An effort to promote the growth of carbon nanotubes and eliminate the formation of carbon nanoparticles is presently underway.

4. CONCLUSIONS

In this article, we have reported the structural, magnetic, and transport properties of bundles of buckytubes produced by an arc discharge. By adjusting the arc mode into a stable glow discharge, evenly spaced and parallel buckybundles with diameters up to 200 μm have been synthesized. The magnetic susceptibility of a bulk sample of buckybundles is -10.75×10^{-6} emu/g for the magnetic field parallel to the bundle axes, which is approximately 1.1 times the perpendicular value and 30 times larger than that of C_{60}. The magnetoresistance (MR) and Hall coefficient measurements on the buckybundles show a negative MR at low temperature, a positive MR at a temperature above 60 K, and a conductivity which increases approximately linearly with temperature. Our results show that a buckybundle may best be described as a semimetal.

Acknowledgements—We are grateful to A. Patashinski for useful discussions. This work was performed under the support of NSF grant #9320520 and DMR-9357513 (NYI award for VPD). The use of MRC central facilities supported by NSF is gratefully acknowledged.

REFERENCES

1. S. Iijima, *Nature* **354**, 56 (1993).
2. S. Iijima, T. Ichihashi, and Y. Ando, *Nature* **356**, 776 (1992).
3. T. W. Ebbesen and P. M. Ajayan, *Nature* **358**, 220 (1992).
4. Y. Ando and S. Iijima, *Jpn. J. Appl. Phys.* **32**, L107 (1993).
5. T. W. Ebbesen, H. Hiura, J. Fujita, Y. Ochiai, S. Matsui, and K. Tanigaki, *Chem. Phys. Lett.* **209**, 83 (1993).
6. Y. Saito, T. Yoshikawa, M. Inagaki, M. Tomita, and T. Nayashi, *Chem. Phys. Lett.* **204**, 277 (1993).

7. Z. Zhang and C. M. Lieber, *Appl. Phys. Lett.* **62**, 2792 (1993).
8. S. N. Song, X. K. Wang, R. P. H. Chang, and J. B. Ketterson, *Phys. Rev. Lett.* **72**, 697 (1994).
9. D. Ugarte, *Nature* **359**, 707 (1992).
10. S. Iijima, P. M. Ajayan, and T. Ichihashi, *Phys. Rev. Lett.* **69**, 3100 (1992).
11. R. A. Jishi, M. S. Dresselhaus, and G. Dresselhaus, *Phys. Rev. B* **48**, 11385 (1993).
12. L. Langer, L. Stockman, J. P. Heremans, V. Bayot, C. H. Olk, C. V. Haesendonck, and Y. Bruynseraede, *J. Mater. Res.* **9**, 927 (1994).
13. J. W. Mintmire, B. I. Dunlap, and C. T. White, *Phys. Rev. Lett.* **68**, 631 (1992).
14. D. H. Robertson, D. W. Brenner, and J. W. Mintmire, *Phys. Rev. B* **45**, 12592 (1992).
15. N. Hamada, S. I. Sawana, and A. Oshiyama, *Phys. Rev. Lett.* **68**, 1579 (1993).
16. R. Saito, M. Fujita, G. Dresselhaus, and M. S. Dresselhaus, *Mater. Res. Soc. Symp. Proc.* **247**, 333 (1992).
17. G. Dresselhaus, M. S. Dresselhaus, and R. Saito, *Phys. Rev.* **B45**, 6234 (1992).
18. H. W. Kroto, J. R. Heath, S. C. O'Brien, R. F. Curl, and R. E. Smalley, *Nature* **318**, 162 (1985).
19. R. Tycko, R. C. Haddon, G. Dabbagh, S. H. Glarum, D. C. Douglass, and A. M. Mujsce, *J. Phys. Chem.* **95**, 518 (1991).
20. S. Iijima and T. Ichihashi, *Nature* **363**, 603 (1993).
21. D. S. Bethune, C. H. Klang, M. S. deVries, G. Gorman, R. Savoy, J. Vazqiez, and R. Beyers, *Nature* **363**, 605 (1993).
22. X. W. Lin, X. K. Wang, V. P. Dravid, R. P. H. Chang, and J. B. Ketterson, *Appl. Phys. Lett.* **64**, 181 (1994).
23. R. E. Haufler, J. Conceicao, L. P. F. Chibante, Y. Chai, N. E. Byrne, S. Flanagan, M. M. Haley, S. C. O'Brien, C. Pan, Z. Xiao, W. E. Billups, M. A. Ciufolini, R. H. Hauge, J. L. Margrave, L. J. Wilson, R. F. Curl, and R. E. Smalley, *J. Phys. Chem.* **94**, 8634 (1990).
24. V. Elser and R. C. Haddon, *Nature* **325**, 793 (1987).
25. R. C. Haddon, L. F. Schneemeyer, J. V. Waszczak, S. H. Glarum, R. Tycko, G. Dabbagh, A. R. Kortan, A. J. Muller, A. M. Mujsce, M. J. Rosseinsky, S. M. Zahurak, A. V. Makhija, F. A. Thiel, K. Raghavachari, E. Cockayne, and V. Elser, *Nature* **350**, 46 (1991).
26. A. Pasquarello, M. Schluter, and R. C. Haddon, *Science* **257**, 1660 (1992).
27. P. W. Fowler, P. Lazzeretti, M. Malagoli, and R. Zanasi, *Chem. Phys. Lett.* **179**, 174 (1991).
28. T. G. Schmalz, *Chem. Phys. Lett.* **175**, 3 (1990).
29. R. S. Ruoff, D. Beach, J. Cuomo, T. McGuire, R. L. Whetten, and F. Diederich, *J. Phys. Chem.* **95**, 3457 (1991).
30. H. Ajiki and T. Ando, *J. Phys. Soc. Japan* **62**, 2470 1993.
31. V. P. Dravid, X. W. Lin, Y. Y. Wang, X. K. Wang, A. Yee, J. B. Ketterson, and H. P. H. Chang, *Science* **259**, 1601 (1993).
32. X. K. Wang, X. W. Lin, V. P. Dravid, J. B. Ketterson, and R. P. H. Chang, *Appl. Phys. Lett.* **62**, 1881 (1993).
33. J. C. Slonczewski and P. R. Weiss, *Phys. Rev.* **109**, 272 (1958).
34. J. W. McClure, *Phys. Rev.* **199**, 606 (1960).
35. R. M. White, *Quantum Theory of Magnetism*, *Springer Series in Solid State Sciences*, Springer-Verlag Berlin, Heidelberg, New York (1983).
36. X. K. Wang, R. P. H. Chang, A. Patashinski, and J. B. Ketterson, *J. Mater. Res.* **9**, 1578 (1994).
37. S. Saito and A. Oshiyama, *Phys. Rev. Lett.* **66**, 2637 (1991).
38. L. Pauling, *J. Chem. Phys.* **4**, 637 (1936).
39. S. N. Song, X. K. Wang, R. P. H. Chang, and J. B. Ketterson, *Phys. Rev. Lett.* **72**, 679 (1994).
40. J. Callaqay, In *Quantum Theory of the Solids*, p. 614, Academic Press, New York (1976).
41. P. Delhaes, P. de Kepper, and M. Uhlrich, *Philos. Mag.* 29, 1301 (1974).
42. M. Ge and K. Sattler, *Appl. Phys. Lett.* **64**, 710 (1994).
43. M. J. Yacaman, M. M. Yoshida, L. Rendon, and J. G. Santiesteban, *Appl. Phys. Lett.* **62**, 657 (1993).
44. M. Endo and H. W. Kroto, *J. Phys. Chem.* **96**, 6941 (1992).
45. J. R. Brock and P. Lim, *Appl. Phys. Lett.* **58**, 1259 (1991).
46. O. Zhou, R. M. Fleming, O. W. Murphy, C. H. Chen, R. C. Haddon, A. P. Ramire, and S. H. Glarum, *Science* **263**, 1774 (1994).
47. R. E. Smalley, *Mater. Sci. Engin.* **B19**, 1 (1993).

ELECTRONIC PROPERTIES OF CARBON NANOTUBES: EXPERIMENTAL RESULTS

J.-P. Issi,[1] L. Langer,[1] J. Heremans,[2] and C. H. Olk[2]
[1]Unité PCPM, Université Catholique de Louvain, Louvain-la-Neuve, Belgium
[2]General Motors Research and Development Center, Warren, MI 48090, U.S.A.

(*Received* 6 *February* 1995, *accepted* 10 *February* 1995)

Abstract—Band structure calculations show that carbon nanotubes exist as either metals or semiconductors, depending on diameter and degree of helicity. When the diameters of the nanotubes become comparable to the electron wavelength, the band structure becomes noticeably one-dimensional. Scanning tunneling microscopy and spectroscopy data on nanotubes with outer diameters from 2 to 10 nm show evidence of one-dimensional behavior: the current-voltage characteristics are consistent with the functional energy dependence of the density-of-states in 1D systems. The measured energy gap values vary linearly with the inverse nanotube diameter. Electrical resistivity and magnetoresistance measurements have been reported for larger bundles, and the temperature dependence of the electrical resistance of a single microbundle was found to be similar to that of graphite and its magnetoresistance was consistent with the formation of Landau levels. Magnetic susceptibility data taken on bundles of similar tubes reveal a mostly diamagnetic behavior. The susceptibility at fields above the value at which the magnetic length equals the tube diameter has a graphite-like dependence on temperature and field. At low fields, where electrons sample the effect of the finite tube diameter, the susceptibility has a much more pronounced temperature dependence.

Key Words—Carbon nanotubes, scanning tunneling microscopy, spectroscopy, magnetoresistance, electrical resistivity, magnetic susceptibility.

1. INTRODUCTION

The existence of carbon nanotubes with diameters small compared to the de Broglie wavelength has been described by Iijima[1,2,3] and others[4,5]. The energy band structures for carbon nanotubes have been calculated by a number of authors and the results are summarized in this issue by M.S. Dresselhaus, G. Dresselhaus, and R. Saito. In short, the tubules can be either metallic or semiconducting, depending on the tubule diameter and chirality[6,7,8]. The calculated density of states[8] shows $1/(E\text{-}E_i)^{1/2}$ singularities characteristic of one-dimensional (1D) systems. The separation between the singularities around the Fermi energy is the energy gap for the tubes that are semiconducting, and scales linearly with the inverse of the tube outer diameter[7,8]. This contrasts with the case of a rod-shaped quantum wire, for which the gap is expected to scale with the inverse square of the diameter. The relevant energy scale for the gap in carbon nanotubes is the nearest-neighbor overlap integral in graphite (3.14 eV)[9]. This makes room-temperature observation of the quantum size effects, in principle, possible in nanotubes with diameters in the nm range, because the sublevel energy separations are of the order of 1 eV.

Experimental measurements to test these remarkable theoretical predictions of the electronic structure of carbon nanotubes are difficult to carry out because of the strong dependence of the predicted properties on tubule diameter and chirality. Ideally, electronic or optical measurements should be made on individual single-wall nanotubes that have been characterized with regard to diameter and chiral angle. Further experimental challenges result from the fact that tubes are often produced in bundles, so that obtaining data on single, well-characterized tubes has not yet been achieved. We review here some experimental observations relevant to the electronic structure of individual nanotubes or on bundles of nanotubes: combined scanning tunneling microscopy and spectroscopy, temperature-dependent resistivity, magnetoresistance (MR), and magnetic susceptibility.

2. SCANNING TUNNELING SPECTROSCOPY STUDIES

Scanning tunneling spectroscopy (STS) can, in principle, probe the electronic density of states of a single-wall nanotube, or the outermost cylinder of a multi-wall tubule, or of a bundle of tubules. With this technique, it is further possible to carry out both STS and scanning tunneling microscopy (STM) measurements at the same location on the same tubule and, therefore, to measure the tubule diameter concurrently with the STS spectrum. No reports have yet been made of a determination of the chiral angle of a tubule with the STM technique. Several groups have, thus far, attempted STS studies of individual tubules.

The first report of current-voltage (I-V) measurements by Zhang and Lieber[10] suggested a gap in the density of states below about 200 MeV and semiconducting behavior in the smallest of their nanotubes (6 nm diameter). The study that provides the most detailed test of the theory for the electronic properties of the 1D carbon nanotubes, thus far, is the combined STM/STS study by Olk and Heremans[11], even though it is still preliminary. In this study, more than nine

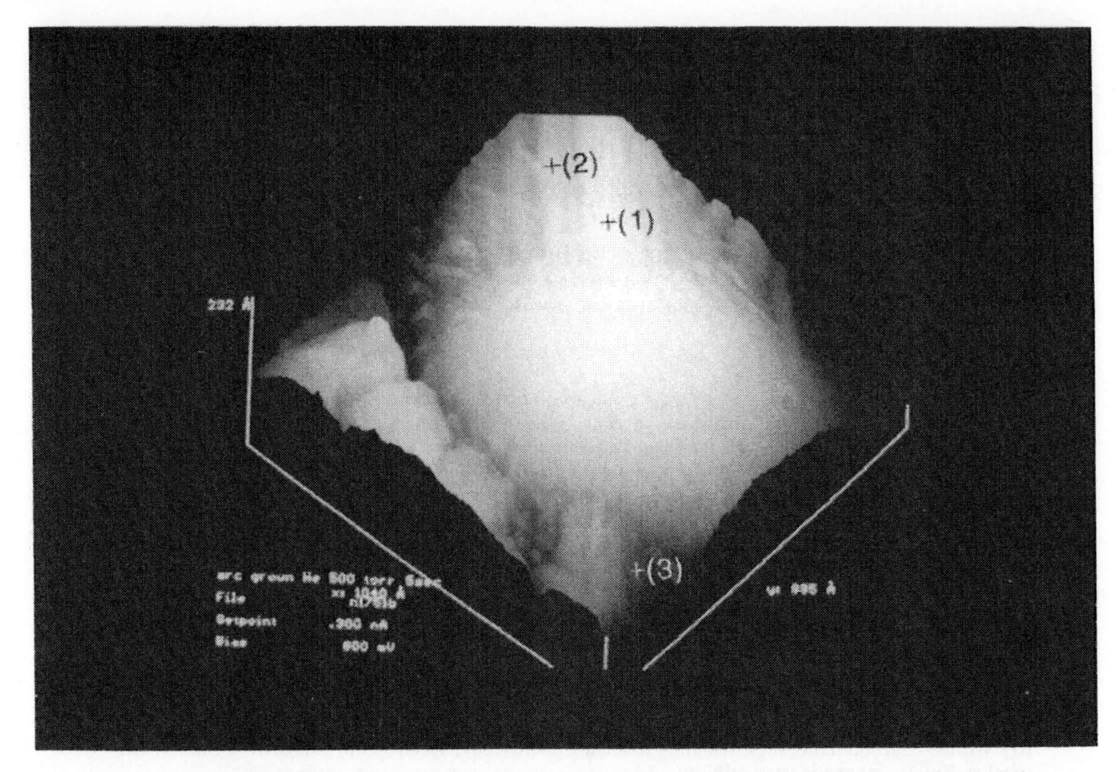

Fig. 1. Topographic STM scan of a bundle of nanotubes. STS data were collected at points (1), (2), and (3), on tubes with diameters of 8.7, 4.0, and 1.7 nm, respectively. The diameters were determined from image cross-sections of the variations in height in a direction perpendicular to the tubes (adapted from Olk *et al.*[11]).

individual multilayer tubules with diameters ranging from 2 to 10 nm, prepared by the standard carbon-arc technique, were examined. STM measurements were taken first, and a topographic STM scan of a bundle of nanotubes is shown in Fig. 1. The exponential relation between the tunneling current and the tip-to-tubule distance was experimentally verified to confirm that the tunneling measurements pertain to the tubule and not to contamination on the tubule surface. From these relations, barrier heights were measured to establish the range in which the current-voltage characteristics can be taken for further STS studies. The image in Fig. 1 is used to determine the diameter of the individual tubule on which the STS scans are carried out. During brief interruptions in the STM scans, the instrument was rapidly switched to the STS mode of operation, and I-V plots were made on the same region of the same tubule as was characterized for its diameter by the STM measurement. The I-V plots for three typical tubules, identified (1)–(3), are shown in Fig. 2. The regions (1)–(3) correspond to interruptions in the STM scans at the locations identified by crosses in the topographic scan, Fig. 1. Although acquisition of spectroscopic data in air can be complicated by contamination-mediated effects on the tunneling gap, several studies on a wide variety of surfaces have been reported[12]. Trace (1) in Fig. 2, taken on a tube with 8.7 nm diameter, has an ohmic behavior, providing evidence for the metallic nature of that tubule. Two tubules (trace 2 for a tubule with diameter = 4.0 nm, and trace 3 for a tubule with diameter = 1.7 nm) show plateaus in the I/V characteristics at zero current. This rectifying behavior is the signature of semiconducting tubules. The dI/dV plot in the inset crudely mimics a 1D density of states, the peaks in the dI/dV plot be-

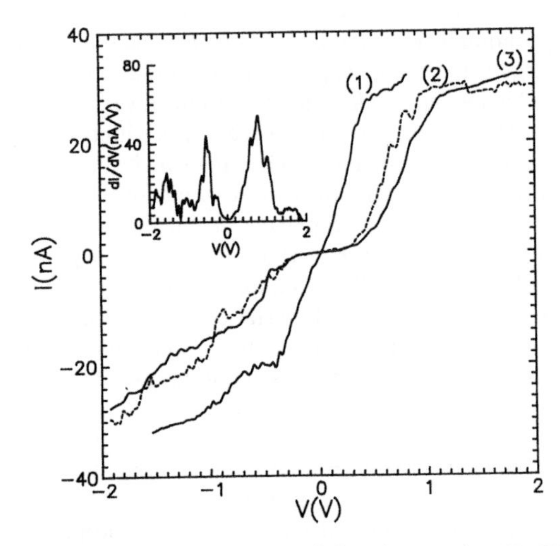

Fig. 2. Current-voltage characteristics taken at points (1), (2), and (3) in Fig. 3. The top insert shows the conductance versus voltage plot, for the data taken at point (3) (adapted from Olk *et al.*[11]).

ing attributed to $[1/(E_0 - E)^{1/2}]$-type singularities in the 1D density of states seen in the density of states versus energy diagrams calculated in[8]. Several I-V curves were collected along the length of each tube. Reproducible spectra were obtained on 9 tubes with different diameters.

The energy gap of the semiconducting tubes was estimated around $V = 0V$ by drawing two tangents at the points of maximum slope nearest zero in the I-V spectra, and measuring the voltage difference between the intercepts of these tangents with the abscissa. A plot of these energy gaps versus inverse tube diameter for all samples studied is shown in Fig. 3[11]. Surface contamination may account for the scatter in the data points, though the correlation between E_g and the inverse diameter shown in Fig. 3 is illustrated by the dashed line. The data in Fig. 3 is consistent with the predicted dependence on the inverse diameter[13]. The experimentally measured values of the bandgaps are, however, about a factor of two greater than the theoretically estimated ones on the basis of a tight binding calculation (full line in Fig. 3)[7,13]. Further experimental and theoretical work is needed to reach a detailed understanding of these phenomena.

3. ELECTRICAL RESISTIVITY AND MAGNETORESISTANCE

The remarkable theoretical predictions mentioned above are even more difficult to verify by experimental measurements in the case of electrical conductivity. Ideally, one has to solve two experimental problems. First, one has to realize a four-point measurement on an individual nanotube. That means four contacts on a sample with typical dimensions of the order of a nm

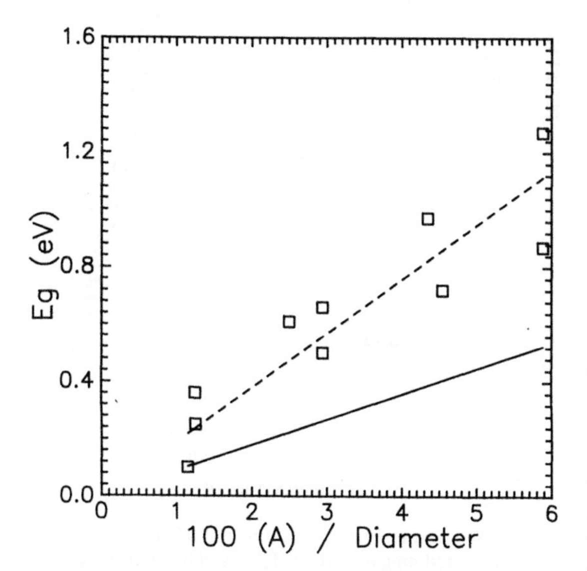

Fig. 3. Energy gap versus inverse nanotube diameter, for the nine nanotubes studied; the dashed line is a regression through the points, the full line is a calculation for semiconducting zigzag nanotubes[7,13] (adapted from Olk *et al.*[11]).

diameter and a few μm length. Second, this sample with its contacts must be characterized to determine its exact diameter and helicity. To take up this challenge it is necessary to resort to nanotechnologies.

Before reviewing the results of different measurements, we need to first briefly describe the nature of the deposit formed during the carbon-arc experiment in a way first proposed by Ebbesen[5]. He suggested that the carbon nanotubes produced by classical carbon arc-discharge present a fractal-like organization. The deposit on the negative electrode consists of a hard gray outer shell and a soft black fibrous core containing the carbon nanotubes. If we examine in detail this core material by scanning electron microscopy, we observe a fractal-like structure. This means that the black core is made of fiber-like entities that are, in reality, bundles of smaller fiber-like systems. These smaller systems are, in turn, formed of smaller bundles, and so on. The micro-bundle, which is the smallest bundle, consists of a few perfectly aligned nanotubes of almost equal lengths. Finally, each of these individual nanotubes is generally formed of several concentric single-shell nanotubes.

The fractal-like organization led, therefore, to conductivity measurements at three different scales: (1) the macroscopic, mm-size core of nanotube containing material, (2) a large (60 μm) bundle of nanotubes and, (3) a single microbundle, 50 nm in diameter. These measurements, though they do not allow direct insights on the electronic properties of an individual tube give, nevertheless, at a different scale and within certain limits fairly useful information on these properties.

Ebbesen[4] was the first to estimate a conductivity of the order of 10^{-4} Ωm for the black core bulk material existing in two thirds of tubes and one third of nanoparticles. From this observation, it may naturally be inferred that the carbon arc deposit must contain material that is electrically conducting. An analysis of the temperature dependence of the zero-field resistivity of similar bulk materials[14,15] indicated that the absolute values of the conductivity were very sample dependent.

Song *et al.* [16] reported results relative to a four-point resistivity measurement on a large bundle of carbon nanotubes (60 μm diameter and 350 μm in length between the two potential contacts). They explained their resistivity, magnetoresistance, and Hall effect results in terms of a conductor that could be modeled as a semimetal. Figures 4 (a) and (b) show the magnetic field dependence they observed on the high- and low-temperature MR, respectively.

At high temperature, the conductivity was found to increase linearly with temperature and the observed high-temperature MR was positive. In fact, by fitting the data using a simple two-band model[17] the authors obtained the theoretical curve in Fig. 4 (a). The fitting parameters showed that the ratio σ_p/σ_e, where σ_p and σ_e are the partial conductivities of holes and electrons, respectively, decreases with increasing tem-

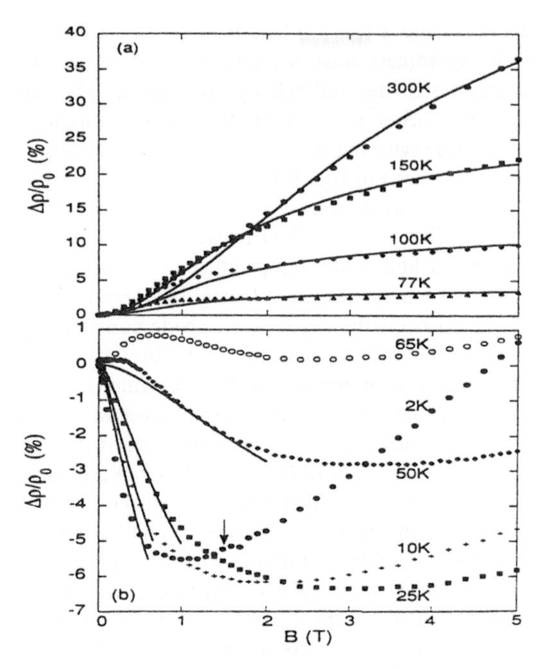

Fig. 4. (a) Magnetic field dependence of the high- and low-temperature MR, respectively. The solid lines are calculated using a simple two-band model for (a) and the 2D weak localization theory for (b) (after Song *et al.*[16]).

perature implying that, as the temperature increases the Fermi level shifts closer to the conduction band. The increase in total conductivity, $\sigma_e + \sigma_p$, with temperature was attributed to an increase in electron concentration. Song *et al.* [16] attributed the fact that above 60 K the conductivity increases almost linearly with temperature, as opposed to an exponential or variable range hopping-type temperature dependence, to a semimetallic behavior.

At low temperature and low field, the observed MR was found to be negative, Fig. 4 (b), while positive MR contributions were increasingly important for higher fields. While 1D weak localization (WL) was found by Song *et al.* [16] to be inadequate to describe the data, they claimed that 2D WL fit the experimental results for the temperature dependence of the resistivity and the MR at low temperature and low fields. They estimated a resistivity of 0.65×10^{-4} Ωm at 300 K and 1.6×10^{-4} Ωm at 5 K. However, the actual resistivity of the metallic tubules along their axis should be much lower than these values for many reasons: (1) as predicted by the theory described in [7], only one third of the tubules are conducting; (2) the filling factor of a large bundle like the measured one is less than 1; (3) when the tubes are not single-walled, the cross-section of the nanotubes where conduction occurs is unknown and affected by their inner structure and, finally, (4) the nanotube is anisotropic.

To avoid the problems described above relative to the interpretation of the results, it is necessary to work at a lower scale. In other words, modern nanotechnology must be used to contact electrically smaller nanotube structures like microbundles, or even better, single nanotubes. A rather sophisticated technique, namely submicronic lithographic patterning of gold films with a scanning tunneling microscope[18], was developed by Langer *et al.* [19] to attach two electrical contacts to a single microbundle. This direct electrical resistance measurement on this quasi-lD system excludes the errors that may result from the situations described in (2) and (4) in the last paragraph. The reported temperature dependence measured from 300 K down to 0.3 K by Langer *et al.* [19] is shown in Fig. 5. Above 2 K, some interesting information was obtained by fitting the experimental zero-field resistance data to a simple two-band (STB) model successfully applied by Klein[20–22] to semimetallic graphite. In this STB-model, the electron and hole densities, *n* and *p*, respectively, can be expressed by:

$$n = C_n k_B T \ln[1 + \exp(E_F/k_B T)] \quad (1)$$

$$p = C_p k_B T \ln[1 + \exp[(\Delta - E_F)/k_B T]] \quad (2)$$

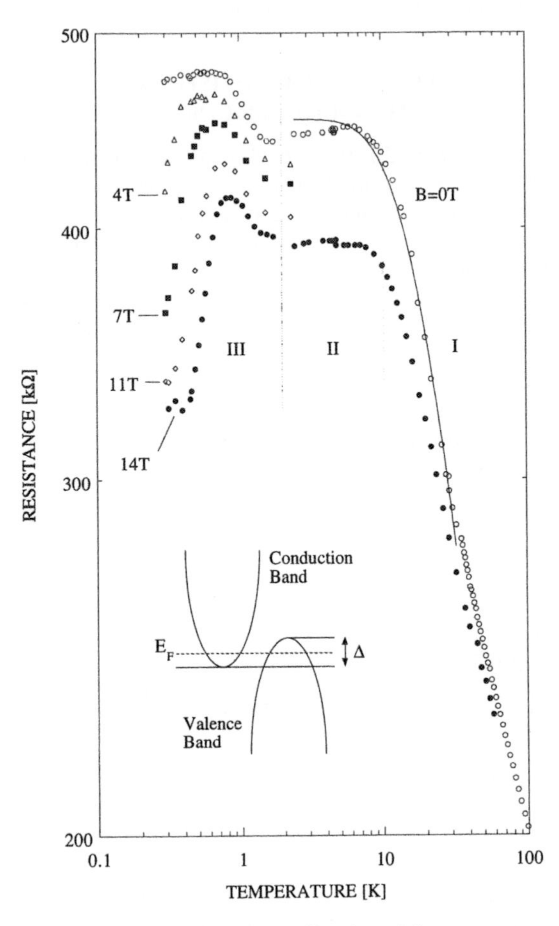

Fig. 5. Electrical resistance as a function of the temperature at the indicated magnetic fields for a single microbundle of carbon nanotubes. The solid line is a fit using the two-band model for graphite (see inset) with an overlap $\Delta = 3.7$ meV and a Fermi level right in the middle of the overlap (after Langer *et al.* [19]).

where E_f is the Fermi energy and Δ is the band overlap. C_n and C_p are the fitting parameters. The results of this fit, which is shown in Fig. 5, is obtained for $\Delta = 3.7$ MeV and with the Fermi energy right in the middle of the overlap. Ideally, an overlap of the order of 40 MeV is expected for the case of multilayered nanotubes[23] with an interlayer configuration similar to that of crystalline graphite. For real nanotubes instead, a turbostratic stacking of the adjacent layers would reduce drastically the interlayer interactions, as in disordered graphite, and so it is not surprising to find a band overlap 10 times smaller than in crystalline graphite. This implies also that, within the frame of the STB model, the carrier density is 10 times smaller. The apparent electrical resistivity (4×10^{-4} Ωm) measured at low temperature is certainly still higher than the resistivity of a single nanotube. In fact, the conductance of the system is dominated by that of the nanotubes with the highest conductance (i.e., the semimetallic nanotubes). All these nanotubes are not necessarily in contact with the measuring probes. Finally, the inner structure of each individual nanotube in this microbundle remains unknown.

By applying a magnetic field normal to the tube axis, Langer *et al.* [19] observed a MR (Fig. 6) which, in contrast to the case of graphite, remains negative at all fields. The negative MR was found consistent with the formation of Landau levels. Ajiki and Ando[24] have predicted that a magnetic field applied perpendicularly to the sample axis should introduce a Landau level at the crossing of the valence and conduction bands. It results in an increase of the density of states at the Fermi level and, hence, a reduction of the resistance, which is in agreement with the experimental data. Moreover, the theory predicts a MR that is temperature independent at low temperature and decreasing when k_BT becomes larger than the Landau level. This is also what was experimentally observed. Thus, in contrast to what was reported by Song *et al.* [16], these results are not consistent with WL. In the framework of WL, the observed temperature independent MR below 10 K and down to 1 K could only be explained in the presence of large amounts of magnetic impurities, as is the case for some pyrocarbons[25]. However, in the present case, the spectrographic analysis performed by Heremans *et al.* [26] excludes the presence of magnetic impurities.

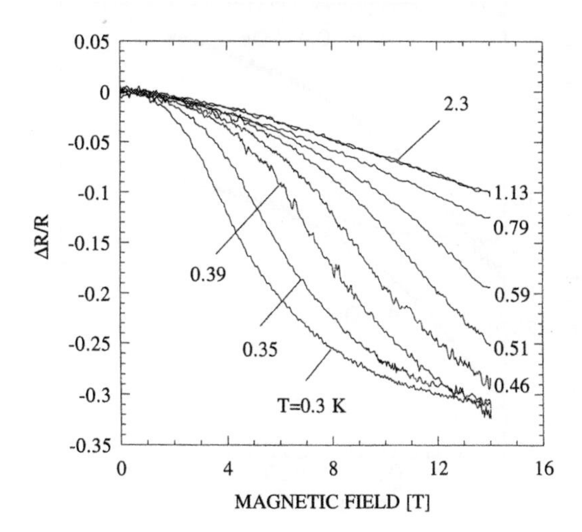

Fig. 6. The magnetic field dependence of the magnetoresistance at different temperature for the same microbundle measured in Fig. 5 (after Langer *et al.*[19]).

Below 2 K, an unexpected temperature dependence of the resistance and MR is observed. As shown in Fig. 5, the resistance presents, after an initial increase, a saturation or a broad maximum and an unexplained sharp drop when a magnetic field is applied (see Fig. 6). Further theoretical work is needed to get a better understanding of these striking physical observations. Finally, it is interesting to note that Whitesides and Weisbecker[27] developed a technique to estimate the conductivity of single nanotubes by dispersing nanotubes onto lithographically defined gold contacts to realize a 'nano-wire' circuit. From this 2-point resistance measurement and, after measuring the diameter of the single nanotubes by non-contact AFM, they estimated the room-temperature electrical resistivity along the nanotube axis to be 9.5×10^{-5} Ωm. This is consistent with the values obtained for a microbundle by Langer *et al.* [19].

The most promising way to study the electrical conductivity of a single nanotube is, thus, tightly dependent on the development or/and the adaptation of modern nanolithographic techniques. The goal to achieve is within reach and a detailed study of the electronic properties with reference to helicity and diameter will provide instrumental information about these fascinating materials.

4. MAGNETIC SUSCEPTIBILITY

The presence of aromatic-like electrons strongly determines the magnetic susceptibility of the diverse forms of carbons. The susceptibility of diamond (-4.9×10^{-7} emu/g) is ascribed to diamagnetic contributions from core and valence electrons, and a Van Vleck paramagnetic term[28]. Graphite has an anisotropic diamagnetic susceptibility[29]. The susceptibility of graphite[29,30] parallel to the planes is about equal to the free atom susceptibility of -5×10^{-7} emu/g, but when the magnetic field is aligned parallel to the c-axis, the susceptibility of graphite (-30×10^{-6} emu/g below 100 K) is due mainly to free electron contributions and is much larger. The magnetic susceptibility of C_{60} (-3.5×10^{-7} emu/g) and C_{70} (-5.9×10^{-7} emu/g) is small again[31,32], as a result of a cancellation between a diamagnetic and a paramagnetic term.

In the light of these results, it is not surprising that a very large anisotropy has been calculated[24] to exist in the magnetic susceptibility of nanotubes. The susceptibility with the field parallel to the tube axis is predicted to be as much as 3 orders of magnitude smaller than that with the field perpendicular. Again, measurements on individual single-wall well-characterized tubes have not been published to date. Experiments usually involve gathering nanotube material from several growths, to obtain quantities of material on the order of tens of mg. A measurement on oriented bundles of tubes[30] at 0.5 T as a function of temperature gives evidence for anisotropy. Such a measurement is, however, easily affected by a small misalignment of the sample. It is, therefore, possible that the data reported for the case where the field is parallel to the tube axis are, in fact, dominated by contributions from the perpendicular susceptibility.

A second study [33] on samples that contain a mixture of nanotubes, together with several percent "buckyonion"-type structures, was carried out at temperatures between 4.5 and 300 K, and fields between 0 and 5.5 T. The moment M is plotted as a function of field in Fig. 7, for the low-field range, and in Fig. 8 for the high-field range. The field dependence is clearly non-linear, unlike that of graphite, in which both the basal plane and the c-axis moments are linear in field, except for the pronounced de Haas-van Alphen oscillations at low temperature.

The a.c. susceptibility ($\chi = dM/dH$, where M is the moment), measured at 5, 0.4, and 0.04 T, is shown as a function of temperature in Fig. 9. Three regimes of magnetic fields are identified in ref. [33]. The high-field susceptibility (at 5 T) has a temperature dependence similar to that of graphite, but a magnitude reduced by a factor of 2. In this regime, the magnetic radius $(hb/eB)^{1/2}$ becomes shorter than the tube diameter. The electronic diamagnetism is then a local probe of the graphene planes, and its value is expected to be the geometrical average of that of rolled-up graphene. Because with the field in the basal plane, χ is much smaller than along the c-axis, this geometrical average comes out to be about $\frac{1}{2}$[30]. The low-field susceptibility (at 0.04 T) is a better probe of the finite size effects of the tubes, because the magnetic length is larger than the tube diameter. The sample, however, consists of a mixture of semiconducting and metallic tubes. The component of χ perpendicular to the tube axis dominates the measured susceptibility, and the value is expected to scale with $1/E_F$[24]. The electrical resistivity data show that, even in the metallic samples, the energy gap is on the order of a few MeV[19]. At room temperature $k_BT > E_F$ for both metallic and semiconducting samples. Thus, a thermally activated behavior is expected for the average susceptibility. Furthermore, carriers may be scattered at a temperature-dependent rate, for instance by acoustic phonons. These two mechanisms are consistent with the much more pronounced temperature dependence of the low-field susceptibility of nanotubes than of graphite. More quantitative models remain to be developed. In the in-

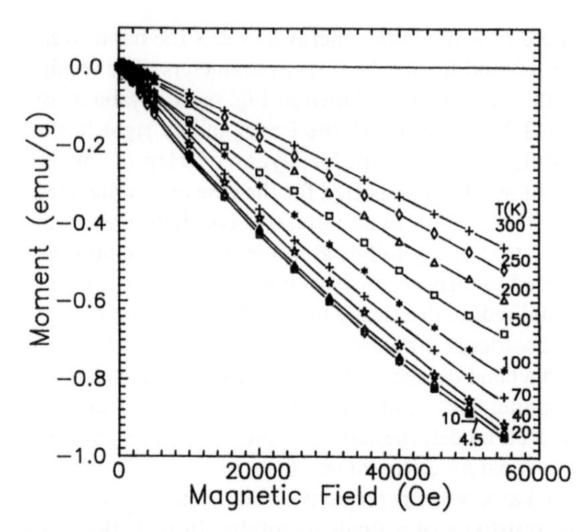

Fig. 8. Field dependence of the moment of carbon nanotubes at the temperatures shown at high magnetic fields (after Heremans et al.[26]).

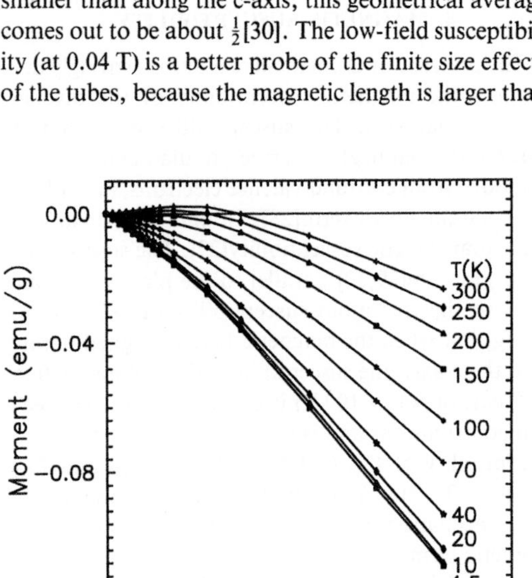

Fig. 7. Field dependence of the moment of carbon nanotubes at the temperatures shown at low magnetic fields (after Heremans et al.[26]).

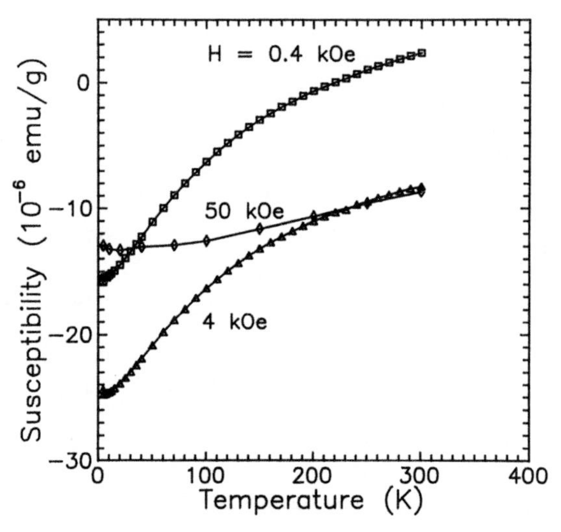

Fig. 9. Susceptibility of carbon nanotubes versus temperature at the different fields identified in the figure (after Heremans et al.[26]).

termediate field regime (around 0.5 T), χ is probably sensitive to the fact that the sample consists of a mixture of semiconducting and metallic tubes. Furthermore, the tubes in that mixture have different diameters and, thus, go from the low-field to the high-field regime at different fields. Figure 10 shows a summary of the a.c. susceptibilities of various forms of carbon, as an easy ready reference. The values for diamond and the basal plane of graphite represent the atomic contributions, and the c-axis of graphite and the susceptibilities of nanotubes are dominated by free carrier contributions.

5. CONCLUSIONS

The purpose of this work is to review experimental data on the electronic properties of carbon nanotubes. Although most of the theoretical work has been focused on single-walled individual tubes, performing transport and other measurements on such entities is extremely difficult and remains unachieved. Combined STM and STS measurements access single tubes within bundles containing a few a tubes. They provide experimental evidence that graphite nanotubes behave as nanowires, with a density of states and an energy gap dependence on inverse diameter as predicted. Studies in ultra-high vacuum are needed to provide more quantitative data on the dependence of the gap on diameter. No electrical resistance measurements are currently available on a well-characterized single nanotube. However, electrical resistivity measurements performed at low temperature on a large nanotube bundle were interpreted in terms of a 2D WL. On the other hand, MR data obtained for a single microbundle were consistent with the formation of Landau levels using the model developed by Ajiki and Ando[24]. From the temperature dependence of the electrical resistance of this microbundle, it appears that the measured nanotubes are semimetallic and behave like rolled-up graphene sheets. Below 1 K, the results are puzzling and yet unexplained. Susceptibility was measured on 20-mg samples containing a large variety of tubes. Free carrier contributes to the diamagnetic behavior of nanotubes, which is similar to that of graphite when the magnetic length is smaller than the tube diameter. The low-field diamagnetism contains more information about the specific band structure of the tubes.

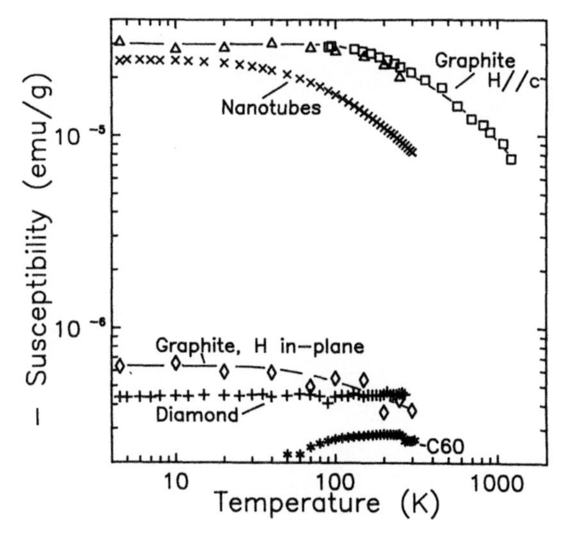

Fig. 10. Temperature dependence of the magnetic susceptibility of various carbon-based materials. The data on HOPG (H//c) are taken at 200 Oe. The data reported for nanotubes, graphite (H in-plane), and diamond, were taken at 4 kOe, those on diamond at 8 kOe. The ordinate axis is negative (after Heremans *et al.*[26]).

Acknowledgements—The authors gratefully acknowledge much help from M.S. Dresselhaus and G. Dresselhaus in the preparation of this article.

REFERENCES

1. S. Iijima, *Nature* (London) **354**, 56 (1991).
2. S. Iijima, *Mater. Sci. Eng.* **B19**, 172 (1993).
3. S. Iijima, T. Ichihashi, and Y. Ando, *Nature* (London) **356**, 776 (1992).
4. T. W. Ebbesen and P. M. Ajayan, *Nature* (London) **358**, 220 (1992).
5. T. W. Ebbesen, H. Hiura, J. Fujita, Y. Ochiai, S. Matsui, and K. Tanigaki, *Chem. Phys. Lett.* **209**, 83 (1993).
6. J. C. Charlier and J. P. Michenaud, *Phys. Rev. Lett.* **70**, 1858 (1993). J. C. Charlier, *Carbon Nanotubes and Fullerenes.* PhD thesis, Catholic University of Louvain, May 1994.
7. C. T. White, D. H. Roberston, and J. W. Mintmire, *Phys. Rev. B* **47**, 5485 (1993).
8. R. Saito, G. Dresselhaus, and M. S. Dresselhaus, *J. Appl. Phys.* **73**, 494 (1993). M. S. Dresselhaus, G. Dresselhaus, and R. Saito, *Solid State Commun.* **84**, 201 (1992).
9. R. Saito, M. Fujita, G. Dresselhaus, and M. S. Dresselhaus, In *Electrical, Optical and Magnetic Properties of Organic Solid State Materials*, MRS Symposia Proceedings, Boston (Edited by L. Y. Chiang, A. F. Garito, and D. J. Sandman), vol. 247, page 333. Materials Research Society Press, Pittsburgh, PA (1992).
10. Z. Zhang and C. M. Lieber, *Appl. Phys. Lett.* **62**, 2792 (1993).
11. C. H. Olk and J. P. Heremans, *J. Mater. Res.* **9**, 259 (1994).
12. N. Venkateswaran, K. Sattler, U. Muller, B. Kaiser, G. Raina, and J. Xhie, *J. Vac. Sci. Technol.* **B9**, 1052 (1991). M. Jobin, R. Emch, F. Zenhausern, S. Steinemann, and P. Descouts, *J. Vac. Sci. Technol.* **B9**, 1263 (1991). Z. Zhang, C. M. Lieber, D. S. Ginley, R. J. Baughman, and B. Morosin, *J. Vac. Sci. Technol.* **B9**, 1009 (1991). H. Enomoto, H. Ozaki, M. Suzuki, T. Fujii, and M. Yamaguchi, *J. Vac. Sci. Technol.* **B9**, 1022 (1991).
13. M. S. Dresselhaus, R. A. Jishi, G. Dresselhaus, and Riichiro Saito, *Fullerenes (1994).* St. Petersburg, Russia Fullerene workshop, October 1993.
14. R. Heyd, A. Charlier, J. F. Marêché, E. McRae, and O. V. Zharikov, *Solid State Commun.* **89**, 989 (1994).
15. R. Seshardi, H. N. Aiyer, A. Govindaraj, and C. N. Rao, *Solid State Commun.* **91**, 195 (1994).
16. S. N. Song, X. K. Wang, R. P. H. Chang, and J. B. Ketterson, *Phys. Rev. Lett.* **72**, 697 (1994).
17. K. Noto and T. Tsuzuku, *Jpn. J. Appl. Phys.* **14**, 46 (1975).
18. L. Stockman, G. Neuttiens, C. Van Haesendonck, and Y. Bruynseraede, *Appl. Phys. Lett.* **62**, 2935 (1993).

19. L. Langer, L. Stockman, J. P. Heremans, V. Bayot, C. H. Olk, C. Van Haesendonck, Y. Bruynseraede, and J. P. Issi, *J. Mat. Res.* **9**, 927 (1994).
20. C. A. Klein, *J. Appl. Phys.* **33**, 3388 (1962).
21. C. A. Klein, *J. Appl. Phys.* **35**, 2947 (1964).
22. C. A. Klein, in *Chemistry and Physics of Carbon* (Edited by P. L. Waiker, Jr.), Vol. 2, page 217. Marcel Dekker, New York (1966).
23. Ph. Lambin, L. Philippe, J. C. Charlier, and J. P. Michenaud, *Comput. Mater. Sci.* **2**, 350 (1994).
24. H. Ajiki and T. Ando, *J. Phys. Soc. Jpn.* **62**, 1255 (1993).
25. V. Bayot, L. Piraux, J.-P. Michenaud, J.-P. Issi, M. Lelaurain, and A. Moore, *Phys. Rev.* **B41**, 11770 (1990).
26. J. Heremans, C. H. Olk, and D. T. Morelli, *Phys. Rev.* **B49**, 15122 (1994).
27. G. M. Whitesides, C. S. Weisbecker, private communication.
28. S. Hudgens, M. Kastner, and H. Fritzsche, *Phys. Rev. Lett.* **33**, 1552 (1974).
29. N. Ganguli and K. S. Krishnan, *Proc. Roy. Soc. London* **177**, 168 (1941).
30. K. S. Krishnan, *Nature* **133**, 174 (1934).
31. R. C. Haddon, L. F. Schneemeyer, J. V. Waszczak, S. H. Glarum, R. Tycko, G. Dabbagh, A. R. Kortan, A. J. Muller, A. M. Musjsce, M. J. Rosseinsky, S. M. Zahurak, A. V. Makhija, F. A. Thiel, K. Raghavachari, E. Cockayne, and V. Elser, *Nature* **350**, 46 (1991).
32. R. S. Ruoff, D. Beach, J. Cuomo, T. McGuire, R. L. Whetten, and F. Diedrich, *J. Phys. Chem* **95**, 3457 (1991).
33. X. K. Wang, R. P. H. Chang, A. Patashinski, and J. B. Ketterson, *J. Mater. Res.* **9**, 1578 (1994).

NOTE ADDED IN PROOF

Since this paper was written, low-temperature measurements on carbon nanotubes revealed the existence of Universal Conductance Fluctuations with magnetic field. These results will be reported elsewhere.

L. Langer, L. Stockman, J. P. Heremans, V. Bayot, C. H. Olk, C. Van Haesendonck, Y. Bruynseraede, and J. P. Issi, to be published.

VIBRATIONAL MODES OF CARBON NANOTUBES; SPECTROSCOPY AND THEORY

P. C. EKLUND,[1] J. M. HOLDEN,[1] and R. A. JISHI[2]
[1]Department of Physics and Astronomy and Center for Applied Energy Research, University of Kentucky, Lexington, KY 40506, U.S.A.
[2]Department of Physics, Massachusetts Institute of Technology, Cambridge, MA 02139, U.S.A.; Department of Physics, California State University, Los Angeles, CA 90032, U.S.A.

(*Received* 9 *February* 1995; *accepted in revised form* 21 *February* 1995)

Abstract—Experimental and theoretical studies of the vibrational modes of carbon nanotubes are reviewed. The closing of a 2D graphene sheet into a tubule is found to lead to several new infrared (IR)- and Raman-active modes. The number of these modes is found to depend on the tubule symmetry and not on the diameter. Their diameter-dependent frequencies are calculated using a zone-folding model. Results of Raman scattering studies on arc-derived carbons containing nested or single-wall nanotubes are discussed. They are compared to theory and to that observed for other sp^2 carbons also present in the sample.

Key Words—Vibrations, infrared, Raman, disordered carbons, carbon nanotubes, normal modes.

1. INTRODUCTION

In this paper, we review progress in the experimental detection and theoretical modeling of the normal modes of vibration of carbon nanotubes. Insofar as the theoretical calculations are concerned, a carbon nanotube is assumed to be an infinitely long cylinder with a monolayer of hexagonally ordered carbon atoms in the tube wall. A carbon nanotube is, therefore, a one-dimensional system in which the cyclic boundary condition around the tube wall, as well as the periodic structure along the tube axis, determine the degeneracies and symmetry classes of the one-dimensional vibrational branches [1–3] and the electronic energy bands[4–12].

Nanotube samples synthesized in the laboratory are typically not this perfect, which has led to some confusion in the interpretation of the experimental vibrational spectra. Unfortunately, other carbonaceous material (e.g., graphitic carbons, carbon nanoparticles, and amorphous carbon coatings on the tubules) are also generally present in the samples, and this material may contribute artifacts to the vibrational spectrum. Defects in the wall (e.g., the inclusion of pentagons and heptagons) should also lead to disorder-induced features in the spectra. Samples containing concentric, coaxial, "nested" nanotubes with inner diameters ~8 nm and outer diameters ~80 nm have been synthesized using carbon arc methods[13,14], combustion flames[15], and using small Ni or Co catalytic particles in hydrocarbon vapors[16–20]. Single-wall nanotubes (diameter 1–2 nm) have been synthesized by adding metal catalysts to the carbon electrodes in a dc arc[21,22]. To date, several Raman scattering studies[23–28] of nested and single-wall carbon nanotube samples have appeared.

2. OVERVIEW OF RAMAN SCATTERING FROM SP² CARBONS

Because a single carbon nanotube may be thought of as a graphene sheet rolled up to form a tube, carbon nanotubes should be expected to have many properties derived from the energy bands and lattice dynamics of graphite. For the very smallest tubule diameters, however, one might anticipate new effects stemming from the curvature of the tube wall and the closing of the graphene sheet into a cylinder. A natural starting point for the discussion of the vibrational modes of carbon nanotubes is, therefore, an overview of the vibrational properties of sp^2 carbons, including carbon nanoparticles, disordered sp^2 carbon, and graphite. This is also important because these forms of carbon are also often present in tubule samples as "impurity phases."

In Fig. 1a, the phonon dispersion relations for 3D graphite calculated from a Born-von Karman lattice-dynamical model are plotted along the high symmetry directions of the Brillouin zone (BZ). For comparison, we show, in Fig. 1b, the results of a similar calculation[29] for a 2D infinite graphene sheet. Interactions up to fourth nearest neighbors were considered, and the force constants were adjusted to fit relevant experimental data in both of these calculations. Note that there is little dispersion in the k_z (Γ to A) direction due to the weak interplanar interaction in 3D graphite (Fig. 1c). To the right of each dispersion plot is the calculated one-phonon density of states. On the energy scale of these plots, very little difference is detected between the structure of the 2D and 3D one-phonon density of states. This is due to the weak interplanar coupling in graphite. The eigenvectors for the optically

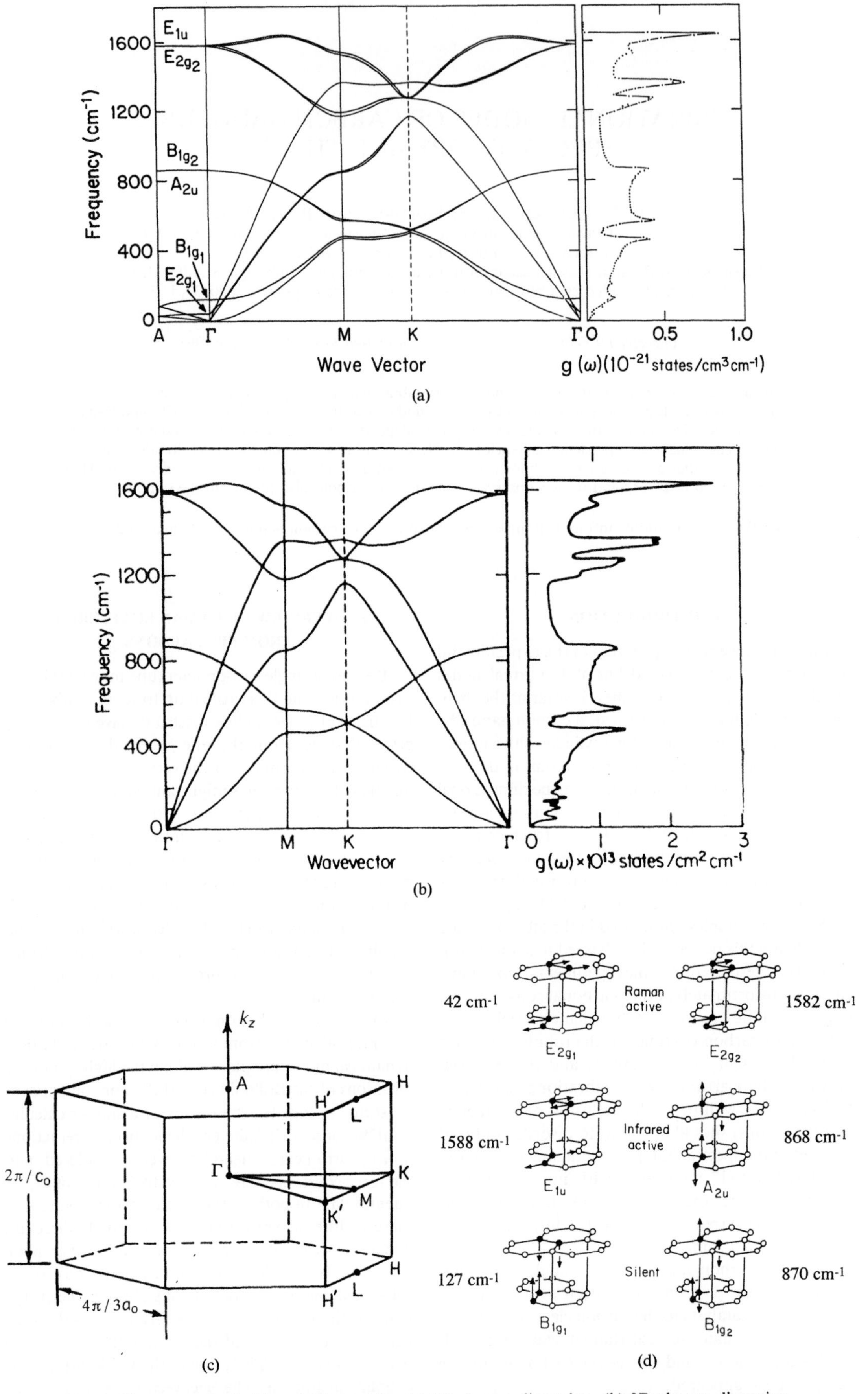

Fig. 1. Phonon modes in 2D and 3D graphite: (a) 3D phonon dispersion, (b) 2D phonon dispersion, (c) 3D Brillouin zone, (d) zone center $q = 0$ modes for 3D graphite.

allowed Γ-point vibrations for graphite (3D) are shown in Fig. 1d, which consist of two, doubly degenerate, Raman-active modes ($E_{2g}^{(1)}$ at 42 cm^{-1}, $E_{2g}^{(2)}$ at 1582 cm^{-1}), a doubly degenerate, infrared-active E_{1u} mode at 1588 cm^{-1}, a nondegenerate, infrared-active A_{2u} mode at 868 cm^{-1}, and two doubly degenerate B_{2g} modes (127 cm^{-1}, 870 cm^{-1}) that are neither Raman- nor infrared-active. The lower frequency $B_{2g}^{(1)}$ mode has been observed by neutron scattering, and the other is predicted at 870 cm^{-1}. Note the Γ-point E_{1u} and $E_{2g}^{(2)}$ modes have the same intralayer motion, but differ in the relative phase of their C-atom displacements in adjacent layers. Thus, it is seen that the interlayer interaction in graphite induces only an ~6 cm^{-1} splitting between these modes ($\omega(E_{1u}) - \omega(E_{2g}^{(2)}) = 6$ cm^{-1}). Furthermore, the frequency of the rigid-layer, shear mode ($\omega(E_{2g}^{(1)}) = 42$ cm^{-1}) provides a second spectroscopic measure of the interlayer interaction because, in the limit of zero interlayer coupling, we must have $\omega(E_{2g}^{(1)}) \rightarrow 0$.

The Raman spectrum (300 cm$^{-1} \leq \omega \leq 3300$ cm^{-1}) for highly oriented pyrolytic graphite (HOPG)[1] is shown in Fig. 2a, together with spectra (Fig. 2b–e) for several other forms of sp^2 bonded carbons with varying degrees of intralayer and interlayer disorder. For HOPG, a sharp first-order line at 1582 cm^{-1} is observed, corresponding to the Raman-active $E_{2g}^{(2)}$ mode observed in single crystal graphite at the same frequency[31]. The first- and second-order mode frequencies of graphite, disordered sp^2 carbons and carbon nanotubes, are collected in Table 1.

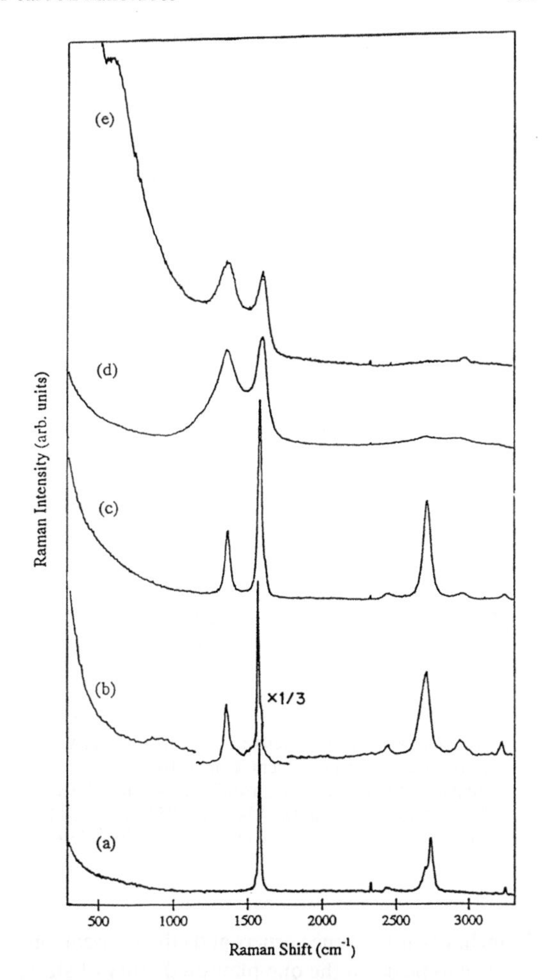

Fig. 2. Raman spectra (T = 300 K) from various sp^2 carbons using Ar-ion laser excitation: (a) highly ordered pyrolytic graphite (HOPG), (b) boron-doped pyrolytic graphite (BHOPG), (c) carbon nanoparticles (dia. 20 nm) derived from the pyrolysis of benzene and graphitized at 2820°C, (d) as-synthesized carbon nanoparticles (~850°C), (e) glassy carbon (after ref. [24]).

Graphite exhibits strong second-order Raman-active features. These features are expected and observed in carbon tubules, as well. Momentum and energy conservation, and the phonon density of states determine, to a large extent, the second-order spectra. By conservation of energy: $\hbar\omega = \hbar\omega_1 + \hbar\omega_2$, where ω and ω_i ($i = 1,2$) are, respectively, the frequencies of the incoming photon and those of the simultaneously excited normal modes. There is also a crystal momentum selection rule: $\hbar\mathbf{k} = \hbar\mathbf{q}_1 + \hbar\mathbf{q}_2$, where $\mathbf{k}$ and $\mathbf{q}_i$ ($i = 1.2$) are, respectively, the wavevectors of the incoming photon and the two simultaneously excited normal modes. Because $\mathbf{k} \ll \mathbf{q}_B$, where $\mathbf{q}_B$ is a typical wavevector on the boundary of the BZ, it follows that $\mathbf{q}_1 \approx -\mathbf{q}_2$. For a second-order process, the strength of the IR lattice absorption or Raman scattering is proportional to $|M(\omega)|^2 g_2(\omega)$, where $g_2(\omega) = g_1(\omega_1) \cdot g_1(\omega_2)$ is the two-phonon density of states subject to the condition that $\mathbf{q}_1 = -\mathbf{q}_2$, and where $g_1(\omega)$ is the one-phonon density of states and $|M(\omega)|^2$ is the effective two-phonon Raman matrix element. In covalently bonded solids, the second-order spectral features are generally broad, consistent with the strong dispersion (or wide bandwidth) of both the optical and acoustic phonon branches.

However, in graphite, consistent with the weak interlayer interaction, the phonon dispersion parallel to the c-axis (i.e., along the k_z direction) is small. Also, there is little in-plane dispersion of the optic branches and acoustic branches near the zone corners and edges (M to K). This low dispersion enhances the peaks in the one-phonon density of states, $g_1(\omega)$ (Fig. 1a). Therefore, relatively sharp second-order features are observed in the Raman spectrum of graphite, which correspond to characteristic combination ($\omega_1 + \omega_2$) and overtone (2ω) frequencies associated with these low-dispersion (high one-phonon density of states) regions in the BZ. For example, a second-order Raman feature is detected at 3248 cm^{-1}, which is close to 2(1582 cm^{-1}) = 3164 cm^{-1}, but significantly upshifted due to the 3D dispersion of the uppermost phonon branch in graphite. The most prominent feature in the graphite second-order spectrum is a peak close to 2(1360 cm^{-1}) = 2720 cm^{-1} with a shoulder at 2698 cm^{-1}, where the lineshape reflects the density of two-phonon states in 3D graphite. Similarly, for a 2D graphene sheet, in-plane dispersion (Fig. 1b) of the optic branches at the zone center and in the acoustic

[1]HOPG is a synthetic polycrystalline form of graphite produced by Union Carbide[30]. The c-axes of each grain (dia. ~1 μm) are aligned to ~1°.

Table 1. Table of frequencies for graphitic carbons and nanotubes

Mode assignment*† (tube dia.)	Planar graphite HOPG[31]	Planar graphite BHOPG[31]	Single-wall tubules: Holden‡ *et al.*[27] (1–2 nm)	Single-wall tubules: Holden *et al.*[28] (1–2 nm)	Nested tubules: Hiura *et al.*[23]	Nested tubules: Chandrabhas *et al.*[24] (15–50 nm)	Nested tubules: Bacsa *et al.*[26] (8–30 nm)	Nested tubules: Kastner *et al.*[25] (20–80 nm)
$E_{2g}^{(1)}$ (R, ⊥)	42[c]					49, 58[c]		
$B_{2g}^{(1)}$ (S, ∥)	127[h]							
A_{2u} (ir, ∥)	868[g]	~900[c]				~700[c]		868[g]
$B_{2g}^{(2)}$ (S, ∥)	870[i]	~900[c]						
$E_{2g}^{(2)}$ (R, ⊥)	1582[c] 1577[e]	1585[c] 1591[e]	1566[c,d] 1592[c,d]	1568[c] 1594[c]	1574[a]	1583[c]	1581[a]	1582[e]
E_{1u} (ir, ⊥)	1588[g]							1575[g]
D (R, ⊥)	1350[c] 1365[e]	1367[c] 1380[e]		1341[c]	1340[a]	1353[c]	1356[a]	varies[f]
E'_{2g} (R, ⊥)		1620[c]						1620[a]
2 × 1220 (R, ⊥)	2441[c]	2450[c] 2440[e]		2450[c]	2455[a]	2455[c]	2450[a]	2455[e]
2 × D (R, ⊥)	2722[c] 2746[e]	2722[c] 2753[e]	2681[c,d]	2680[c]	2687[a]	2709[c]		2734[c]
$E_{2g}^{(2)} + D$ (R, ⊥)		2950[c] 2974[e]		2925[c]				2925[e]
$2E'_{2g}$ (R, ⊥)	3247[c] 3246[e]	3240[c] 3242[e]	3180[c,d]	3180[c]		3250[a]	3250[a]	3252[e]

*Activity: R = Raman-active, ir = infrared-active, S = optically silent, observed in neutron scattering.
†Carbon atom displacement ∥ or ⊥ to $\hat{C}$.
‡Peaks in "difference spectrum" (see section 4.3).
[a–e]Excitation wavelength: [a]742 nm, [b]532 nm, [c]514 nm, [d]488 nm, [e]458 nm; [f] resonance Raman scattering study; [g]ir-absorption study; [h]from neutron scattering; [i]predicted.

branches near the zone corners and edges is weak, giving rise to peaks in the one-phonon density of states. One anticipates, therefore, that similar second-order features will also be observed in carbon nanotubes. This is because the zone folding (c.f., section 4) preserves in the tubule the essential character of the in-plane dispersion of a graphene sheet for **q** parallel to the tube axis. However, in small-diameter carbon nanotubes, the cyclic boundary conditions around the tube wall activate many new first-order Raman- and IR-active modes, as discussed below.

Figure 2b shows the Raman spectrum of Boron-doped, highly oriented pyrolytic-graphite (BHOPG) according to Wang *et. al*[32]. Although the BHOPG spectrum is similar to that of HOPG, the effect of the 0.5% substitutional boron doping is to create in-plane disorder, without disrupting the overall AB stacking of the layers or the honeycomb arrangement of the remaining C-atoms in the graphitic planes. However, the boron doping relaxes the $\mathbf{q} \approx 0$ optical selection rule for single-phonon scattering, enhancing the Raman activity of the graphitic one- and two-phonon density of states. Values for the peak frequencies of the first- and second-order bands in BHOPG are tabulated in Table 1. Significant disorder-induced Raman activity in the graphitic one-phonon density of states is observed near 1367 cm^{-1}, similar to that observed in other disordered sp^2 bonded carbons, where features in the range ~1360–1365 cm^{-1} are detected. This band is referred to in the literature as the "D-band," and the position of this band has been shown to depend weakly on the laser excitation wavelength[32]. This unusual effect arises from a resonant coupling of the excitation laser with electronic states associated with the disordered graphitic material. Small basal plane crystallite size (L_a) has also been shown[33] to activate disorder-induced scattering in the D-band. The high frequency $E_{2g}^{(2)}(q = 0)$ mode has also been investigated in a wide variety of graphitic materials that have various degrees of in-plane and stacking disorder[32]. The frequency, strength, and line-width of this mode is also found to be a function of the degree of the disorder, but the peak position depends much less strongly on the excitation frequency.

The Raman spectrum of a strongly disordered sp^2 carbon material, "glassy" carbon, is shown in Fig. 2e. The $E_{2g}^{(2)}$-derived band is observed at 1600 cm^{-1} and is broadened along with the D-band at 1359 cm^{-1}. The similarity of the spectrum of glassy carbon (Fig. 2e) to the one-phonon density of states of graphite (Fig. 1a) is apparent, indicating that despite the disorder, there is still a significant degree of sp^2 short-range order in the glassy carbon. The strongest second-order feature is located at 2973 cm^{-1}, near a combination band ($\omega_1 + \omega_2$) expected in graphite at D (1359 cm^{-1}) + E'_{2g} (1620 cm^{-1}) = 2979 cm^{-1}, where the E'_{2g} (1620 cm^{-1}) frequency is associated with a mid-zone maximum of the uppermost optical branch in graphite (Fig. 1a).

The carbon black studied here was prepared by a CO_2 laser-driven pyrolysis of a mixture of benzene, ethylene, and iron carbonyl[34]. As synthesized, TEM images show that this carbon nanosoot consists of disordered sp^2 carbon particles with an average particle diameter of ~200 Å. The Raman spectrum (Fig. 2d) of the "as synthesized" carbon black is very similar to that of glassy carbon (Fig. 2e) and has broad disorder-induced peaks in the first-order Raman spectrum at 1359 and 1600 cm^{-1}, and a broad second-order feature near 2950 cm^{-1}. Additional weak features are observed in the second-order spectrum at 2711 and 3200 cm^{-1}, similar to values in HOPG, but appearing closer to 2(1359 cm^{-1}) = 2718 cm^{-1} and 2(1600 cm^{-1}) = 3200 cm^{-1}, indicative of somewhat weaker 3D phonon dispersion, perhaps due to weaker coupling between planes in the nanoparticles than found in HOPG. TEM images[34] show that the heat treatment of the laser pyrolysis-derived carbon nanosoot to a temperature $T_{HT} = 2820°C$ graphitizes the nanoparticles (i.e., carbon layers spaced by ~3.5 Å are aligned parallel to facets on hollow polygonal particles). As indicated in Fig. 2c, the Raman spectrum of this heat-treated carbon black is much more "graphitic" (similar to Fig. 2a) and, therefore, a decrease in the integrated intensity of the disorder-induced band at 1360 cm^{-1} and a narrowing of the 1580 cm^{-1} band is observed. Note that heat treatment allows a shoulder associated with a maximum in the mid-BZ density of states to be resolved at 1620 cm^{-1}, and dramatically enhances and sharpens the second-order features.

3. THEORY OF VIBRATIONS IN CARBON NANOTUBES

A single-wall carbon nanotube can be visualized by referring to Fig. 3, which shows a 2D graphene sheet with lattice vectors $\mathbf{a}_1$ and $\mathbf{a}_2$, and a vector $\mathbf{C}$ given by

$$\mathbf{C} = n\mathbf{a}_1 + m\mathbf{a}_2, \tag{1}$$

where n and m are integers. By rolling the sheet such that the tip and tail of $\mathbf{C}$ coincide, a cylindrical nanotube specified by (n,m) is obtained. If $n = m$, the resulting nanotube is referred to as an "armchair" tubule, while if $n = 0$ or $m = 0$, it is referred to as a "zigzag" tubule; otherwise ($n \neq m \neq 0$) it is known as a "chiral" tubule. There is no loss of generality if it is assumed that $n > m$.

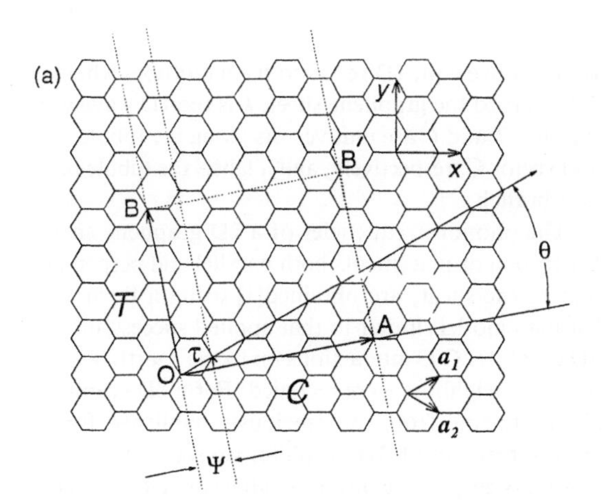

Fig. 3. Translation vectors used to define the symmetry of a carbon nanotube (see text). The vectors $\mathbf{a}_1$ and $\mathbf{a}_2$ define the 2D primitive cell.

The electronic properties of single-walled carbon nanotubes have been studied theoretically using different methods[4–12]. It is found that if $n - m$ is a multiple of 3, the nanotube will be metallic; otherwise, it will exhibit a semiconducting behavior. Calculations on a 2D array of identical armchair nanotubes with parallel tube axes within the local density approximation framework indicate that a crystal with a hexagonal packing of the tubes is most stable, and that intertubule interactions render the system semiconducting with a zero energy gap[35].

3.1 *Symmetry groups of nanotubes*

A cylindrical carbon nanotube, specified by (n,m), can be considered a one-dimensional crystal with a fundamental lattice vector $\mathbf{T}$, along the direction of the tube axis, of length given by[1,3]

$$T = \sqrt{3}C/d_R \tag{2}$$

where

$$d_R = d \quad \text{if } n - m \neq 3dr$$
$$= 3d \quad \text{if } n - m = 3dr \tag{3}$$

where C is the length of the vector in eqn (1), d is the greatest common divisor of n and m, and r is any integer. The number of atoms per unit cell is $2N$ such that

$$N = 2(n^2 + m^2 + nm)/d_R. \tag{4}$$

For a chiral nanotube specified by (n,m), the cylinder is divided into d identical sections; consequently a rotation about the tube axis by the angle $2\pi/d$ constitutes a symmetry operation. Another symmetry operation, $R = (\psi, \boldsymbol{\tau})$ consists of a rotation by an angle ψ given by

$$\psi = 2\pi \frac{\Omega}{Nd} \tag{5}$$

followed by a translation $\boldsymbol{\tau}$, along the direction of the tube axis, given by

$$\boldsymbol{\tau} = \mathbf{T}\frac{d}{N}. \tag{6}$$

The quantity Ω that appears in eqn (5) is expressed in terms of n and m by the relation

$$\Omega = \{p(m + 2n) + q(n + 2m)\}(d/d_R) \tag{7}$$

where p and q are integers that are uniquely determined by the eqn

$$mp - nq = d, \tag{8}$$

subject to the conditions $q < m/d$ and $p < n/d$.

For the case $d = 1$, the symmetry group of a chiral nanotube specified by (n,m) is a cyclic group of order N given by

$$\mathcal{C}_{N/\Omega} = \{\mathcal{R}_{N/\Omega}, \mathcal{R}^2_{N/\Omega}, \ldots, \mathcal{R}^{N-1}_{N/\Omega}, \mathcal{R}^N_{N/\Omega} = E\} \tag{9}$$

where E is the identity element, and $\mathcal{R}_{N/\Omega} = (2\pi(\Omega/N), \mathbf{T}/N))$. For the general case when $d \neq 1$, the cylinder is divided into d equivalent sections. Consequently, it follows that the symmetry group of the nanotube is given by

$$\mathcal{C} = \mathcal{C}_d \otimes \mathcal{C}'_{Nd/\Omega} \tag{10}$$

where

$$\mathcal{C}_d = \{C_d, C_d^2, \ldots, C_d^d = E\} \tag{11}$$

and

$$\mathcal{C}'_{Nd/\Omega} = \{\mathcal{R}_{Nd/\Omega}, \mathcal{R}^2_{Nd/\Omega}, \ldots, \mathcal{R}^{N/d}_{ND/\Omega} = E\}. \tag{12}$$

Here the operation $\mathcal{C}_d$ represents a rotation by $2\pi/d$ about the tube axis; the angles of rotation in $\mathcal{C}'_{Nd/\Omega}$ are defined modulo $2\pi/d$, and the symmetry element $\mathcal{R}_{Nd/\Omega} = (2\pi(\Omega/Nd), \mathbf{T}d/N))$.

The irreducible representations of the symmetry group $\mathcal{C}$ are given by $A, B, E_1, E_2, \ldots, E_{N/2-1}$. The A representation is completely symmetric, while in the B representation, the characters for the operations C_d and $\mathcal{R}_{Nd/\Omega}$ are

$$\chi(C_d) = +1, \tag{13}$$

and

$$\chi(\mathcal{R}_{Nd/\Omega}) = -1. \tag{14}$$

In the E_n irreducible representation, the character of any symmetry operation corresponding to a rotation by an angle ξ is given by

$$\chi(C) = \begin{cases} e^{i2\pi\xi n} \\ e^{-i2\pi\xi n}. \end{cases} \tag{15}$$

Equations (13–15) completely determine the character table of the symmetry group $\mathcal{C}$ for a chiral nanotube.

Applying the above symmetry formulation to armchair $(n = m)$ and zigzag $(m = 0)$ nanotubes, we find that such nanotubes have a symmetry group given by the product of the cyclic group $\mathcal{C}_n$ and $\mathcal{C}'_{2n}$, where $\mathcal{C}'_{2n}$ consists of only two symmetry operations: the identity, and a rotation by $2\pi/2n$ about the tube axis followed by a translation by $\mathbf{T}/2$. Armchair and zigzag nanotubes, however, have other symmetry operations, such as inversion and reflection in planes parallel to the tube axis. Thus, the symmetry group, assuming an infinitely long nanotube with no caps, is given by

$$\mathcal{C} = \begin{cases} \mathfrak{D}_{nd} \otimes \mathcal{C}'_{2n} & n \text{ odd} \\ \mathfrak{D}_{nh} \otimes \mathcal{C}'_{2n} & n \text{ even} \end{cases} \tag{16}$$

Thus $\mathcal{C} = \mathfrak{D}_{2nh}$ in these cases. The choice of $\mathfrak{D}_{nd}$ or $\mathfrak{D}_{nh}$ in eqn (16) is made to insure that inversion is a symmetry operation of the nanotube. Even though we neglect the caps in calculating the vibrational frequencies, their existence, nevertheless, reduces the symmetry to either $\mathfrak{D}_{nd}$ or $\mathfrak{D}_{nh}$.

Of course, whether the symmetry groups for armchair and zigzag tubules are taken to be $\mathfrak{D}_{nd}$ (or $\mathfrak{D}_{nh}$) or $\mathfrak{D}_{2nh}$, the calculated vibrational frequencies will be the same; the symmetry assignments for these modes, however, will be different. It is, thus, expected that modes that are Raman or IR-active under $\mathfrak{D}_{nd}$ or $\mathfrak{D}_{nh}$ but are optically silent under $\mathfrak{D}_{2nh}$ will only show a weak activity resulting from the fact that the existence of caps lowers the symmetry that would exist for a nanotube of infinite length.

3.2 *Model calculations of phonon modes*

The BZ of a nanotube is a line segment along the tube direction, of length $2\pi/T$. The rectangle formed by vectors $\mathbf{C}$ and $\mathbf{T}$, in Fig. 3, has an area N times larger than the area of the unit cell of a graphene sheet formed by vectors $\mathbf{a}_1$ and $\mathbf{a}_2$, and gives rise to a rectangular BZ than is N times smaller than the hexagonal BZ of a graphene sheet. Approximate values for the vibrational frequencies of the nanotubes can be obtained from those of a graphene sheet by the method of zone folding, which in this case implies that

$$\omega_{1\mathrm{D}}(k;\mu) = \omega_{2\mathrm{D}}\left(k\hat{T} + \frac{2\pi\mu}{C}\hat{C}\right), \quad \mu = 0, 1, \ldots N-1. \tag{17}$$

In the above eqn, 1D refers to the nanotubes whereas 2D refers to the graphene sheet, k is the 1D wave vector, and $\hat{T}$ and $\hat{C}$ are unit vectors along the tubule axis and vector $\mathbf{C}$, respectively, and μ labels the tubule phonon branch.

The phonon frequencies of a 2D graphene sheet, for carbon displacements both parallel and perpendicular to the sheet, are obtained[1] using a Born-Von Karman model similar to that applied successfully to 3D graphite. C-C interactions up to the fourth nearest in-plane neighbors were included. For a 2D graphene sheet, starting from the previously published force-constant model of 3D graphite, we set all the force constants connecting atoms in adjacent layers to zero, and we modified the in-plane force constants slightly to describe accurately the results of electron energy loss spectroscopic measurements, which yield the phonon dispersion curves along the M direction in the BZ. The dispersion curves are somewhat different near M, and along M–K, than the 2D calculations shown in Fig. 1b. The lattice dynamical model for 3D graphite produces dispersion curves $\omega_i(q)$ that are in good agreement with experimental results from inelastic neutron scattering, Raman scattering, and IR spectroscopy.

The zone-folding scheme has two shortcomings. First, in a 2D graphene sheet, there are three modes with vanishing frequencies as $q \to 0$; they correspond to two translational modes with in-plane C-atom displacements and one mode with out-of-plane C-atom displacements. Upon rolling the sheet into a cylinder, the translational mode in which atoms move perpendicular to the plane will now correspond to the breathing mode of the cylinder for which the atoms vibrate along the radial direction. This breathing mode has a nonzero frequency, but the value cannot be obtained by zone folding; rather, it must be calculated analytically. The frequency of the breathing mode ω_{rad} is readily calculated and is found to be[1,2]

$$\omega_{\text{rad}} = \frac{3a}{4\sqrt{m_C}\, r_0} \left[\phi_r^{(1)} + 6\phi_r^{(2)} + 4\phi_r^{(3)} + 14\phi_r^{(4)}\right]^{1/2} \tag{18}$$

where $a = 2.46$ Å is the lattice constant of a graphene sheet, r_0 is the tubule radius, m_C is the mass of a carbon atom, and $\phi_r^{(i)}$ is the bond stretching force constant between an atom and its ith nearest neighbor. It should be noted that the breathing mode frequency is found to be independent of n and m, and that it is inversely proportional to the tubule radius. The value of $\omega_{\text{rad}} = 300\ \text{cm}^{-1}$ for $r_0 = 3.5$ Å, the radius that corresponds to a nanotube capped by a C_{60} hemisphere.

Second, the zone-folding scheme cannot give rise to the two zero-frequency tubule modes that correspond to the translational motion of the atoms in the two directions perpendicular to the tubule axis. That is to say, there are no normal modes in the 2D graphene sheet for which the atomic displacements are such that if the sheet is rolled into a cylinder, these displacements would then correspond to either of the rigid tubule translations in the directions perpendicular to the cylinder axis. To convert these two translational modes into eigenvectors of the tubule dynamical matrix, a perturbation matrix must be added to the dynamical matrix. As will be discussed later, these translational modes transform according to the E_1 irreducible representation; consequently, the perturbation should be constructed so that it will cause a mixing of the E_1 modes, but should have no effect in first order on modes with other symmetries. The perturbation matrix turns out to cause the frequencies of the E_1 modes with lowest frequency to vanish, affecting the other E_1 modes only slightly.

Finally, it should be noted that in the *zone-folding* scheme, the effect of curvature on the force constants has been neglected. We make this approximation under the assumption that the hybridization between the sp^2 and p_z orbitals is small. For example, in the armchair nanotube based on C_{60}, with a diameter of approximately 0.7 nm, the three bond angles are readily calculated and they are found to be 120.00°, 118.35°, and 118.35°. Because the deviation of these angles from 120° is very small, the effect of curvature on the force constants might be expected to be small. Based on a calculation using the semi-empirical interatomic Tersoff potential, Bacsa *et al.* [26,36] estimate considerable mode softening with decreasing diameter. For tubes of diameter greater than ~10 nm, however, they predict tube wall curvature has negligible effect on the mode frequencies.

3.3 *Raman- and infrared-active modes*

The frequencies of the tubule phonon modes at the Γ-point, or BZ center, are obtained from eqn (17) by setting $k = 0$. At this point, we can classify the modes according to the irreducible representations of the symmetry group that describes the nanotube. We begin by showing how the classification works in the case of chiral tubules. The nanotube modes obtained from the zone-folding eqn by setting $\mu = 0$ correspond to the Γ-point modes of the 2D graphene sheet. For these modes, atoms connected by any lattice vector of the 2D sheet have the same displacement. Such atoms, under the symmetry operations of the nanotubes, transform into each other; consequently, the nanotubes modes obtained by setting $\mu = 0$ are completely symmetric and they transform according to the A irreducible representation.

Next, we consider the Γ-point nanotube modes obtained by setting $k = 0$ and $\mu = N/2$ in eqn (17). The modes correspond to 2D graphene sheet modes at the point $\mathbf{k} = (N\pi/C)\hat{C}$ in the hexagonal BZ. We consider how such modes transform under the symmetry operations of the groups $\mathcal{C}_d$ and $\mathcal{C}'_{Nd/\Omega}$. Under the action of the symmetry element C_d, an atom in the 2D graphene sheet is carried into another atom separated from it by the vector

$$\mathbf{r}_1 = (n/d)\mathbf{a}_1 + (m/d)\mathbf{a}_2. \tag{19}$$

The displacements of two such atoms at the point $\mathbf{k} = (N\pi/C)\hat{C}$ have a phase difference given by

$$\frac{N}{2}\,\mathbf{k}\cdot\mathbf{r}_1 = 2\pi(n^2 + m^2 + nm)/(dd_R) \tag{20}$$

which is an integral multiple of 2π. Thus, the displacements of the two atoms are equal and it follows that

$$\chi(C_d) = 1. \tag{21}$$

The symmetry operation $R_{Nd/\Omega}$ carries an atom into another one separated from it by the vector

$$\mathbf{r}_2 = p\mathbf{a}_1 + q\mathbf{a}_2, \tag{22}$$

where p and q are the integers uniquely determined by eqn (8). The atoms in the 2D graphene sheet have displacements, at the point $\mathbf{k} = (N\pi/C)\hat{C}$, that are completely out of phase. This follows from the observation that

$$\left(\frac{N\pi}{C}\hat{C}\right)\cdot \mathbf{r}_2 = \frac{\pi\Omega}{d}, \tag{23}$$

and that Ω/d is an odd integer; consequently

$$\chi(R_{Nd/\Omega}) = -1. \tag{24}$$

From the above, we therefore conclude that the nanotube modes obtained by setting $\mu = N/2$, transform according to the B irreducible representation of the chiral symmetry group $\mathcal{C}$.

Similarly, it can be shown that the nanotube modes at the Γ-point obtained from the zone-folding eqn by setting $\mu = \eta$, where $0 < \eta < N/2$, transform according to the E_η irreducible representation of the symmetry group $\mathcal{C}$. Thus, the vibrational modes at the Γ-point of a chiral nanotube can be decomposed according to the following eqn

$$\Gamma^{vib} = 6A + 6B6E_1 + 6E_2 + \cdots + 6E_{N/2-1} \tag{25}$$

Modes with A, E_1, or E_2 symmetry are Raman active, while only A and E_1 modes are infrared active. The A modes are nondegenerate and the E modes are doubly degenerate. According to the discussion in the previous section, two A modes and one of the E_1 modes have vanishing frequencies; consequently, for a chiral nanotube there are 15 Raman- and 9 IR-active modes, the IR-active modes being also Raman-active. It should be noted that the number of Raman- and IR-active modes is independent of the nanotube diameter. For a given chirality, as the diameter of the nanotube increases, the number of phonon modes at the BZ center also increases. Nevertheless, the number of the modes that transform according to the A, E_1, or E_2 irreducible representations does not change. Since only modes with these symmetries will exhibit optical activity, the number of Raman or IR modes does not increase with increasing diameter. This, perhaps unanticipated, result greatly simplifies the data analysis. The symmetry classification of the phonon modes in armchair and zigzag tubules have been studied in ref. [2,3] under the assumption that the symmetry group of these tubules is isomorphic with either $\mathfrak{D}_{nd}$ or $\mathfrak{D}_{nh}$, depending on whether n is odd or even. As noted earlier, if one considers an infinite tubule with no caps, the relevant symmetry group for armchair and zigzag tubules would be the group $\mathfrak{D}_{2nh}$. For armchair tubules described by the $\mathfrak{D}_{nd}$ group there are, among others, $3A_{1g}$, $6E_{1g}$, $6E_{2g}$, $2A_{2u}$, and $5E_{1u}$ optically active modes with nonzero frequencies; consequently, there are 15 Raman- and 7 IR-active modes. All zigzag tubules, under $\mathfrak{D}_{nd}$ or $\mathfrak{D}_{nh}$ symmetry group have, among others, $3A_{1g}$, $6E_{1g}$, $6E_{2g}$, $2A_{2u}$, and $5E_{1u}$ optically active modes with nonzero frequencies; thus there are 15 Raman- and 7 IR-active modes.

3.4 *Mode frequency dependence on tubule diameter*

In Figs. 4–6, we display the calculated tubule frequencies as a function of tubule diameter. The results are based on the zone-folding model of a 2D graphene sheet, discussed above. IR-active (a) and Raman-active (b) modes appear separately for chiral tubules (Fig. 4), armchair tubules (Fig. 5) and zig-zag tubules (Fig. 6). For the chiral tubules, results for the representative (n, m), indicated to the left in the figure, are displaced vertically according to their calculated diameter, which is indicated on the right. Similar to modes in a C_{60} molecule, the lower and higher frequency modes are expected, respectively, to have radial and tangential character. By comparison of the model calculation results in Figs. 4–6 for the three tube types (armchair, chiral, and zig-zag) a common general behavior is observed for both the IR-active (a) and Raman-active (b) modes. The highest frequency modes exhibit much less frequency dependence on diameter than the lowest frequency modes. Taking the large-diameter tube frequencies as our reference, we see that the four lowest modes stiffen dramatically (150–400 cm^{-1}) as the tube diameter approaches ~1 nm. Conversely, the modes above ~800 cm^{-1} in the large-diameter tubules are seen to be relatively less sensitive to tube diameter: one Raman-active mode stiffens with increasing tubule diameter (armchair), and a few modes in all the three tube types soften (100–200 cm^{-1}), with decreasing tube diameter. It should also be noted that, in contrast to armchair and zig-zag tubules, the mode frequencies in chiral tubules are grouped near 850 cm^{-1} and 1590 cm^{-1}.

All carbon nanotube samples studied to date have been undoubtedly composed of tubules with a distribution of diameters and chiralities. Therefore, whether one is referring to nanotube samples comprised of single-wall tubules or nested tubules, the results in Figs. 4–6 indicate one should expect inhomogeneous broadening of the IR- and Raman-active bands, particularly if the range of tube diameter encompasses the 1–2 nm range. Nested tubule samples must have a broad diameter distribution and, so, they should exhibit broader spectral features due to inhomogeneous broadening.

4. SYNTHESIS AND RAMAN SPECTROSCOPY OF CARBON NANOTUBES

We next address selected Raman scattering data collected on nanotubes, both in our laboratory and elsewhere. The particular method of tubule synthesis may also produce other carbonceous matter that is both difficult to separate from the tubules and also exhibits potentially interfering spectral features. With this in mind, we first digress briefly to discuss synthesis and purification techniques used to prepare nanotube samples.

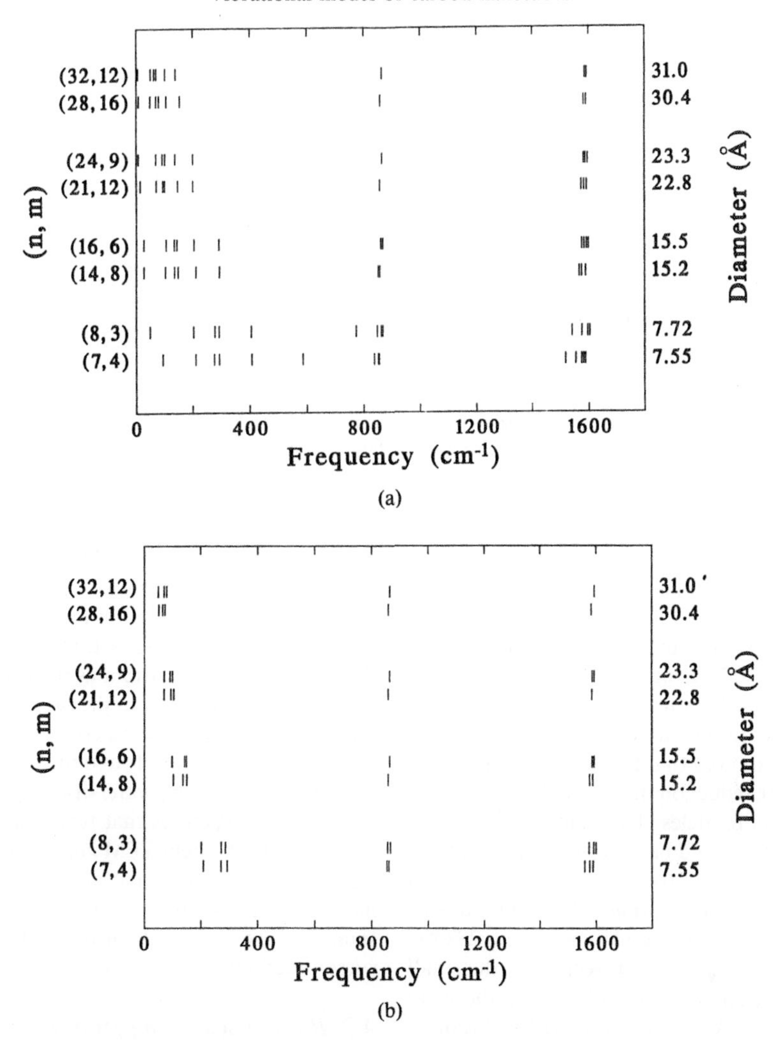

Fig. 4. Diameter dependence of the first order (a) IR-active and, (b) Raman-active mode frequencies for "chiral" nanotubes.

4.1 *Synthesis and purification*

Nested carbon nanotubes, consisting of closed concentric, cylindrical tubes were first observed by Iijima by TEM[37]. Later TEM studies[38] showed that the tubule ends were capped by the inclusion of pentagons and that the tube walls were separated by ~3.4 Å. A dc carbon-arc discharge technique for large-scale synthesis of nested nanotubes was subsequently reported [39]. In this technique, a dc arc is struck between two graphite electrodes under an inert helium atmosphere, as is done in fullerene generation. Carbon vaporized from the anode condenses on the cathode to form a hard, glassy outer core of fused carbon and a soft, black inner core containing a high concentration of nanotubes and nanoparticles. Each nanotube typically contains between 10 and 100 concentric tubes that are grouped in "microbundles" oriented axially within the core[14].

These nested nanotubes may be harvested from the core by grinding and sonication; nevertheless, substantial fractions of other types of carbon remain, all of which are capable of producing strong Raman bands as discussed in section 2. It is very desirable, therefore, to remove as much of these impurity carbon phases as possible. Successful purification schemes that exploit the greater oxidation resistance of carbon nanotubes have been investigated[40–42]. Thermogravimetric analyses reveal weight loss rate maxima at 420°C, 585°C, and 645°C associated with oxidation (in air) of fullerenes, amorphous carbon soot, and graphite, respectively, to form volatile CO and/or CO_2. Nanotubes and onion-like nanoparticles were found to lose weight rapidly at higher temperatures around 695°C. Evidently, the concentration of these other forms of carbon can be lowered by oxidation. However, the abundant carbon nanoparticles, which are expected to have a Raman spectrum similar to that shown in Figs. 1d or 1c are more difficult to remove in this way. Nevertheless, Ebbesen *et al.* [43] found that, by heating core material to 700°C in air until more than 99% of the starting material had been removed by oxidation, the remaining material consisted solely of open-ended, nested nanotubes. The oxidation was found to initiate at the reactive end caps and progress toward the cen-

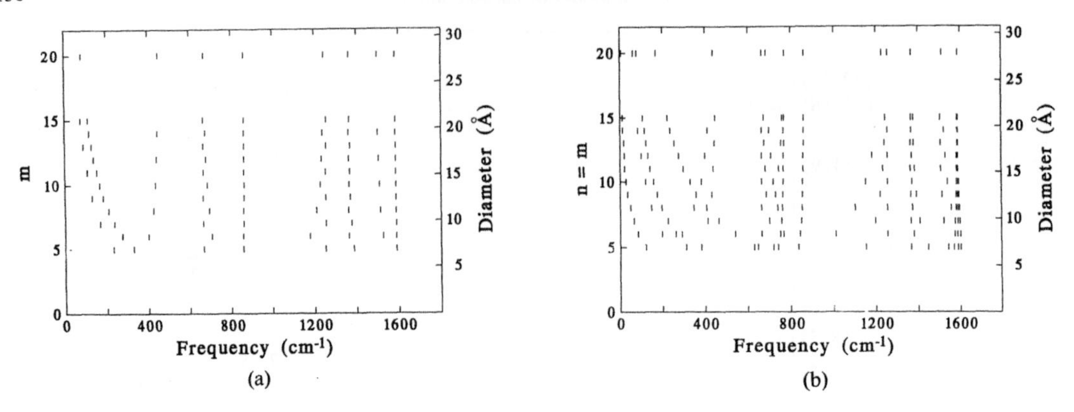

Fig. 5. Diameter dependence of the first order (a) IR-active and, (b) Raman-active mode frequencies for "armchair" nanotubes.

ter only at the open ends[42]. The inner tubules appear to be protected by the outer layers and survive the purification process.

Similar results were found by Bacsa *et al.* [26] for cathode core material. Raman scattering spectra were reported by these authors for material shown in these figures, and these results are discussed below. Their HRTEM images showed that heating core material in air induces a clear reduction in the relative abundance of the carbon nanoparticles. The Raman spectrum of these nanoparticles would be expected to resemble an intermediate between a strongly disordered carbon black synthesized at ~850°C (Fig. 2d) and that of carbon black graphitized in an inert atmosphere at 2820°C (Fig. 2c). As discussed above in section 2, the small particle size, as well as structural disorder in the small particles (dia. ~200 Å), activates the D-band Raman scattering near 1350 cm^{-1}.

Small diameter, *single-wall* nanotubes have been synthesized with metal catalysts by maintaining a dc arc (30 V, 95 A) between two electrodes in ~300 Torr of He gas.[21,22] The metal catalyst (cobalt[22] or nickel and iron[21]) is introduced into the arc synthesis as a mixture of graphite and pure metal powders pressed into a hole bored in the center of the graphite anode. The cathode is translated to maintain a fixed gap and stable current as the anode is vaporized in a helium atmosphere. In the case of nickel and iron, methane is added to the otherwise inert helium atmosphere. Nanotubes are found in carbonaceous material condensed on the water-cooled walls and also in cobweb-like structures that form throughout the arc chamber. Bright-field TEM images (100,000×) of the Co-catalyzed, arc-derived carbon material reveal numerous narrow-diameter single-wall nanotubes and small Co particles with diameters in the range 10–50 nm surrounded by a thick (~50 nm) carbon coating[27].

4.2 *Raman scattering from nested carbon nanotubes*

Several Raman studies have been carried out on *nested* nanotubes[23–26]. The first report was by Hiura *et al.* [23], who observed a strong first-order band at 1574 cm^{-1} and a weaker, broader D-band at 1346

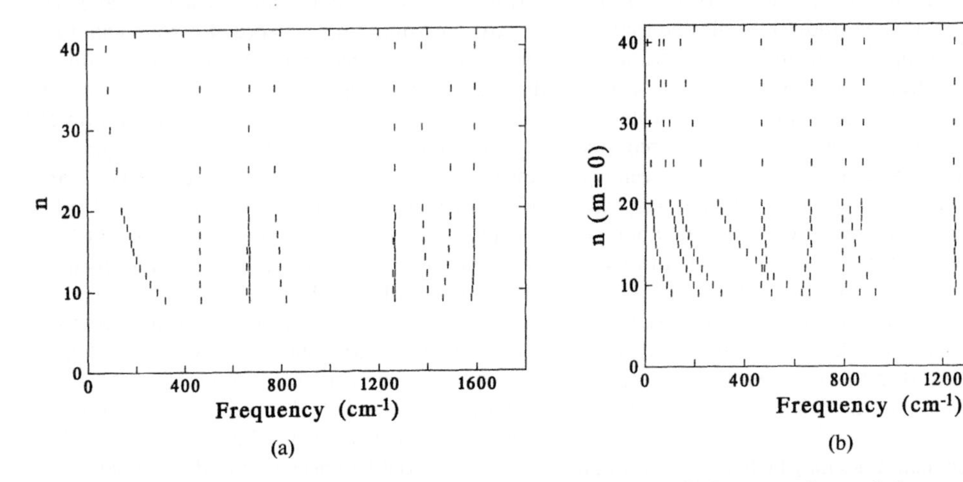

Fig. 6. Diameter dependence of the first order (a) IR-active and, (b) Raman-active mode frequencies for "zig-zag" nanotubes.

cm^{-1}. The feature at 1574 cm^{-1} is strongly downshifted relative to the 1582 cm^{-1} mode observed in HOPG, possibly a result of curvature and closure of the tube wall. These authors also observe reasonably sharp second order Raman bands at 2687 cm^{-1} and 2455 cm^{-1}.

Other Raman studies of cathode *core* material grown by the same method, and also shown by TEM to contain nested nanotubes as well as carbon nanoparticles, have reported slightly different results (Figs. 7, 8). Chandrabhas *et al.* [24] report a first-order Raman spectrum, Fig. 7, curve (b), for the cathode core material similar to that of polycrystalline graphite, Fig. 7, curve (a), with a strong, disorder-broadened band at 1583 cm^{-1}, and a weaker, D-band at 1353 cm^{-1}. For comparison, the Raman spectrum for the outer shell material from the cathode, Fig. 7, curve (c), is also shown. The spectrum for the outer shell exhibits the character of a disordered sp^2 carbon (i.e., carbon black or glassy carbon, c.f. Figs. 2d and 2e). Additionally, weak Raman features were observed at very low frequencies, 49 cm^{-1} and 58 cm^{-1}, which are upshifted, respectively, by 7 cm^{-1} and 16 cm^{-1} from the $E_{2g}^{(1)}$ shear mode observed in graphite at 42 cm^{-1} (Fig. 1d). The authors attributed this upshifting to defects in the tubule walls, such as inclusion of pentagons and heptagons. However, two shear modes are consistent with the cylindrical symmetry, as the planar $E_{2g}^{(1)}$ shear modes should split into a rotary and a telescope mode, as shown schematically in Fig. 9. The second-order Raman spectrum of Chandrabhas *et al.*, Fig. 8, curve (b), shows a strong line at 2709 cm^{-1} downshifted and narrower than it's counterpart in polycrystalline graphite at 2716 cm^{-1}, Fig. 8, curve (a). Thus, although the first-order mode (1583 cm^{-1}) in the core material is broader than in graphite, indicating some disorder in the tubule wall, the 2709 cm^{-1} feature is actually narrower than its graphitic counterpart, suggesting a reduction in the phonon dispersion in tubules relative to that in graphite.

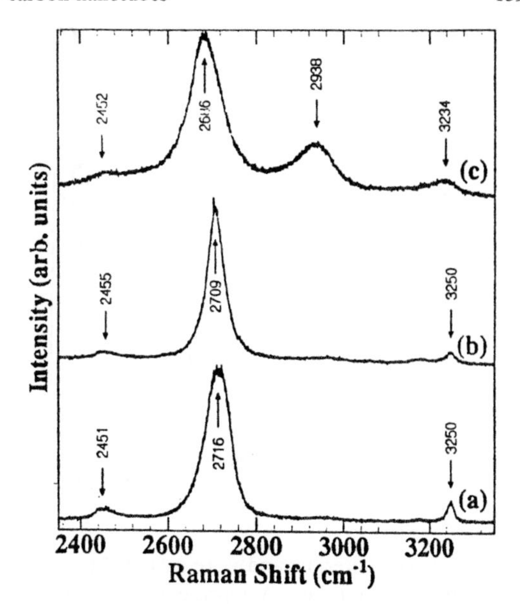

Fig. 8. Second-order Raman spectra of (a) graphite, (b) inner core material containing nested nanotubes, (c) outer shell of cathode (after ref. [24]).

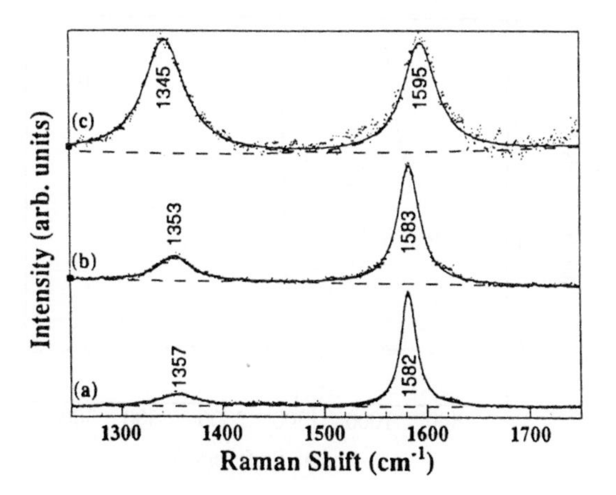

Fig. 7. First-order Raman spectra of (a) graphite, (b) inner core material containing nested nanotubes, (c) outer shell of carbonaceous cathode deposit (after ref. [24]).

Kastner *et al.* [25] also reported Raman spectra of cathode core material containing nested tubules. The spectral features were all identified with tubules, including weak D-band scattering for which the laser excitation frequency dependence was studied. The authors attribute some of the D-band scattering to curvature in the tube walls. As discussed above, Bacsa *et al.* [26] reported recently the results of Raman studies on oxidatively purified tubes. Their spectrum is similar to that of Hiura *et al.* [23], in that it shows very weak D-band scattering. Values for the frequencies of all the first- and second-order Raman features reported for these nested tubule studies are also collected in Table 1.

4.3 *Small diameter single-wall nanotubes*

Recently, Bethune *et al.* [22] reported that single-wall carbon nanotubes with diameters approaching the diameter of a C_{60} fullerene (7 Å) are produced when cobalt is added to the dc arc plasma, as observed in TEM. Concurrently, Iijima *et al.* [21] described a similar route incorporating iron, methane, and argon in the dc arc plasma. These single-wall tubule samples provided the prospect of observing experimentally the many intriguing properties predicted theoretically for small-diameter carbon nanotubes.

Holden *et al.* [27] reported the first Raman results on nanotubes produced from a Co-catalyzed carbon arc. Thread-like material removed from the chamber was encapsulated in a Pyrex ampoule in ~500 Torr of He gas for Raman scattering measurements. Sharp first-order lines were observed at 1566 and 1592 cm^{-1} and second-order lines at 2681 and 3180 cm^{-1}, but *only when cobalt was present in the core of the anode.* These sharp lines had not been observed previously in

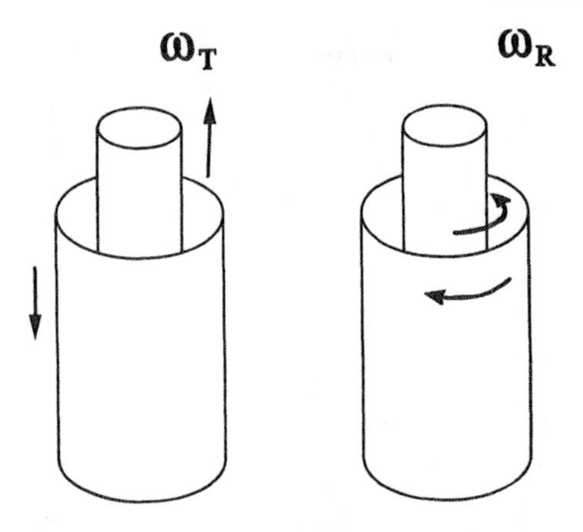

Fig. 9. Schematic view of cylindrical shear modes for a nested tubule: telescope mode (ω_T) and rotary mode (ω_R).

carbonaceous materials and were assigned to single-wall carbon nanotubes. A representative spectrum, $I_{Co}(\omega)$, is shown in Fig. 10a for *Co-catalyzed, arc-derived carbons* (solid line) over the frequency range 300–3300 cm^{-1}. This sample also contained a large fraction of other sp^2 carbonaceous material, so a subtraction scheme was devised to remove the spectral contributions from these carbons. The dashed line in the figure represents the spectrum $I_0(\omega)$ obtained from thread-like carbon removed from the chamber *when cobalt was not present* in the carbon anode. All other sample preparation conditions were identical to those used to prepare the Co-catalyzed carbons. $I_0(\omega)$ was scaled by a factor $\alpha = 0.85$ to superimpose with $I_{Co}(\omega)$ in the region near 1590 cm^{-1}.

Prominent in both first-order Raman spectra Fig. 10a is the broad D-band centered at 1341 cm^{-1}. Two second-order features, one at 2681 cm^{-1} $\approx$ 2(1341 cm^{-1}) and 3180 the other at cm^{-1} $\approx$ 2(1592 cm^{-1}) are apparent in the Co-catalyzed carbon. Weak features near 1460 cm^{-1} were identified with fullerenes because they were not present after boiling the thread-like material in toluene. The overall strength and relative intensities of the sharp peaks at 1566, 1592, 2681, and 3180 cm^{-1} remained the same, implying that these features were not related to fullerenes or other toluene soluble impurities, such as polyaromatic hydrocarbons. The significant strength of the 1592 cm^{-1} line suggests that a resonant Raman scattering process may be involved[27]. Importantly, $I_0(\omega)$ shows no evidence for any of the sharp first- or second-order features and is very similar to that of Fig. 2d (disordered carbon black). Noting that these disordered sp^2 carbons likely contribute to both $I_0(\omega)$ and $I_{Co}(\omega)$, Holden *et al.* [27] compute the "difference spectrum"; $I_{diff}(\omega) = I_{Co}(\omega) - \alpha I_0(\omega)$, which is shown in Fig. 10b. $I_{diff}(\omega)$ was constructed to emphasize contributions from new carbonaceous material(s) (e.g., carbon nanotubes), which form only when Co is present in the plasma. This difference spectrum has a fairly flat baseline with sharp first-order lines at 1566 and 1592 cm^{-1}. The inset shows a Lorentzian lineshape fit to the first-order spectrum. Sharp second-order features at 2681 and 3180 cm^{-1} are also observed.

Hiura *et al.* [23] observed two Raman lines in their spectrum of *nested* carbon nanotubes at 1574 (FWHM = 23 cm^{-1}) and at 2687 cm^{-1}. It is interesting to note that their first-order peak at 1574 cm^{-1} lies between, and is more than twice as broad, as either of the two first-order lines in $I_{diff}(\omega)$ identified[27] with single-wall nanotubes. These two observations may be consistent if an inhomogeneous broadening mechanism, originating from a distribution of tubule diameters and chiralities is active. Also, the second-order feature of Hiura *et al.* [23] at 2687 cm^{-1} is slightly broader than, and upshifted from, the second-order feature at 2681 cm^{-1} in $I_{diff}(\omega)$. It should also be noted that the second-order features in $I_{diff}(\omega)$ are downshifted significantly relative to other sp^2 carbons (see Table 1).

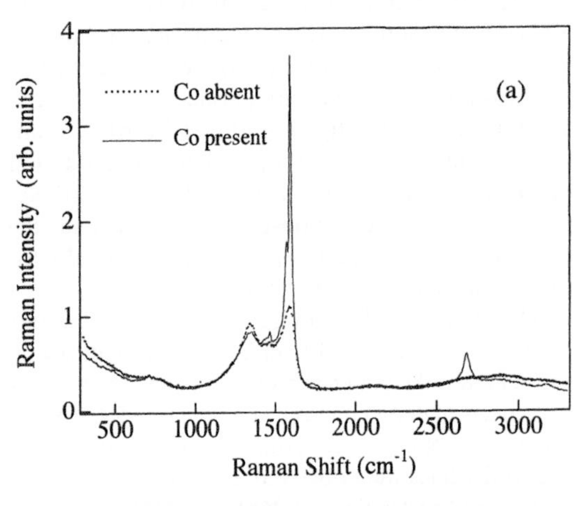

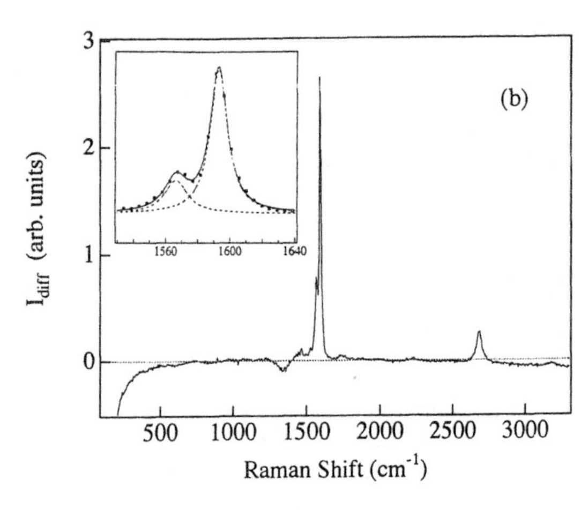

Fig. 10. (a) Raman spectra (T = 300 K) of arc-derived carbons from a dc arc: cobalt was absent (dotted line) and cobalt was present (solid line) in the carbon anode, (b) the difference spectrum calculated from (a), emphasizing the contribution from Co-catalyzed nanotubes, the inset to (b) depicts a Lorentzian fit to the first-order spectrum (after ref. [27]).

In Fig. 11 we show the Raman spectrum of carbonaceous soot containing ~1–2 nm diameter, single-wall nanotubes produced from Co/Ni-catalyzed carbon plasma[28]. These samples were prepared at MER, Inc. The sharp line components in the spectrum are quite similar to that from the Co-catalyzed carbons. Sharp, first-order peaks at 1568 cm^{-1} and 1594 cm^{-1}, and second-order peaks at ~2680 cm^{-1} and ~3180 cm^{-1} are observed, and identified with single-wall nanotubes. Superimposed on this spectrum is the contribution from disordered sp^2 carbon. A narrowed, disorder-induced D-band and an increased intensity in the second-order features of this sample indicate that these impurity carbons have been partially graphitized (i.e., compare the spectrum of carbon black prepared at 850°C, Fig. 1d, to that which has been heat treated at 2820°C, Fig. 1c).

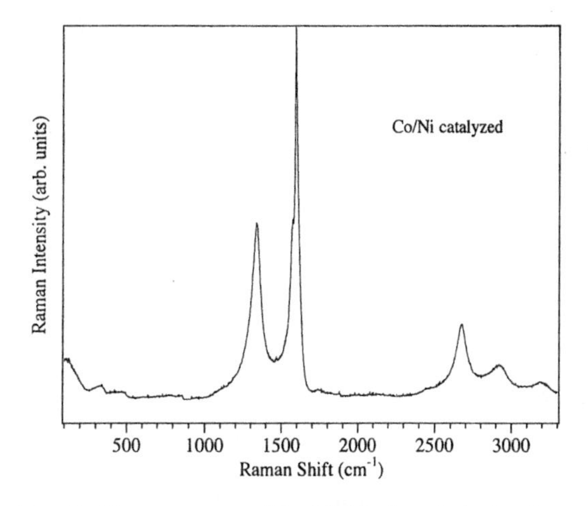

Fig. 11. Raman spectrum (T = 300 K) of arc-derived carbons containing single-wall nanotubes generated in a Ni/Co-catalyzed dc arc (after ref. [42]).

5. CONCLUSIONS

It is instructive to compare results from the various Raman scattering studies discussed in sections 4.2 (nested nanotubes) and 4.3 (single-wall nanotubes). Ignoring small changes in eigenmode frequencies, due to curvature of the tube walls, and the weak van der Waals interaction between nested nanotubes, the zone-folding model should provide reasonable predictions for trends in the Raman data. Of course, the low-frequency telescope and rotary, shear-type modes anticipated in the range ~30–50 cm^{-1} (Fig. 9) are outside the scope of the single sheet, zone-folding model.

Considering all the spectra from nested tubule samples first, it is clear from Table 1 that the data from four different research groups are in reasonable agreement. The spectral features identified with tubules appear very similar to that of graphite with sample-dependent variation in the intensity in the "D" (disorder-induced) band near 1350 cm^{-1} and also in the second-order features associated with the D-band (i.e., $2 \times D \approx 2722$ cm^{-1}) and $E_{2g}^{(2)} + D \approx 2950$ cm^{-1}. Sample-dependent D-band scattering may stem from the relative admixture of nanoparticles and nanotubes, or defects in the nanotube wall.

The zone-folding model calculations predict ~14 new, first-order Raman-active modes activated by the closing of the graphene sheet into a tube. The Raman activity (i.e., spectral strength) of these additional modes has not been addressed theoretically, and it must be a function of tubule diameter, decreasing with increasing tubule diameter. Thus, although numerous first-order modes are predicted by group theoretical arguments in the range from 200 to 1600 cm^{-1}, their Raman activity may be too small to be observed in the larger diameter, nested nanotube samples. As reported by Bacsa *et al.* [26], their nested tubule diameter distribution peaked near 10 nm and extended from ~8–40 nm, and the Raman spectrum for this closely resembled graphite. No zone-folded modes were resolved in their study. Importantly, they oxidatively purified their sample to enhance the concentration of tubules and observed no significant change in the spectrum other than a new peak at ~2900 cm^{-1}, which they attributed to C-H stretching modes.[2] We can, then, be reasonably certain that their spectrum is primarily associated with large-diameter carbon nanotubes, and not nanoparticles. In addition, they observed a very weak D-band, suggesting the tubes were fairly defect-free or that D-band scattering stems only from nanoparticles or other disordered sp^2 carbons. We can conclude that tubules with diameters greater than ~8 nm will have a Raman spectrum very similar to graphite, and that the Raman activity for the zone-folded modes may be too small to be detected experimentally. The tube diameter distributions in two other nested-tube studies[24,25] reviewed here (see Table 1) were somewhat larger than reported by Bacsa *et al.* [26]. In both these cases[24,25], the Raman spectra were very similar to disordered graphite. Interestingly, the spectra of Hiura *et al.* [23], although appearing nearly identical to other nested tubule spectra, exhibit a significantly lower first-order mode frequency (1574 cm^{-1}).

Metal-catalyzed, single-wall tubes, by comparison, are found by high-resolution TEM to have much smaller diameters (1 to 2 nm)[44], which is in the range where the zone-folding model predicts noticeable mode frequency dependence on tubule diameter[27]. This is the case for the single-wall tube samples whose data appear in columns 4 and 5 in Table 1. Sharp line contributions to the Raman spectra for single-wall tubule samples produced by Co[27] and Ni/Co[28] are also found, and they exhibit frequencies in very good agreement with one another. Using the difference spectrum of Holden *et al.* [27] to enhance the contribution from the nanotubes results in the first- and second-order frequencies found in column 4 of Table 1. As can be seen in the table, the single-wall tube frequencies are noticeably different from those reported for larger diameter (nested) tubules. For example, in

[2]The source of the hydrogen in their air treatment is not mentioned; presumably, it is from H_2O in the air.

first-order, sharp modes are observed[27] at 1566 and 1594 cm^{-1}, one downshifted and one upshifted from the value of the E_{2g}^2 mode at 1582 cm^{-1} for graphite. From zone-folding results, the near degeneracy of the highest frequency Raman modes is removed with decreasing tubule diameter; the mode frequencies spread, some upshifting and some downshifting relative to their common large-diameter values. Thus, the observation of the sharp modes at 1566 cm^{-1} and 1592 cm^{-1} for 1-2 nm tubules is consistent with this theoretical result.

Finally, in second order, the Raman feature at ~3180 cm^{-1} observed in Co- and Ni/Co-catalyzed single-wall nanotube corresponds to a significantly downshifted $2 \times E'_{2g}$ mode, where E'_{2g} represents the mid-zone (see Figs. 1a and 1b) frequency maximum of the uppermost optic branch seen in graphite at 3250 cm^{-1}.

Acknowledgement—We gratefully acknowledge valuable discussions with M. S. Dresselhaus and G. Dresselhaus, and Y. F. Balkis for help with computations. One of the authors (RAJ) acknowledges AFOSR Grant No. F49620-92-J-0401. The other author (PCE) acknowledges support from University of Kentucky Center for Applied Energy Research and the NSF Grant No. EHR-91-08764.

REFERENCES

1. R. A. Jishi, L. Venkataraman, M. S. Dresselhaus, and G. Dresselhaus, *Chem. Phys. Lett.* **209**, 77 (1993).
2. R. A. Jishi, L. Venkataraman, M. S. Dresselhaus, and G. Dresselhaus, *Chem. Phys. Lett.* **209**, 77 (1993).
3. R. A. Jishi, M. S. Dresselhaus, and G. Dresselhaus, *Phys. Rev. B* **47**, 16671 (1993).
4. E. G. Gal'pern, I. V. Stankevich, A. L. Christyakov, and L. A. Chernozatonskii, *JETP Lett.* (*Pis'ma Zh. Eksp. Teor.*) **55**, 483 (1992).
5. N. Hamada, S.-I. Sawada, and A. Oshiyama, *Phys. Rev. Lett.* **68**, 1579 (1992).
6. R. Saito, G. Dresselhaus, and M. S. Dresselhaus, *Chem. Phys. Lett.* **195**, 537 (1992).
7. J. W. Mintmire, B. I. Dunalp, and C. T. White, *Phys. Rev. Lett.* **68**, 631 (1992).
8. K. Harigaya, *Chem. Phys. Lett.* **189**, 79 (1992).
9. K. Tanaka *et al.*, *Phys. Lett.* **A164**, 221 (1992).
10. J. W. Mintmire, D. H. Robertson, and C. T. White, *J. Phys. Chem. Solids* **54**, 1835 (1993).
11. P. W. Fowler, *J. Phys. Chem. Solids* **54**, 1825 (1993).
12. C. T. White, D. H. Roberston, and J. W. Mintmire, *Phys. Rev. B* **47**, 5485 (1993).
13. T. W. Ebbesen and P. M. Ajayan, *Nature* (London) **358**, 220 (1992).
14. T. W. Ebbesen *et al.*, *Chem. Phys. Lett.* **209**, 83 (1993).
15. H. M. Duan and J. T. McKinnon, *J. Phys. Chem.* **49**, 12815 (1994).
16. M. Endo, Ph.D. thesis (in French), University of Orleans, Orleans, France (1975).
17. M. Endo, Ph.D. thesis (in Japanese), Nagoya University, Japan (1978).
18. M. Endo and H. W. Kroto, *J. Phys. Chem.* **96**, 6941 (1992).
19. X. F. Zhang *et al.*, *J. Cryst. Growth* **130**, 368 (1993).
20. S. Amelinckx *et al.*, personal communication.
21. S. Iijima and T. Ichihashi, *Nature* (London) **363**, 603 (1993).
22. D. S. Bethune *et al.*, *Nature* (London) **363**, 605 (1993).
23. H. Hiura, T. W. Ebbesen, K. Tanigaki, and H. Takahashi, *Chem. Phys. Lett.* **202**, 509 (1993).
24. N. Chandrabhas *et al.*, *PRAMA-J. Physics* **42**, 375 (1994).
25. J. Kastner *et al.*, *Chem. Phys. Lett.* **221**, 53 (1994).
26. W. S. Bacsa, D. U. A., Châtelain, and W. A. de Heer, *Phys. Rev. B* **50**, 15473 (1994).
27. J. M. Holden *et al.*, *Chem. Phys. Lett.* **220**, 186 (1994).
28. J. M. Holden, R. A. Loufty, and P. C. Eklund (unpublished).
29. P. Lespade, R. Al-Jishi, and M. S. Dresselhaus, *Carbon* **20**, 427 (1982).
30. A. W. Moore, In *Chemistry and Physics of Carbon* (Edited by P. L. Walker and P. A. Thrower), Vol. 11, p. 69, Marcel Dekker, New York (1973).
31. L. J. Brillson, E. Burnstein, A. A. Maradudin, and T. Stark, In *Proceedings of the International Conference on Semimetals and Narrow Gap Semiconductors* (Edited by D. L. Carter and R. T. Bate), p. 187, Pergamon Press, New York (1971).
32. Y. Wang, D. C. Alsmeyer, and R. L. McCreery, *Chem. Matter.* **2**, 557 (1990).
33. F. Tunistra and J. L. Koenig, *J. Chem. Phys.* **53**, 1126 (1970).
34. X.-X. Bi *et al.*, *J. Mat. Res.* (1994), submitted.
35. J. C. Charlier, Ph.D. thesis (unpublished), Universite Chatholique De Louvain (1994).
36. W. S. Bacsa, W. A. de Heer, D. Ugarte, and A. Châtelain, *Chem. Phys. Lett.* **211**, 346 (1993).
37. S. Iijima, *Nature* (London) **354**, 56 (1991).
38. S. Iijima, T. Ichihashi, and Y. Ando, *Nature* (London) **356**, 776 (1992).
39. P. M. Ajayan and S. Iijima, *Nature* (London) **361**, 333 (1993).
40. L. S. K. Pang, J. D. Saxby, and S. P. Chatfield, *J. Phys. Chem.* **97**, 6941 (1993).
41. S. C. Tsang, P. J. F. Harris, and M. L. H. Green, *Nature* (London) **362**, 520 (1993).
42. P. M. Ajayan *et al.*, *Nature* (London) **362**, 522 (1993).
43. T. W. Ebbesen, P. M. Ajayan, H. Hiura, and K. Tanigaki, *Nature* (London) **367**, 519 (1994).
44. C. H. Kiang *et al.*, *J. Phys. Chem.* **98**, 6612 (1994).

MECHANICAL AND THERMAL PROPERTIES OF CARBON NANOTUBES

RODNEY S. RUOFF and DONALD C. LORENTS
Molecular Physics Laboratory, SRI International, Menlo Park, CA 94025, U.S.A.

(*Received* 10 *January* 1995; *accepted* 10 *February* 1995)

Abstract—This chapter discusses some aspects of the mechanical and thermal properties of carbon nanotubes. The tensile and bending stiffness constants of ideal multi-walled and single-walled carbon nanotubes are derived in terms of the known elastic properties of graphite. Tensile strengths are estimated by scaling the 20 GPa tensile strength of Bacon's graphite whiskers. The natural resonance (fundamental vibrational frequency) of a cantilevered single-wall nanotube of length 1 micron is shown to be about 12 MHz. It is suggested that the thermal expansion of carbon nanotubes will be essentially isotropic, which can be contrasted with the strongly anisotropic expansion in "conventional" (large diameter) carbon fibers and in graphite. In contrast, the thermal conductivity may be highly anisotropic and (along the long axis) perhaps higher than any other material. A short discussion of topological constraints to surface chemistry in idealized multi-walled nanotubes is presented, and the importance of a strong interface between nanotube and matrix for formation of high strength nanotube-reinforced composites is highlighted.

Key Words—Nanotubes, mechanical properties, thermal properties, fiber-reinforced composites, stiffness constant, natural resonance.

1. INTRODUCTION

The discovery of multi-walled carbon nanotubes (MWNTs), with their nearly perfect cylindrical structure of seamless graphite, together with the equally remarkable high aspect ratio single-walled nanotubes (SWNTs) has led to intense interest in these remarkable structures[1]. Work is progressing rapidly on the production and isolation of pure bulk quantities of both MWNT and SWNT, which will soon enable their mechanical, thermal, and electrical properties to be measured[2]. Until that happens, we can speculate about the properties of these unique one-dimensional carbon structures. A preview of the mechanical properties that might be expected from such structures was established in the 1960s by Bacon[3], who grew carbon fibers with a scroll structure that had nearly the tensile mechanical properties expected from ideal graphene sheets.

The mechanical and thermal properties of nanotubes (NTs) have not yet been measured, mainly because of the difficulties of obtaining pure homogeneous and uniform samples of tubes. As a result we must rely, for the moment, on *ab initio* calculations or on continuum calculations based on the known properties of graphite. Fortunately, several theoretical investigations already indicate that the classical continuum theory applied to nanotubes is quite reliable for predicting the mechanical and some thermal properties of these tubes[4,5]. Of course, care must be taken in using such approximations in the limit of very small tubes or when quantum effects are likely to be important. The fact that both MWNTs and SWNTs are simple single or multilayered cylinders of graphene sheets gives confidence that the in-plane properties of the graphene sheet can be used to predict thermal and mechanical properties of these tubes.

2. MECHANICAL PROPERTIES

2.1 *Tensile strength and yield strength*

Tersoff[4] has argued convincingly that the elastic properties of the graphene sheet can be used to predict the stain energy of fullerenes and nanotubes. Indeed, the elastic strain energy that results from simple calculations based on continuum elastic deformation of a planar sheet compares very favorably with the more sophisticated *ab initio* results. The result has been confirmed by *ab initio* calculations of Mintmire *et al.*[6] This suggests that the mechanical properties of nanotubes can be predicted with some confidence from the known properties of single crystal graphite.

We consider the case of defect-free nanotubes, both single-walled and multi-walled (SWNT and MWNT). The stiffness constant for a SWNT can be calculated in a straightforward way by using the elastic moduli of graphite[7] because the mechanical properties of single-crystal graphite are well understood. To good approximation, the in-plane elastic modulus of graphite, C_{11}, which is 1060 GPa, gives directly the on-axis Young's modulus for a homogeneous SWNT. To obtain the stiffness constant, one must scale the Young's modulus with the cross-sectional area of the tube, which gives the scaling relation

$$K = E_0(A_0 - A_1)/A_0 = E_0(r_0^2 - r_1^2)/r_0^2 \qquad (1)$$

where A_0 is the cross-sectional area of the nanotube, and A_1 is the cross-sectional area of the hole. Because we derive the tensile stiffness constant from the material properties of graphite, each cylinder has a wall thickness equivalent to that of a single graphene sheet in graphite, namely, 0.34 nm. We can, thus, use this relationship to calculate the tensile stiffness of a SWNT,

which for a typical 1.0-nm tube is about 75% of the ideal, or about 800 GPa. To calculate K for MWNTs we can, in principle, use the scaling relation given by eqn (1), where it is assumed that the layered tubes have a homogeneous cross-section. For MWNTs, however, an important issue in the utilization of the high strength of the tubes is connected with the question of the binding of the tubes to each other. For ideal MWNTs, that interact with each other only through weak van der Waals forces, the stiffness constant K of the individual tubes cannot be realized by simply attaching a load to the outer cylinder of the tube because each tube acts independently of its neighbors, so that ideal tubes can readily slide within one another.

For ideal tubes, calculations[8] support that tubes can translate with respect to one another with low energy barriers. Such tube slippage may have been observed by Ge and Sattler in STM studies of MWNTs[9]. To realize the full tensile strength of a MWNT, it may be necessary to open the tube and secure the load to each of the individual nanotubes. Capped MWNTs, where only the outer tube is available for contact with a surface, are not likely to have high tensile stiffness or high yield strength. Because the strength of composite materials fabricated using NTs will depend mainly on the surface contact between the matrix and the tube walls, it appears that composites made from small-diameter SWNTs are more likely to utilize the high strength potential of NTs than those made from MWNTs.

A milestone measurement in carbon science was Bacon's production of graphite whiskers. These were grown in a DC arc under conditions near the triple point of carbon and had a Young's modulus of 800 GPa and a yield strength of 20 GPa. If we assume that these whiskers, which Bacon considered to be a scroll-like structure, had no hollow core in the center, then the same scaling rule, eqn (1), can be used for the yield strength of carbon nanotubes. As a practical means of estimating yield strengths, it is usually assumed that the yield strength is proportional to the Youngs modulus (i.e., $Y_{max} = \beta E$), where β ranges from 0.05 to 0.1[10]. Using Bacon's data, $\beta = 0.025$, which may indicate the presence of defects in the whisker. Ideally, one would like to know the in-plane yield strength of graphite, or directly know the yield strengths of a variety of nanotubes (whose geometries are well known) so that the intrinsic yield strength of a graphene sheet, whether flat or rolled into a scroll, could be determined. This is fundamentally important, and we call attention to Coulson's statement that "the C-C bond in graphite is the strongest bond in nature[11]." This statement highlights the importance of Bacon's determination of the yield strength of the scroll structures: it is the only available number for estimating the yield strength of a graphene sheet.

The yield strengths of defect-free SWNTs may be higher than that measured for Bacon's scroll structures, and measurements on defect-free carbon nanotubes may allow the prediction of the yield strength of a single, defect-free graphene sheet. Also, the yield strengths of MWNTs are subject to the same limitations discussed above with respect to tube slippage. All the discussion here relates to ideal nanotubes; real carbon nanotubes may contain faults of various types that will influence their properties and require experimental measurements of their mechanical constants.

2.2 *Bending of tubes*

Due to the high in-plane tensile strength of graphite, we can expect SW and MW nanotubes to have large bending constants because these depend, for small deflections, only on the Young's modulus. Indeed, the TEM photos of MWNTs show them to be very straight, which indicates that they are very rigid. In the few observed examples of sharply bent MWNTs, they appear to be buckled on the inner radius of the bend as shown in Fig. 1. Sharp bends can also be produced in NTs by introducing faults, such as pentagon-heptagon pairs as suggested by theorists[12], and these are occasionally also seen in TEM photos. On the other hand, TEM photos of SWNTs show them to be much more pliable, and high curvature bends without buckling are seen in many photos of web material contain-

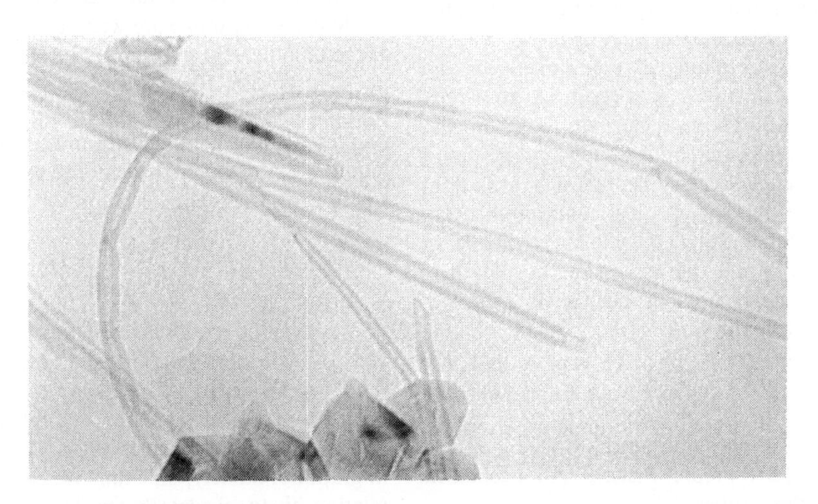

Fig. 1a. Low-resolution TEM photograph of a bent MWNT showing kinks along the inner radius of the bend resulting from bending stress that exceeds the elastic limit of the tube.

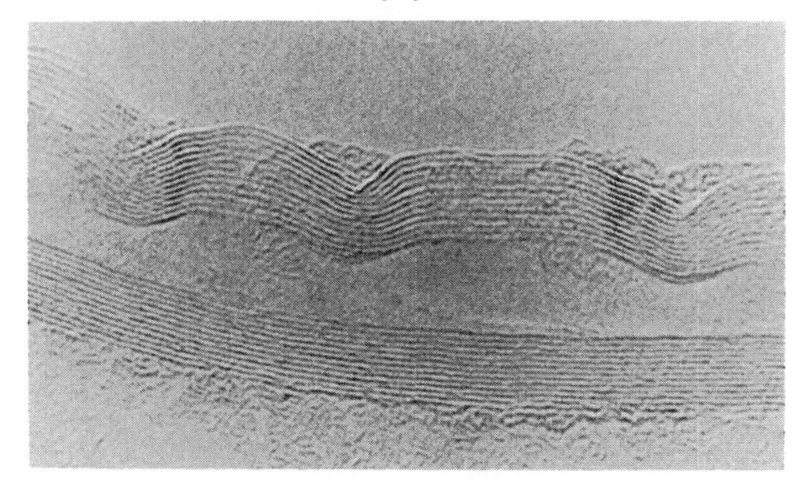

Fig. 1b. HRTEM of a bent tube (not the same as 1a) showing the strain in the region of the kinks, including a stress fracture; note the compression of the layers at the kinks and their expansion in the regions between kinks.

ing SWNTs. Alignment of tubes in a composite matrix caused by slicing of the matrix has indicated that the thinner MW tubes are also quite flexible[13].

Considering a nanotube to be a graphite cylinder means that the extremely high elastic constant of in-plane graphite (C_{11} = 1060 GPa) can be used as the Young's modulus for calculating both the elastic bending and the extension of NTs. Thus, one can use the standard beam deflection formula[14] to calculate the bending of a tube under an applied force. For example, the deflection of a cantilever beam of length l with a force f exerted at its free end is given by

$$d = fl^3/(3EI) \quad (2)$$

where E is the Young's modulus and I is the areal moment of inertia of the cross-section of the tube about its central axis, $I = \pi(r_2^4 - r_1^4)/4$. For a typical 10-layer MWNT with an inner diameter of 3 nm, an outer diameter of 6.5 nm, and length of 1 μm, the deflection would be 2.3 nm/μdyne. This calculation assumes that the 10 SWNTs that make up this MWNT act as a single, uniform, homogeneous medium.

Overney *et al.*[15] calculated the rigidity of short SW tubes using *ab initio* local density calculations to determine the parameters in a Keating potential. The Young's modulus resulting from this calculation is about 1500 GPa, which is in very good agreement with the continuum value of 1060 GPa. Again, it appears that use of the continuum model of MWNTs and SWNTs based on the properties of the graphene sheet is well justified. It is important to recognize that in calculating the moment of inertia of a single walled tube, one must consider the wall thickness of the tube to be 0.34 nm (i.e., the normal graphite layer separation). Thus, a typical 1 μm long single wall tube with a diameter of 1.1 nm will deflect 16 nm/ndyne; indeed, SWNTs are much more flexible than the thicker MWNTs, an observation that is well documented by the TEM photos of these tubes.

One can calculate the vibrational frequency of a cantilevered SWNT of length 1 μm with the bending force constant. The fundamental vibrational frequency in this case is about 12 MHz, in a range that is easily observable by electrical methods. This range suggests a possible means of measuring the mechanical properties when individual isolated tubes are cantilever-mounted to a larger body and can be readily manipulated.

The mechanical properties of the NTs have not as yet been experimentally studied because the difficulty of getting pure samples free of amorphous, graphitic, and polyhedral carbon particles and the need to characterize the tubules (e.g., their size and number of layers). However, rapid progress is being made on the production, purification, and isolation of nanotubes so that it is likely that some definitive measurements will appear in the near future. Recent demonstrations of alignment of nanotubes using polymer matrices are showing promise as a method for alignment and separation and may provide a means to investigate the mechanical properties of individual, as well as assemblies of, SWNTs and MWNTs[13,16].

Work on the production and oxidation of SWNT samples at SRI and other laboratories has led to the observation of very long bundles of these tubes, as can be seen in Fig. 2. In the cleanup and removal of the amorphous carbon in the original sample, the SWNTs self-assemble into aligned cable structures due to van der Waals forces. These structures are akin to the SW nanotube crystals discussed by Tersoff and Ruoff; they show that van der Waals forces can flatten tubes of diameter larger than 2.5 nm into a hexagonal cross-sectional lattice or honeycomb structure[17].

Since most SWNTs have diameters in the range of 1–2 nm, we can expect them to remain cylindrical when they form cables. The stiffness constant of the cable structures will then be the sum of the stiffness constants of the SWNTs. However, just as with MWNTs, the van der Waals binding between the tubes limits tensile strength unless the ends of all the tubes can be fused to a load. In the case of bending, a more exact

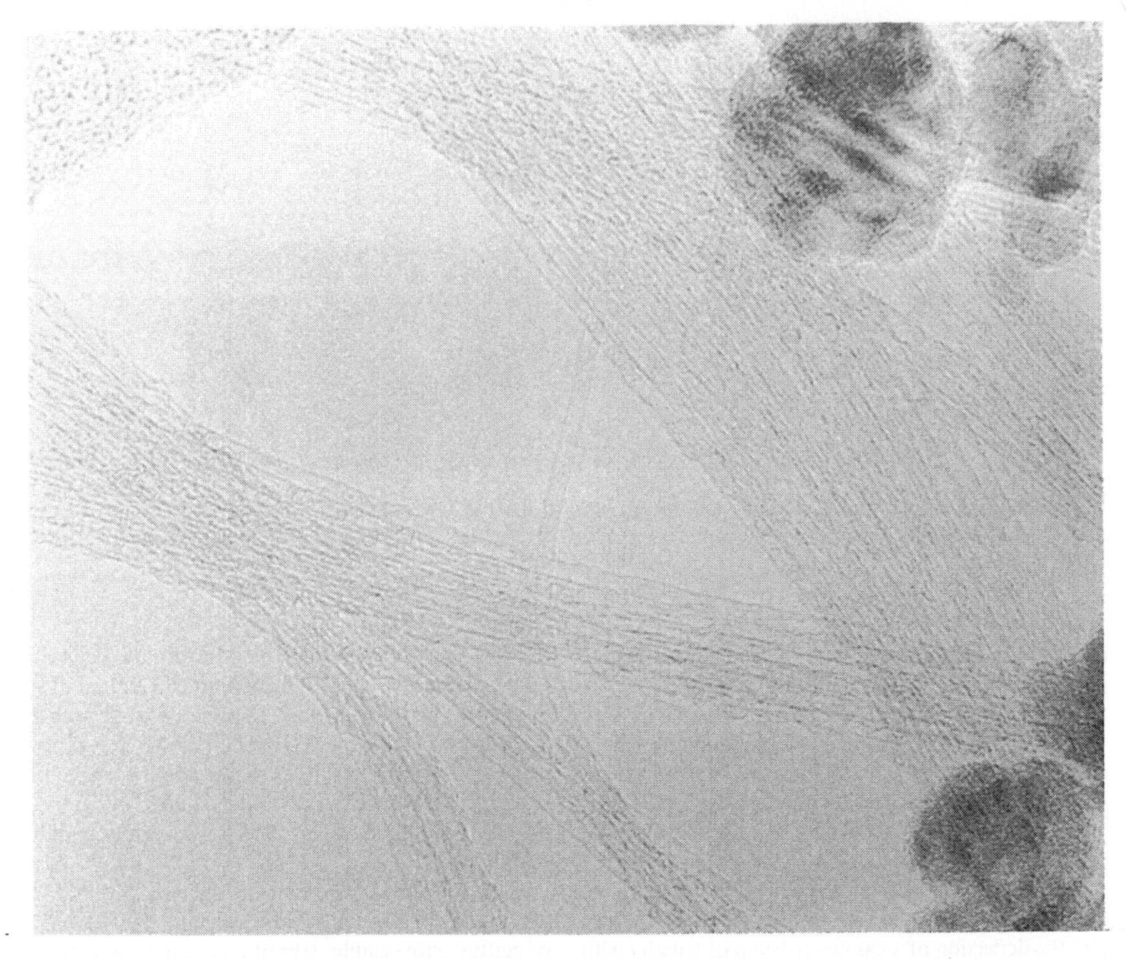

Fig. 2. Cables of parallel SWNTs that have self-assembled during oxidative cleanup of arc-produced soot composed of randomly oriented SWNTs imbedded in amorphous carbon. Note the large cable consisting of several tens of SWNTs, triple and single strand tubes bent without kinks, and another bent cable consisting of 6 to 8 SWNTs.

treatment of such cables will need to account for the slippage of individual tubes along one another as they bend. However, the bending moment induced by transverse force will be less influenced by the tube-tube binding and, thus, be more closely determined by the sum of the individual bending constants.

2.3 *Bulk modulus*

The bulk modulus of an ideal SWNT crystal in the plane perpendicular to the axis of the tubes can also be calculated as shown by Tersoff and Ruoff and is proportional to $D^{1/2}$ for tubes of less than 1.0 nm diameter[17]. For larger diameters, where tube deformation is important, the bulk modulus becomes independent of D and is quite low. Since modulus is independent of D, close-packed large D tubes will provide a very low density material without change of the bulk modulus. However, since the modulus is highly nonlinear, the modulus rapidly increases with increasing pressure. These quantities need to be measured in the near future.

3. THERMAL PROPERTIES

The thermal conductivity and thermal expansion of carbon nanotubes are also fundamentally interesting and technologically important properties. At this stage, we can infer possible behavior from the known in-plane properties of graphite.

The in-plane thermal conductivity of pyrolytic graphite is very high, second only to type II-a diamond, which has the highest measured thermal conductivity of any material[18]. The c-axis thermal conductivity of graphite is, as one might expect, very low due to the weakly bound layers which are attracted to each other only by van der Waals forces. Contributions to a finite in-plane thermal conductivity in graphite have been discussed by several authors[7,19]. At low temperature ($<$140 K), the main scattering mechanism is phonon scattering from the edges of the finite crystallites[19].

Unlike materials such as mica, extremely large *single crystal* graphite has not been possible to grow. Even in highly oriented pyrolytic graphite (HOPG), the in-plane coherence length is typically $<$1000 Å and, at low temperatures, the phonon free path is controlled mainly by boundary scattering; at temperatures above 140 K, phonon-phonon (umklapp processes) dominate [20]. TEM images suggest that defect-free tubes exist with lengths exceeding several microns, which is significantly longer than the typical crystallite diameter present in pyrolytic graphite. Therefore, it is possible that the on-axis thermal conductivity of carbon nanotubes could exceed that of type II-a diamond.

Because direct calculation of thermal conductivity is difficult[21], experimental measurements on composites with nanotubes aligned in the matrix could be a first step for addressing the thermal conductivity of carbon nanotubes. High on-axis thermal conductivities for CCVD high-temperature treated carbon fibers have been obtained, but have not reached the in-plane thermal conductivity of graphite (ref. [3], Fig. 5.11, p. 115). We expect that the radial thermal conductivity in MWNTs will be very low, perhaps even lower than the c-axis thermal conductivity of graphite.

The thermal expansion of carbon nanotubes will differ in a fundamental way from carbon fibers and from graphite as well. Ruoff[5] has shown that the radial thermal expansion coefficient of MWNTs will be essentially identical to the on-axis thermal expansion coefficient, even though the nested nanotubes in a MWNT are separated by distances similar to the interplanar separation in graphite and the forces between nested tubes are also only van der Waals forces. The explanation is simple and based on topology: unlike graphene sheets in graphite, the nanotube sheet is *wrapped onto itself* so that radial expansion is governed entirely by the carbon covalent bonding network; the van der Waals interaction between nested cylinders is, therefore, incidental to the radial thermal expansion. We, therefore, expect that the thermal coefficient of expansion will be isotropic, in a defect-free SWNT or MWNT.

Stress patterns can develop between fibers and matrix in fiber-matrix composites, as a result of differential thermal expansion during composite production. An isotropic thermal coefficient of expansion for carbon nanotubes may be advantageous in carbon-carbon composites, where stress fields often result when commercial high-temperature treated carbon fibers expand (and contract) significantly more radially than longitudinally on heating (and cooling)[22]. The carbon matrix can have a thermal expansion similar to the in-plane thermal expansion of graphite (it is graphitized), and undesirable stress-induced fracture can result; this problem may disappear with NTs substituted for the carbon fibers. However, the very low thermal expansion coefficient expected for defect-free nanotubes may be a problem when bonding to a higher thermal expansion matrix, such as may be the case for various plastics or epoxies, and may cause undesirable stresses to develop.

3.1 *Application of carbon nanotubes for high strength composite materials*

It is widely perceived that carbon nanotubes will allow construction of composites with extraordinary strength:weight ratios, due to the inherent strength of the nanotubes. Several "rules of thumb" have been developed in the study of fiber/matrix composites. Close inspection of these shows that carbon nanotubes satisfy several criteria, but that others remain untested (and therefore unsatisfied to date). High-strength composites involving carbon nanotubes and plastic, epoxy, metal, or carbon matrices remain on the horizon at the time of this review.

The ultimate tensile strength of a uniaxially aligned fiber-reinforced composite is given to reasonable accuracy by the rule of mixtures relation:

$$\sigma_c = \sigma_F V_F + \sigma'_m(1 - V_F), \quad (3)$$

where σ_c is the composite tensile strength, σ_F is the ultimate tensile strength of the fibers, σ'_m is the matrix stress at the breaking strain of the fibers, and V_F is the volume fraction of fibers in the composite. This rule holds, provided that

1. The "critical volume fraction" is exceeded,
2. the strength distribution or average strength of the fibers is known,
3. the dispersion of fibers in the matrix is free of nonuniformities that are a consequence of the fabrication process and that would give rise to stress-concentrating effects,
4. the aspect ratio of the fibers is sufficient for the matrix type,
5. the fiber is bound to the matrix with a high-strength, continuous interface.

The five factors mentioned above are discussed in detail in ref. [23] and we mention only briefly factors 2 through 4 here, and then discuss factor 5 at some length.

The strength distribution of carbon nanotubes, factor 2, could be estimated by a statistical fit to the inner and outer diameter of many (typically 100 or more nanotubes imaged in TEM micrographs) nanotubes in a sample. From such a statistical distribution of nanotube geometries, a strength distribution can be calculated from eqns. discussed above. Factor 3 is a fabrication issue, which does not pose a serious problem and will be addressed in the future by experiments. TEM micrographs have shown SWNTs with aspect ratios exceeding 1000, and a typical number for nanotubes would be 100 to 300. In this range of aspect ratios, the composite strength could approach that of a composite filled with continuous filaments, whose volume fraction is given by eqn (3), factor 4.

Factor 5 is an important issue for future experiments, and binding to a nominally smooth hexagonal bonding network in a nanotube could be a challenging endeavor. We suggest preliminary experiments to see if it is possible to convert some or all of the 3-coordinated C atoms in carbon nanotubes to tetravalent C atoms (e.g., by fluorination or oxidation). By analogy, fluorinated and oxygenated graphites have been made[24]. However, nanotubes may provide a strong topological constraint to chemical functionalization *due to the graphene sheet being wrapped onto itself*. The planes in graphite can "buckle" at a local level, with every neighboring pair of C atoms projecting up and then down due to conversion to sp^3 bonding. Can such buckling be accommodated on the surface of a carbon nanotube? For a MWNT, it seems very unlikely that the outer tube can buckle in this way, because of the geometric constraint that the neighboring tube offers; in graphite, expansion in the c direction occurs readily, as has been shown by intercalation of a wide range of atoms and molecules, such as potassium. However, Tanaka *et al.*[25] have shown that samples of MWNTs purified by extensive oxidation (and removal of other carbon types present, such as carbon polyhedra), do not intercalate *K* because sufficient expansion of the interlayer separation in the radial direction is impossible in a nested MWNT.

Achieving a *continuous* high strength bonding of defect-free MWNTs at their interface to the matrix, as in the discussion above, may simply be impossible. If our argument holds true, efforts for high-strength composites with nanotubes might better be concentrated on SWNTs with open ends. The SWNTs made recently are of small diameter, and some of the strain at each C atom could be released by local conversion to tetravalent bonding. This conversion might be achieved either by exposing both the inner and outer surfaces to a gas such as F_2(g) or through reaction with a suitable solvent that can enter the tube by wetting and capillary action[26–28]. The appropriately pretreated SWNTs might then react with the matrix to form a strong, continuous interface. However, the tensile strength of the chemically modified SWNT might differ substantially from the untreated SWNT.

The above considerations suggest caution in use of the *rules of mixtures*, eqn (3), to suggest that ultrastrong composites will form just because carbon nanotube samples distributions are now available with favorable strength and aspect ratio distributions. *Achieving a high strength, continuous interface between nanotube and matrix may be a high technological hurdle to leap*. On the other hand, other applications where reactivity should be minimized may be favored by the geometric constraints mentioned above. For example, contemplate the oxidation resistance of carbon nanotubes whose ends are in some way terminated with a special oxidation resistant cap, and compare this possibility with the oxidation resistance of graphite. The oxidation resistance of such capped nanotubes could far exceed that of graphite. Very low chemical reactivities for carbon materials are desirable in some circumstances, including use in electrodes in harsh electrochemical environments, and in high-temperature applications.

Acknowledgements—The authors are indebted to S. Subramoney for the TEM photographs. Part of this work was conducted in the program, "Advanced Chemical Processing Technology," consigned to the Advanced Chemical Processing Technology Research Association from the New Energy and Industrial Technology Development Organization, which is carried out under the Industrial Science and Technology Frontier Program enforced by the Agency of Industrial Science and Technology, The Ministry of International Trade and Industry, Japan.

REFERENCES

1. References to other papers in this issue.
2. T. W. Ebbesen, P. M. Ajayan, H. Hiura, and K. Tanigaki, *Nature* **367**, 519 (1994). K. Uchida, M. Yumura, S. Oshima, Y. Kuriki, K. Yase, and F. Ikazaki, *Proceedings 5th General Symp. on C_{60}*, to be published in *Jpn. J. Appl. Phys.*
3. R. Bacon, *J. Appl. Phys.* **31**, 283 (1960).
4. J. Tersoff, *Phys. Rev. B.* **46**, 15546 (1992).
5. R. S. Ruoff, *SRI Report* #MP 92-263, Menlo Park, CA (1992).
6. J. W. Mintmire, D. H. Robertson, and C. T. White, In *Fullerenes: Recent Advances in the Chemistry and Physics of Fullerenes and Related Materials*, (Edited by K. Kadish and R. S. Ruoff), p. 286. The Electrochemical Society, Pennington, NJ (1994).
7. B. T. Kelly, *Physics of Graphite*. Applied Science, London (1981).
8. J. C. Charlier and J. P. Michenaud, *Phys. Rev. Lett.* **70**, 1858 (1993).
9. M. Ge and K. Sattler, *J. Phys. Chem. Solids* **54**, 1871 (1993).
10. M. S. Dresselhaus, G. Dresselhaus, K. Sugihara, I. L. Spain, and H. A. Goldberg, In *Graphite Fibers and Filaments* p. 120. (Springer Verlag (1988).
11. C. A. Coulson, *Valence*. Oxford University Press, Oxford (1952).
12. B. Dunlap, In *Fullerenes: Recent Advances in the Chemistry and Physics of Fullerenes and Related Materials*, (Edited by K. Kádish and R. S. Ruoff), p. 226. The Electrochemical Society, Pennington, NJ (1994).
13. P. M. Ajayan, O. Stephan, C. Colliex, and D. Trauth, *Science* **265**, 1212 (1994).
14. R. A. Beth, Statics of Elastic Bodies, In *Handbook of Physics*, (Edited by E. U. Condon and H. Odishaw). McGraw-Hill, New York (1958).
15. G. Overney, W. Zhong, and D. Tomanek, *Zeit. Physik D* **27**, 93 (1993).
16. M. Yumura, MRS Conference, Boston, December 1994, private communication.
17. J. Tersoff and R. S. Ruoff, *Phys. Rev. Lett.* **73**, 676 (1994).
18. *CRC Handbook of Chemistry and Physics* (Edited by David R. Lide) 73rd edition, p. 4–146. CRC Press, Boca Raton (1993).
19. M. S. Dresselhaus, G. Dresselhaus, K. Sugihara, I. L. Spain, and H. A. Goldberg, *Graphite Fibers and Filaments*, Springer Series in Materials Science, Vol. 5 p. 117. Springer Verlag, Berlin (1988).
20. J. Heremans, I. Rahim, and M. S. Dresselhaus, *Phys. Rev. B* **32**, 6742 (1985).
21. R. O. Pohl, private communication.
22. G. Rellick, private communication.
23. *Whisker Technology* (Edited by A. P. Levitt), Chap. 11, Wiley-Interscience, New York (1970).
24. F. A. Cotton, and G. Wilkinson, *Advanced Inorganic Chemistry* (2nd edition, Chap. 11, John Wiley & Sons, New York (1966).
25. K. Tanaka, T. Sato, T. Yamake, K. Okahara, K. Vehida, M. Yumura, N. Niino, S. Ohshima, Y. Kuriki, K. Yase, and F. Ikazaki. *Chem. Phys. Lett.* **223**, 65 (1994).
26. E. Dujardin, T. W. Ebbesen, H. Hiura, and K. Tanigaki, *Science* **265**, 1850 (1994).
27. S. C. Tsang, Y. K. Chen, P. J. F. Harris, and M. L. H. Green, *Nature* **372** 159 (1994).
28. K. C. Hwang, *J. Chem. Soc. Chem. Comm.* 173 (1995).

Flexibility of graphene layers in carbon nanotubes

J.F. DESPRES and E. DAGUERRE
Laboratoire Marcel Mathieu, 2, avenue du President Pierre Angot
64000 Pau, France

K. LAFDI
Materials Technology Center, Southern Illinois University at Carbondale,
Carbondale, IL 62901-4303

(Received 16 September 1994; accepted in revised form 9 November 1994)

Key Words - Buckeytubes; nanotubes; graphene layers

The Kratschmer-Huffman technique [1] has been widely used to synthesize fullerenes. In this technique, graphite rods serve as electrodes in the production of a continuous dc electric arc discharge within an inert environment. When the arc is present, carbon evaporates from the anode and a carbon slag is deposited on the cathode. In 1991, Ijima *et al.* [2] examined samples of this slag. They observed a new form of carbon which has a tubular structure. These structures, called nanotubes, are empty tubes made of perfectly coaxial graphite sheets and generally have closed ends. The number of sheets may vary from a single sheet to as many as one hundred sheets. The tube length can also vary; and the diameters can be several nanometers. The tube ends are either spherical or polyhedral. The smallest nanotube ever observed consisted of a single graphite sheet with a 0.75 nm diameter [2].

Electron diffraction studies [3] have revealed that hexagons within the sheets are helically wrapped along the axis of the nanotubes. The interlayer spacing between sheets is 0.34 nm which is slightly larger than that of graphite (0.3354 nm). It was also reported [2] that the helicity aspect may vary from one nanotube to another. Ijima *et al.* [2] also reported that in addition to nanotubes, polyhedral particles consisting of concentric carbon sheets were also observed.

An important question relating to the structure of nanotubes is: Are nanotubes made of embedded closed tubes, like "Russian dolls," or are they composed of a single graphene layer which is spirally wound, like a roll of paper? Ijima *et al.* [2] espouse the "Russian doll" model based on TEM work which shows that the same number of sheets appear on each side of the central channel. Dravid et al. [4], however, support a "paper roll" structural model for nanotubes.

Determination of the structure of nanotubes is crucial for two reasons: (1) to aid understanding the nanotube growth mechanism and (2) to anticipate whether intercalation can occur. Of the two models, only the paper roll structure can be intercalated.

The closure of the graphite sheets can be explained by the substitution of pentagons for hexagons in the nanotube sheets. Six pentagons are necessary to close a tube (and Euler's Rule is not violated). Hexagon formation requires a two-atom addition to the graphitic sheet while a pentagon formation requires only one. Pentagon formation may be explained by a temporary reduction in carbon during current fluctuations of the arc discharge. More complex defaults (beyond isolated pentagons and hexagons) may be possible. Macroscopic models have been constructed by Conard *et al.* [5] to determine the angles that would be created by such defaults.

To construct a nanotube growth theory, a new approach, including some new properties of nanotubes, must be taken. The purpose of this work is to present graphene layer flexibility as a new property of graphitic materials. In previous work, the TEM characterization of nanotubes consists of preparing the sample by dispersing the particles in alcohol (ultrasonic preparation). When the particles are dispersed in this manner, individual nanotubes are observed in a stress-free state, i.e. without the stresses that would be present due to other particles in an agglomeration. If one carefully prepares a sample without using the dispersion technique, we expect that a larger variety of configurations may be observed.

Several carbon shapes are presented in Figure 1 in which the sample has been prepared without using ultrasonic preparation. In this figure, there are three polyhedral entities (in which the two largest entities belong to the same family) and a nanotube. The bending of the tube occurs over a length of several hundred nanometers and results in a 60° directional change. Also, the general condition of the tube walls has been modified by local buckling, particularly in compressed areas. Figure 2 is a magnification of this compressed area. A contrast intensification in the tensile area near the compression can be observed in this unmodified photograph. The inset in Figure 2 is a drawing which illustrates the compression of a plastic tube. If the tube is initially straight, buckling occurs on the concave side of the nanotubes as it is bent. As shown in Figure 3, this fact is related to the degree of curvature of the nanotube at a given location. Buckling is not observed in areas where the radius of curvature is large, but a large degree of buckling is observed in severely bent regions.

These TEM photographs are interpreted as

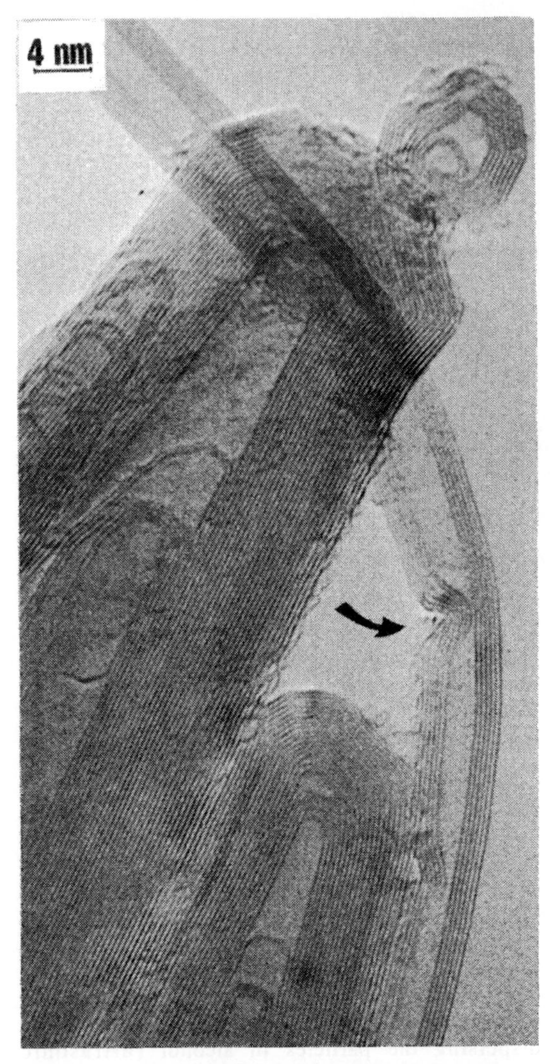

Figure 1. Lattice fringes LF 002 of nanotube particles.

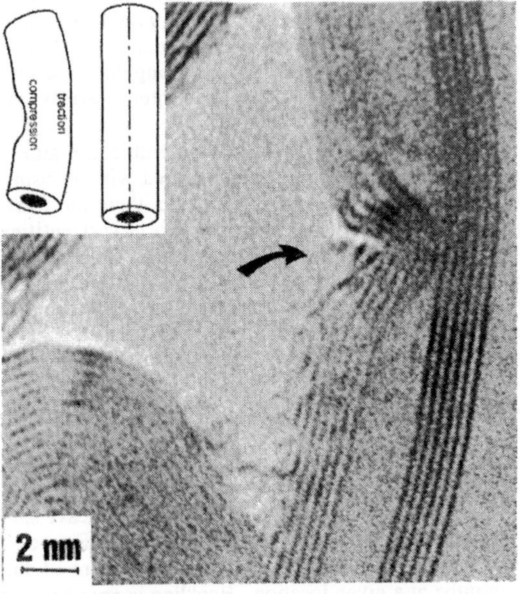

Figure 2. Details of Figure 2 and an inset sketch illustrating what happens before and after traction.

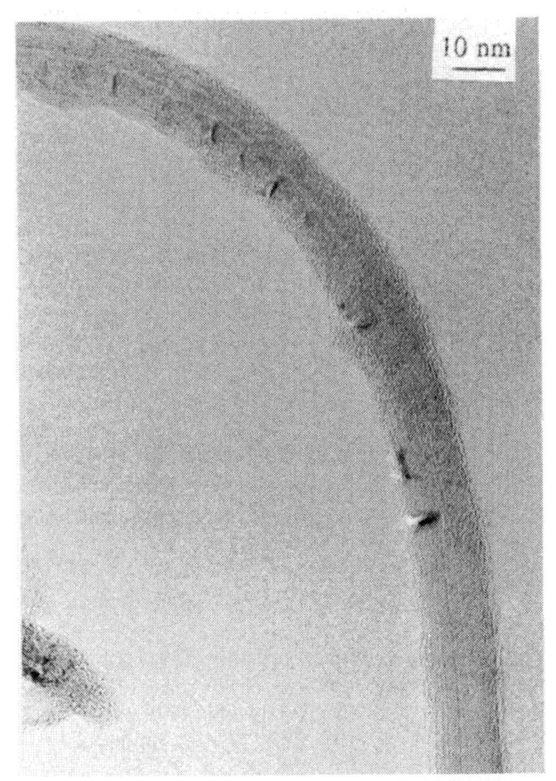

Figure 3. Lattice fringes LF 002 of buckled nanotube particles.

follows: the tube, which is initially straight, is subjected to bending during the preparation of the TEM grid. The stress on the concave side of the tube results in buckling. The buckling extends into the tube until the effect of the stress on the tube is minimized. The effect of this buckling on the graphene layers on the convex side is that they are stretched and become flattened because this is the only way to minimize damage. This extension results in a large coherent volume which causes the observed increase in contrast. On the concave side of the tube, damage is minimized by shortening the graphene layer length in the formation of a buckling location.

We observe that compression and its associated buckling instability only on the concave side of the tube, but never on the convex side. This result suggests that it is only necessary to consider the flexibility of the graphene layers; and, thus, there is no need to invoke the notion of defects due to the substitution of pentagons and hexagons. In the latter case, we would expect to observe the buckling phenomenon on both sides of the nanotube upon bending. Thus, it is clear that further work must be undertaken to study the flexibility of graphene layers since, from the above results, it is possible to conclude that graphene layers are not necessarily rigid and flat entities. These entities do not present undulations or various forms only as a result of the existence of atomic and/or structural defects. The time has come to discontinue the use of the description of graphene layers based on rigid, coplanar chemical bonds (with 120° angles)! A model of graphene layers which under mechanical stress, for example, results in the modification of bond angles and bond length values induce observed curvature effects (without using any structural modifications such as pentagon substitution for hexagons) may be more appropriate.

Acknowledgments - Stimulating discussions with Dr. H. Marsh, M. Wright and D. Marx are acknowledged.

REFERENCES

1. W. Kratshmer and D.R. Huffman, *Chem. Phys. Letter*, **170**, 167 (1990).
2. S. Ijima and P. Ajayan, *Physical Review Letters*, **69**, 3010 (1992).
3. C.T. White, *Physical Review B*, **479**, 5488.
4. V. Dravid and X. Lin, *Science*, **259**, 1601 (1993).
5. C. Clinard, J.N. Rouzaud, S. Delpeux, F. Beguin and J. Conard, *J. Phys. Chem. Solids*, **55**, 651 (1994).

NANOPARTICLES AND FILLED NANOCAPSULES

YAHACHI SAITO
Department of Electrical and Electronic Engineering, Mie University, Tsu 514 Japan

(*Received* 11 *October* 1994; *accepted in revised form* 10 *February* 1995)

Abstract—Encapsulation of foreign materials within a hollow graphitic cage was carried out for rare-earth and iron-group metals by using an electric arc discharge. The rare-earth metals with low vapor pressures, Sc, Y, La, Ce, Pr, Nd, Gd, Tb, Dy, Ho, Er, Tm, and Lu, were encapsulated in the form of carbides, whereas volatile Sm, Eu, and Yb metals were not. For iron-group metals, particles in metallic phases (α-Fe, γ-Fe; hcp-Co, fcc-Co; fcc-Ni) and in a carbide phase (M_3C, M = Fe, Co, Ni) were wrapped in graphitic carbon. The excellent protective nature of the outer graphitic cages against oxidation of the inner materials was demonstrated. In addition to the wrapped nanoparticles, exotic carbon materials with hollow structures, such as single-wall nanotubes, bamboo-shaped tubes, and nanochains, were produced by using transition metals as catalysts.

Key Words—Nanoparticles, nanocapsules, rare-earth elements, iron, cobalt, nickel.

1. INTRODUCTION

The carbon-arc plasma of extremely high temperatures and the presence of an electric field near the electrodes play important roles in the formation of nanotubes[1,2] and nanoparticles[3]. A nanoparticle is made up of concentric layers of closed graphitic sheets, leaving a nanoscale cavity in its center. Nanoparticles are also called nanopolyhedra because of their polyhedral shape, and are sometimes dubbed as nanoballs because of their hollow structure.

When metal-loaded graphite is evaporated by arc discharge under an inactive gas atmosphere, a wide range of composite materials (e.g., filled nanocapsules, single-wall tubes, and metallofullerenes, $R@C_{82}$, where R = La, Y, Sc,[4–6]) are synthesized. Nanocapsules filled with LaC_2 crystallites were discovered in carbonaceous deposits grown on an electrode by Ruoff *et al.*[7] and Tomita *et al.*[8]. Although rare-earth carbides are hygroscopic and readily hydrolyze in air, the carbides nesting in the capsules did not degrade even after a year of exposure to air. Not only rare-earth elements but also 3*d*-transition metals, such as iron, cobalt, and nickel, have been encapsulated by the arc method. Elements that are found, so far, to be incapsulated in graphitic cages are shown in Table 1.

In addition to nanocapsules filled with metals and carbides, various exotic carbon materials with hollow structures, such as single-wall (SW) tubes[9,10], bamboo-shaped tubes, and nanochains[11], are produced by using transition metals as catalysts.

In this paper, our present knowledge and understanding with regard to nanoparticles, filled nanocapsules, and the related carbon materials are described.

2. PREPARATION PROCEDURES

Filled nanocapsules, as well as hollow nanoparticles, are synthesized by the dc arc-evaporation method that is commonly used to synthesize fullerenes and nanotubes. When a pure graphite rod (anode) is evaporated in an atmosphere of noble gas, macroscopic quantities of hollow nanoparticles and multi-wall nanotubes are produced on the top end of a cathode. When a metal-packed graphite anode is evaporated, filled nanocapsules and other exotic carbon materials with hollow structures (e.g., "bamboo"-shaped tubes, nanochains, and single-wall (SW) tubes) are also synthesized. Details of the preparation procedures are described elsewhere[8,11,12].

3. NANOPARTICLES

Nanoparticles grow together with multi-wall nanotubes in the inner core of a carbonaceous deposit formed on the top of the cathode. The size of nanoparticles falls in a range from a few to several tens of nanometers, being roughly the same as the outer diameters of multi-wall nanotubes. High-resolution TEM (transmission electron microscopy) observations reveal that polyhedral particles are made up of concentric graphitic sheets, as shown in Fig. 1. The closed polyhedral morphology is brought about by well-developed graphitic layers that are flat except at the corners and edges of the polyhedra. When a pentagon is introduced into a graphene sheet, the sheet curves positively and the strain in the network structure is localized around the pentagon. The closed graphitic cages produced by the introduction of 12 pentagons will exhibit polyhedral shapes, at the corners of which the pentagons are located. The overall shapes of the polyhedra depend on how the 12 pentagons are located. Carbon nanoparticles actually synthesized are multilayered, like a Russian doll. Consequently, nanoparticles may also be called gigantic multilayered fullerenes or gigantic hyper-fullerenes[13].

The spacings between the layers (d_{002}) measured by selected area electron diffraction were in a range of 0.34 to 0.35 nm[3]. X-ray diffraction (XRD) of the cathode deposit, including nanoparticles and nano-

Table 1. Formation of filled nanocapsules. Elements in shadowed boxes are those which were encapsulated so far. M and C under the chemical symbols represent that the trapped elements are in metallic and carbide phases, respectively. Numbers above the symbols show references.

Li	Be											B	C	N	O	F	Ne
Na	Mg											Al	Si	P	S	Cl	Ar
K	Ca	11, 21 Sc C	Ti	47 V C	Cr C	Mn	11, 29 Fe M,C	11 Co M,C	11, 34 Ni M,C	Cu M	Zn	Ga	Ge	As	Se	Br	Kr
Rb	Sr	12, 23 Y C	47 Zr C	Nb	Mo C	Tc	Ru	Rh	Pd	Ag	Cd	In	Sn	Sb	Te	I	Xe
Cs	Ba	lantha-nides	Hf	48 Ta C	W	Re	Os	Ir	Pt	49 Au M	Hg	Tl	Pb	Bi	Po	At	Rn
Fr	Ra	Ac	25 Th C	Pa	26 U C	Np	Pu	Am	Cm	Bk	Cf	Es	Fm	Md	No	Lr	

7, 8 La C	11, 12 Ce C	11, 12 Pr C	11, 12 Nd C	Pm	Sm	Eu	11, 12 Gd C	11, 12 Tb C	11, 12 Dy C	11, 12 Ho C	11, 12 Er C	12 Tm C	Yb	11, 12 Lu C

tubes, gave $d_{002} = 0.344$ nm[14], being consistent with the result of electron diffraction. The interlayer spacing is wider by a few percent than that of the ideal graphite crystal (0.3354 nm). The wide interplanar spacing is characteristic of the turbostratic graphite[15].

Figure 2 illustrates a proposed growth process[3] of a polyhedral nanoparticle, along with a nanotube. First, carbon neutrals (C and C_2) and ions (C^+)[16] deposit, and then coagulate with each other to form small clusters on the surface of the cathode. Through an accretion of carbon atoms and coalescence between clusters, clusters grow up to particles with the size finally observed. The structure of the particles at this stage may be "quasi-liquid" or amorphous with high structural fluidity because of the high temperature ($\approx$3500 K)[17] of the electrode and ion bombardment. Ion bombardment onto the electrode surface seems to be important for the growth of nanoparticles, as well as tubes. The voltage applied between the electrodes is concentrated within thin layers just above the surface of the respective electrodes because the arc plasma is electrically conductive, and thereby little drop in voltage occurs in a plasma pillar. Near a cathode, the voltage drop of approximately 10 V occurs in a thin layer of 10^{-3} to 10^{-4} cm from the electrode surface[18]. Therefore, C^+ ions with an average kinetic energy of ~10 eV bombard the carbon particles and enhance the fluidity of particles. The kinetic energy of the carbon ions seems to affect the structure of deposited carbon. It is reported that tetrahedrally coordinated amorphous carbon films, exhibiting mechanical properties similar to diamond, have been grown by deposition of carbon ions with energies between 15 and 70 eV[19]. This energy is slightly higher than the present case, indicating that the structure of the deposited material is sensitive to the energy of the impinging carbon ions.

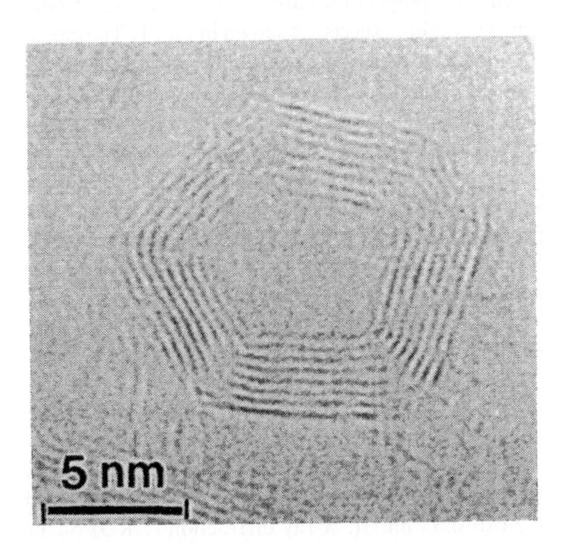

Fig. 1. TEM picture of a typical nanoparticle.

The vapor deposition and ion bombardment onto quasi-liquid particles will continue until the particles are shadowed by the growth of tubes and other particles surrounding them and, then, graphitization occurs. Because the cooling goes on from the surface to the center of the particle, the graphitization initiates on the external surface of the particle and progresses toward its center. The internal layers grow, keeping

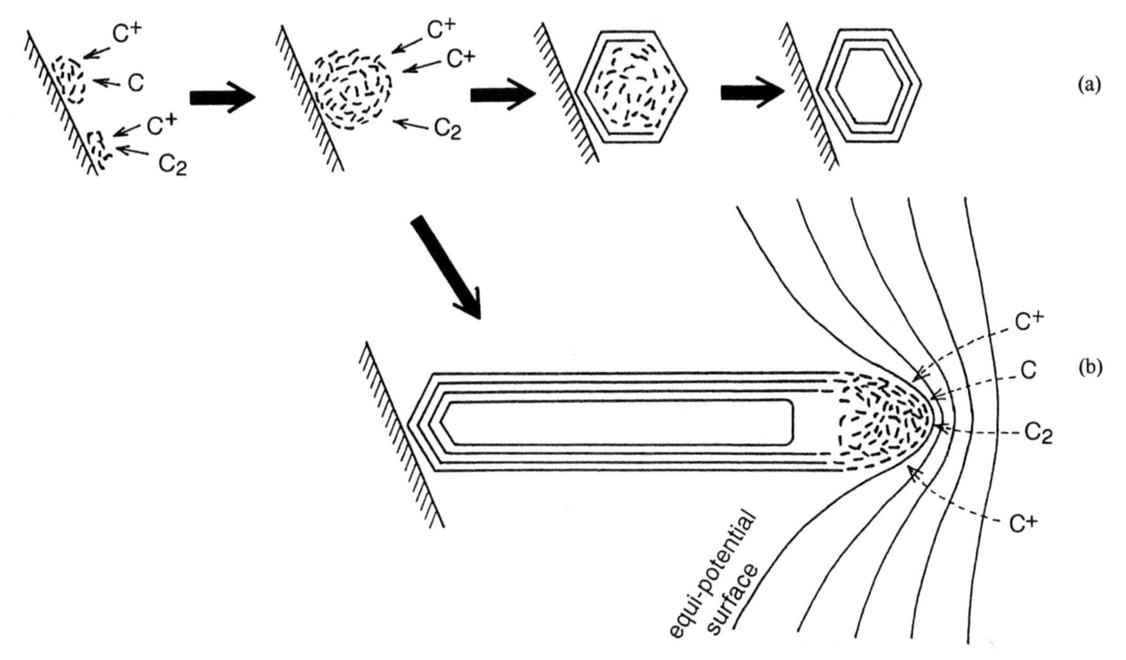

Fig. 2. A model of growth processes for (a) a hollow nanoparticle and, (b) a nanotube; curved lines depicted around the tube tip show schematically equal potential surfaces.

their planes parallel to the external layer. The flat planes of the particle consist of nets of six-member rings, while five-member rings may be located at the corners of the polyhedra. The closed structure containing pentagonal rings diminishes dangling bonds and lowers the total energy of a particle. Because the density of highly graphitized carbon (≈ 2.2 g/cm^3) is higher than that of amorphous carbon (1.3–1.5 g/cm^3), a pore will be left inevitably in the center of a particle after graphitization. In fact, the corresponding cavities are observed in the centers of nanoparticles.

4. FILLED NANOCAPSULES

4.1 *Rare earths*

4.1.1 *Structure and morphology.* Most of the rare-earth elements were encapsulated in multilayered graphitic cages, being in the form of single-domain carbides. The carbides encapsulated were in the phase of RC_2 (R stands for rare-earth elements) except for Sc, for which Sc_3C_4[20] was encapsulated[21].

A high-resolution TEM image of a nanocapsule encaging a single-domain YC_2 crystallite is shown in Fig. 3. In the outer shell, (002) fringes of graphitic layers with 0.34 nm spacing are observed and, in the core crystallite, lattice fringes with 0.307-nm spacing due to (002) planes of YC_2 are observed. The YC_2 nanocrystal partially fills the inner space of the nanocapsule, leaving a cavity inside. No intermediate phase was observed between the core crystallite and the graphitic shell. The external shapes of nanocapsules were polyhedral, like the nanoparticles discussed above, while the volume ratio of the inner space (including the volume of a core crystallite and a cavity) to the whole particle is greater for the stuffed nanocapsules than that for hollow nanoparticles. While the inner space within a hollow nanoparticle is only ~1% of the whole volume of the particle, that for a filled nanocapsule is 10 to 80% of the whole volume.

The lanthanides (from La to Lu) and yttrium form isomorphous dicarbides with a structure of the CaC_2 type (body-centered tetragonal). These lanthanide carbides are known to have conduction electrons (one

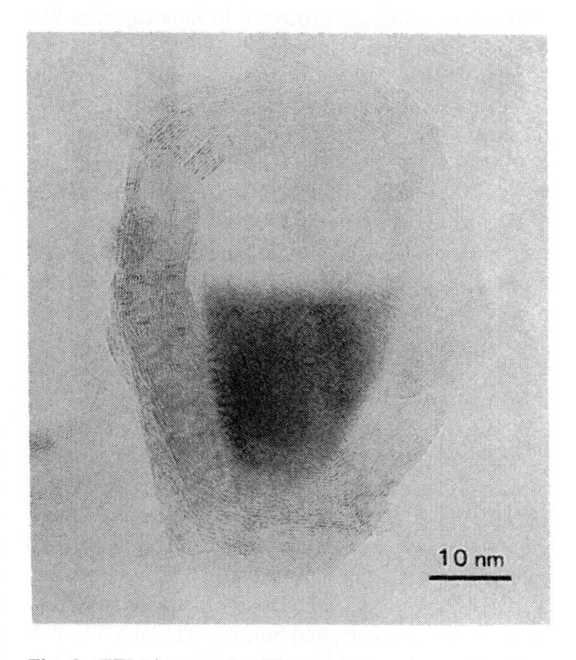

Fig. 3. TEM image of a YC_2 crystallite encapsulated in a nanocapsule.

electron per formula unit, RC_2)[22] (i.e., metallic electrical properties) though they are carbides. All the lanthanide carbides including YC_2 and Sc_3C_4 are hygroscopic; they quickly react with water in air and hydrolyze, emanating hydrogen and acetylene. Therefore, they usually have to be treated and stored in an inactive gas atmosphere or oil to avoid hydrolysis. However, the observation of intact dicarbides, even after exposure to air for over a year, shows the excellent airtight nature of nanocapsules, and supports the hypothesis that their structure is completely closed by introducing pentagons into graphitic sheets like fullerenes[23].

4.1.2 *Correlation between metal volatility and encapsulation.* A glance at Table 1 shows us that carbon nanocapsules stuffed with metal carbides are formed for most of the rare-earth metals, Sc, Y, La, Ce, Pr, Nd, Gd, Tb, Dy, Ho, Er, and Lu. Both TEM and XRD confirm the formation of encapsulated carbides for all the above elements. The structural and morphological features described above for Y are common to all the stuffed nanocapsules: the outer shell, being made up of concentric multilayered graphitic sheets, is polyhedral, and the inner space is partially filled with a single-crystalline carbide. It should be noted that the carbides entrapped in nanocapsules are those that have the highest content of carbon among the known carbides for the respective metal. This finding provides an important clue to understanding the growth mechanism of the filled nanocapsules (see below).

In an XRD profile from a Tm-C deposit, a few faint reflections that correspond to reflections from TmC_2 were observed[12]. Owing to the scarcity of TmC_2 particles, we have not yet obtained any TEM images of nanocapsules containing TmC_2. However, the observation of intact TmC_2 by XRD suggests that TmC_2 crystallites are protected in nanocapsules like the other rare-earth carbides.

For Sm, Eu, and Yb, on the other hand, nanocapsules containing carbides were not found in the cathode deposit by either TEM or XRD. To see where these elements went, the soot particles deposited on the walls of the reaction chamber was investigated for Sm. XRD of the soot produced from Sm_2O_3/C composite anodes showed the presence of oxide (Sm_2O_3) and a small amount of carbide (SmC_2). TEM, on the other hand, revealed that Sm oxides were naked, while Sm carbides were embedded in flocks of amorphous carbon[12]. The size of these compound particles was in a range from 10 to 50 nm. However, no polyhedral nanocapsules encaging Sm carbides were found so far.

Figure 4 shows vapor pressure curves of rare-earth metals[24], clearly showing that there is a wide gap between Tm and Dy in the vapor pressure-temperature curves and that the rare-earth elements are classified into two groups according to their volatility (viz., Sc, Y, La, Ce, Pr, Nd, Gd, Tb, Dy, Ho, Er, and Lu, non-volatile elements, and Sm, Eu, Tm, and Yb, volatile elements). Good correlation between the volatility and the encapsulation of metals was recently

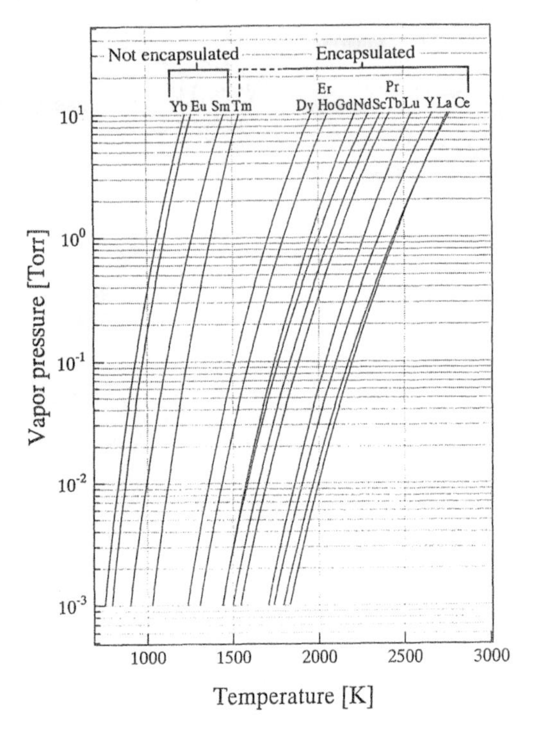

Fig. 4. Vapor pressure curves of rare-earth metals reproduced from the report of Honig[24]. Elements are distinguished by their vapor pressures. Sm, Eu, Tm, and Yb are volatile, and Sc, Y, La, Ce, Pr, Nd, Gd, Tb, Dy, Ho, Er, and Lu are non-volatile.

pointed out[12]; all the encapsulated elements belong to the group of non-volatile metals, and those not encapsulated, to the group of volatile ones with only one exception, Tm.

Although Tm is classified into the group of volatile metals, it has the lowest vapor pressure within this group and is next to the non-volatile group. This intermediary property of Tm in volatility may be responsible for the observation of trace amount of TmC_2. The vapor pressure of Tm suggests the upper limit of volatility of metals that can be encapsulated.

This correlation of volatility with encapsulation suggests the importance of the vapor pressure of metals for their encapsulation. In the synthesis of the stuffed nanocapsules, a metal-graphite composite was evaporated by arc heating, and the vapor was found to deposit on the cathode surface. A growth mechanism for the stuffed nanocapsules (see Fig. 5) has been proposed by Saito *et al.*[23] that explains the observed features of the capsules. According to the model, particles of metal-carbon alloy in a liquid state are first formed, and then the graphitic carbon segregates on the surface of the particles with the decrease of temperature. The outer graphitic carbon traps the metal-carbon alloy inside. The segregation of carbon continues until the composition of alloy reaches RC_2 (R = Y, La, ..., Lu) or Sc_2C_3, which equilibrates with graphite. The co-deposition of metal and carbon atoms on the cathode surface is indispensable for the formation of the stuffed nanocapsules. However, because the

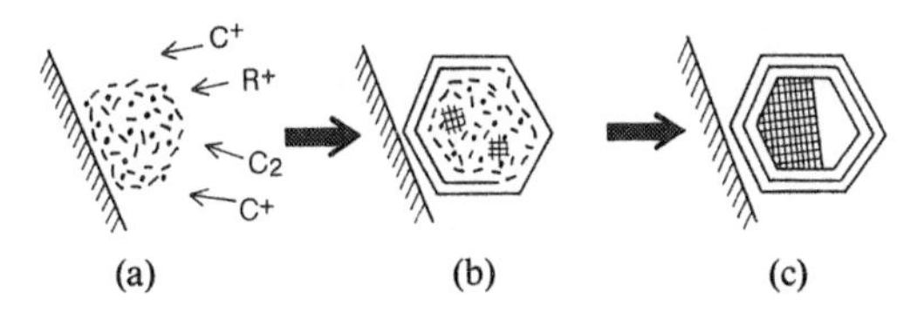

Fig. 5. A growth model of a nanocapsule partially filled with a crystallite of rare-earth carbide (RC_2 for R = Y, La, . . . , Lu; R_3C_4 for R = Sc): (a) R-C alloy particles, which may be in a liquid or quasi-liquid phase, are formed on the surface of a cathode; (b) solidification (graphitization) begins from the surface of a particle, and R-enriched liquid is left inside; (c) graphite cage outside equilibrates with RC_2 (or R_3C_4 for R = Sc) inside.

temperature of the cathode surface is as high as 3500 K, volatile metals do not deposit on a surface of such a high temperature, or else they re-evaporate immediately after they deposit. Alternatively, since the shank of an anode (away from the arc gap) is heated to a rather high temperature (e.g., 2000 K), volatile metals packed in the anode rod may evaporate from the shank into a gas phase before the metals are exposed to the high-temperature arc. For Sm, which was not encapsulated, its vapor pressure reaches as high as 1 atmosphere at 2000 K (see Fig. 4).

The criterion based on the vapor pressure holds for actinide; Th and U, being non-volatile (their vapor pressures are much lower than La), were recently found to be encapsulated in a form of dicarbide, ThC_2[25] and UC_2[26], like lanthanide.

It should be noted that rare-earth elements that form metallofullerenes[27] coincide with those that are encapsulated in nanocapsules. At present, it is not clear whether the good correlation between the metal volatility and the encapsulation found for both nanocapsules and metallofullerenes is simply a result of kinetics of vapor condensation, or reflects thermodynamic stability. From the viewpoint of formation kinetics, to form precursor clusters (transient clusters comprising carbon and metal atoms) of filled nanocapsules or metallofullerenes, metal and carbon have to condense simultaneously in a spatial region within an arc-reactor vessel (i.e., the two regions where metal and carbon condense have to overlap with each other spatially and chronologically). If a metal is volatile and its vapor pressure is too high compared with that of carbon, the metal vapor hardly condenses on the cathode or near the arc plasma region. Instead, it diffuses far away from the region where carbon condenses and, thereby, the formation of mixed precursor clusters scarcely occurs.

4.2 *Iron-group metals (Fe, Co, Ni)*

4.2.1 *Wrapped nanocrystals.* Metal crystallites covered with well-developed graphitic layers are found in soot-like material deposited on the outer surface of a cathode slag. Figure 6 shows a TEM picture of an α(bcc)-Fe particle grown in the cathode soot. Generally, iron crystallites in the α-Fe phase are faceted. The outer shell is uniform in thickness, and it usually consists of several to about 30 graphene layers[28]. Nanocapsules of the iron-group metals (Fe, Co, Ni) show structures and morphology different from those of rare-earth elements in the following ways. First, most of the core crystallites are in ordinary metallic phases (i.e., carbides are minor). The α-Fe, β(fcc)-Co and fcc-Ni are the major phases for the respective metals, and small amounts of γ(fcc)-Fe and α(hcp)-Co are also formed[11]. Carbides formed for the three metals were of the cementite phase (viz., Fe_3C, Co_3C, and Ni_3C). The quantity of carbides formed depends on the affinity of the metal toward carbon; iron forms the carbide most abundantly (about 20% of metal atoms are in the carbide phase)[29], nickel forms the least amount (on the order of 1%), and cobalt, intermediate between iron and nickel.

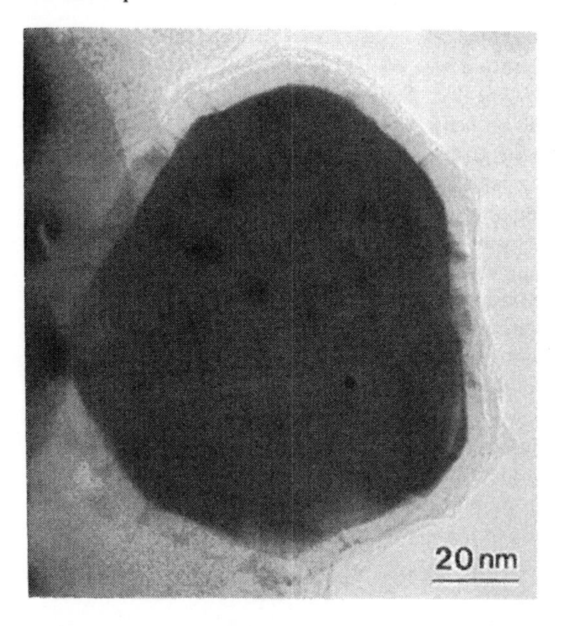

Fig. 6. TEM picture of an α-Fe particle grown in the cathode soot; the core crystallite is wrapped in graphitic carbon.

Secondly, the outer graphitic layers tightly surround the core crystallites without a gap for most of the particles, in contrast to the nanocapsules of rare-earth carbides, for which the capsules are polyhedral and have a cavity inside. The graphite layers wrapping iron (cobalt and nickel) particles bend to follow the curvature of the surface of a core crystallite. The graphitic sheets, for the most part, seem to be stacked parallel to each other one by one, but defect-like contrast suggesting dislocations, was observed[28], indicating that the outer carbon shell is made up of small domains of graphitic carbon stacked parallel to the surface of the core particle. The structure may be similar to that of graphitized carbon blacks, being composed of small segments of graphitic sheets stacked roughly parallel to the particle surface[30].

Magnetic properties of iron nanocrystals nested in carbon cages, which grew on the cathode deposit, have been studied by Hiura *et al.*[29]. Magnetization (*M-H*) curves showed that the coercive force, H_c, of this nanoscale encapsulated material at room temperature is about 80 Oe, larger than that of bulk α-Fe ($H_c \approx 1$ Oe). The particle sizes of iron studied (10 to 100 nm) and the large coercive force suggests that the magnetization process is dominated by the cooperative spin rotation of single domains. The saturation magnetization was 25 emu/g, being smaller than that for pure α-Fe (221.7 emu/g) because the measured sample contains a large amount of free carbon as well as wrapping graphite. For application, magnetic particles have to be extracted, and it is also necessary to control the size and composition of iron particles to obtain larger coercive force and magnetization.

Iron, cobalt, and nickel particles also grow in soot deposited on the chamber walls, but graphitic layers wrapping the metals are not so well-developed as those grown in the cathode soot. Figure 7 shows a TEM picture of iron particles grown in the chamber soot. They are nearly spherical in shape and are embedded in amorphous carbon globules. For some iron particles, lattice fringes (0.34–0.35 nm spacing) suggesting the presence of a few layers of graphene sheets between their surface and the outer amorphous carbon are observed, as indicated by arrows. The iron is predominantly in α-Fe phase, and minorities are in γ-Fe and Fe_3C phases. The iron particles in the chamber soot are smaller (3–10 nm) than those in the cathode soot. Much higher coercive force of 380 Oe and superparamagnetism were observed for the smaller iron nanocrystals grown in the chamber soot[31].

Majetich and coworkers have studied magnetic properties of carbon-coated Co[32], Gd_2C_3, and Ho_2C_3 nanocrystals[33] formed in the chamber soot. A brief account on the coated Co nanocrystals is given here. They extracted magnetic nanocrystals from the crude soot with a magnetic gradient field technique.

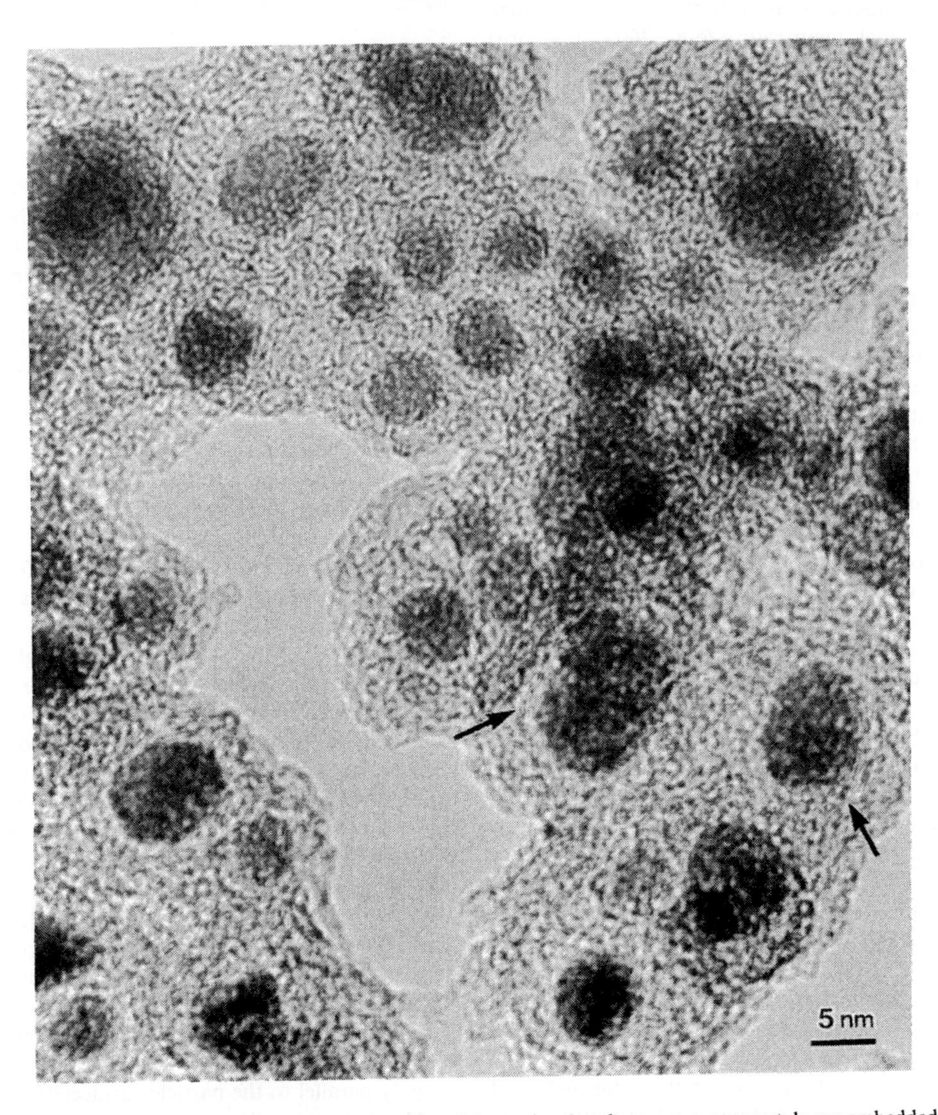

Fig. 7. TEM picture of iron nanocrystals collected from the chamber soot; nanocrystals are embedded in amorphous carbon globules. On the surface of some core crystals, a few fringes with 0.34–0.35 nm spacing suggesting the presence of graphitic layers are observed, as indicated by arrows.

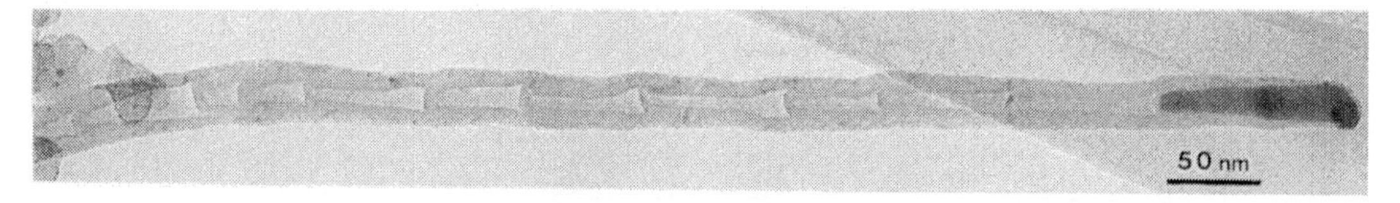

Fig. 8. TEM picture of a bamboo-shaped carbon tube.

The majority of Co nanocrystals (in the fcc phase) exist as nominally spherical particles with a 0.5–5 nm in radius. Hysteretic and temperature-dependent magnetic response, in randomly and magnetically aligned powder samples frozen in epoxy, revealed fine particle magnetism associated with single-domain Co particles. These single-domain particles exhibited superparamagnetic response with magnetic hysteresis observed only at temperatures below T_B (blocking temperature) ≈ 160 K.

4.2.2 *Bamboo-shaped tubes.* A carbon tube with a peculiar shape looking like "bamboo," produced by the arc evaporation of nickel-loaded graphite, is shown in Fig. 8. The tube consists of a linear chain of hollow compartments that are spaced at nearly equal separation from 50 to 100 nm. The outer diameter of the bamboo tubes is about 40 nm, and the length typically several μm. One end of the tube is capped with a needle-shaped nickel particle which is in the normal fcc phase, and the other end is empty. Walls of each compartment are made up by about 20 graphitic layers[34]. The shape of each compartment is quite similar to the needle-shape of the Ni particle at the tip, suggesting that the Ni particle was once at the cavities.

Figure 9 illustrates a growth process of the bamboo tubes. It is not clear whether the Ni particle at the tip was liquid or solid during the growth of the tube. We infer that the cone-shaped Ni was always at the tip and it was absorbing carbon vapor. The dissolved carbon diffused into the bottom of the Ni needle, and carbon segregated as graphite at the bottom and the side of the needle. After graphitic layers (about 20 layers) were formed, the Ni particle probably jumped out of the graphitic sheath to the top of the tube. The motive force of pushing out the Ni needle may be a stress accumulated in the graphitic sheath due to the segregation of carbon from the inside of the sheath.

The segregation process of graphite on the surface of a metal particle is similar to that proposed by Oberlin and Endo[35] for carbon fibers prepared by thermal decomposition of hydrocarbons. However, the lengthening of tubes goes on intermittently for the bamboo-shaped tubes, while the pyrolytic fibers grow continuously.

A piece of Ni metal was sometimes left in a compartment located in the middle of a bamboo tube, as shown in Fig. 10. The shape of the trapped metal is reminiscent of a drop of mercury left inside a glass capillary. The contact angle between the Ni metal and the inner wall of graphite is larger than 90° (measured angle is about 140°), indicating that the metal poorly wets the tube walls. Strong capillary action, anticipated in nanometer-sized cavities[36,37], does not seem to be enough to suck the metal into the tubes, at least for the present system[11,38].

4.2.3 *Nanochains.* Figure 11 shows a TEM picture of nanochains produced from a Ni/C composite anode[11]. The nanochains consist of spherical, hollow graphitic particles with outer diameters of 10–20 nm and inner diameters of several to 10 nm. A nanochain comprises a few tens of hollow particles that are linearly connected with each other. The inside of some particles is filled with a Ni particle, as seen in the figure. The morphology of each particle resembles that of graphitized carbon blacks, which are made up of many patches of small graphitic sheets piling up to form spherical shapes.

The chains of hollow carbon may be initially chains consisting of Ni (or carbide) particles covered with graphitic carbon. The chains lying on the hot surface of the cathode are heated, and Ni atoms evaporate through defects of the outer graphitic carbon because the vapor pressure of Ni is much higher than carbon. Thus, the carbon left forms hollow graphitic layers.

Seraphin *et al.*[39] reported that an arc evaporation of Fe/C composite anode also generated nanochains with similar morphology, described above.

4.2.4 *Single-wall tubes.* Following the synthesis studies of stuffed nanocapsules, single-wall (SW) tubes were discovered in 1993[9,10]. SW tubes are found in chamber soot when iron[9] and cobalt[10] were used as catalysts, and for nickel[11,40] they grow on the surface of the cathode slag. For iron catalyst,

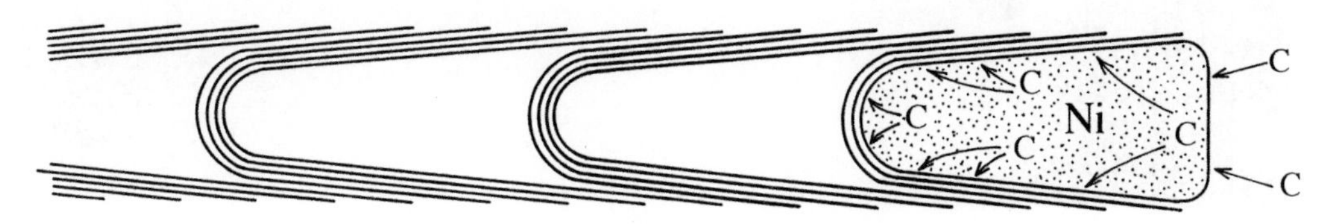

Fig. 9. Growth model of a bamboo tube.

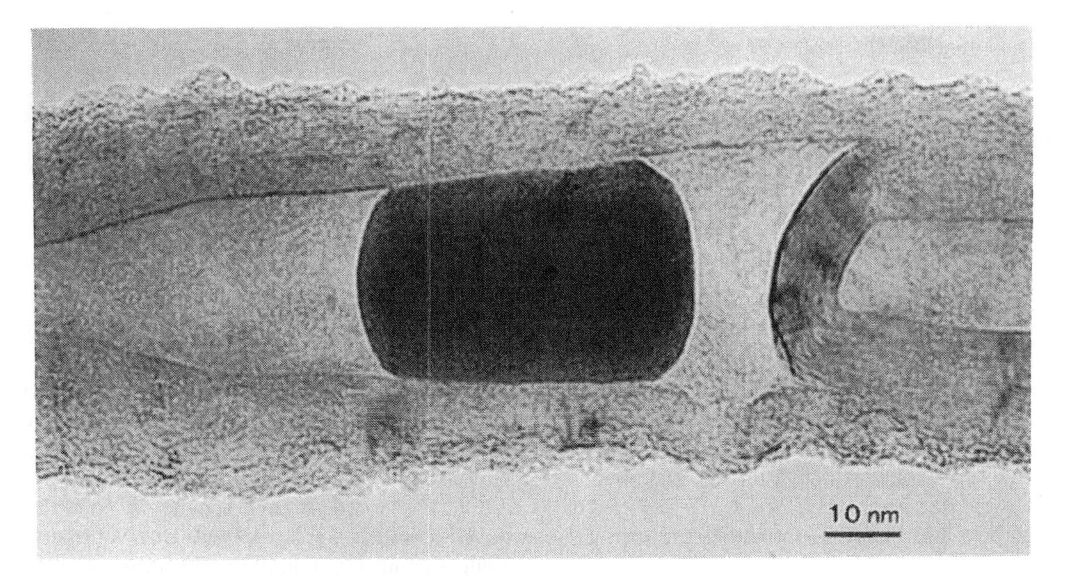

Fig. 10. TEM picture of a Ni metal left in the capillary of a graphite tube. Contact angle of the Ni particle on graphite surface (angle between the Ni/graphite interface and the Ni free surface) is larger than 90° (measured angle is about 140°), indicating poor wetting of Ni on the inner wall of a graphite tube.

methane is reportedly an indispensable ingredient to be added to an inactive gas (Ar for iron)[9]. For cobalt and nickel, on the other hand, no additives are necessary; the arc evaporation of metal-loaded graphite in a pure inactive gas (usually He) produces SW tubes. Figure 12 shows a TEM image of bundles of SW tubes growing radially from a Ni-carbide particle. The diameter of tubes are mostly in a range from 1.0 nm to 1.3 nm. Tips of SW tubes are capped and hollow inside. No contrast suggesting the presence of Ni clusters or particles is observed at the tips.

Effects of various combinations of $3d$-transition metals on the formation of SW tubes have been studied by Seraphin and Zhou[41]. They reported that mixed metals enhanced the production of SW tubes; in particular, a 50% Fe + 50% Ni combination performed much better than Fe, Co, or Ni alone. It was also shown that the addition of some metals, such as Cu, to these metals poisoned their catalytic action.

Catalysts for SW tube formation are not confined to the iron-group metals. Some elements of the lanthanide series can catalyze the formation of SW tubes,

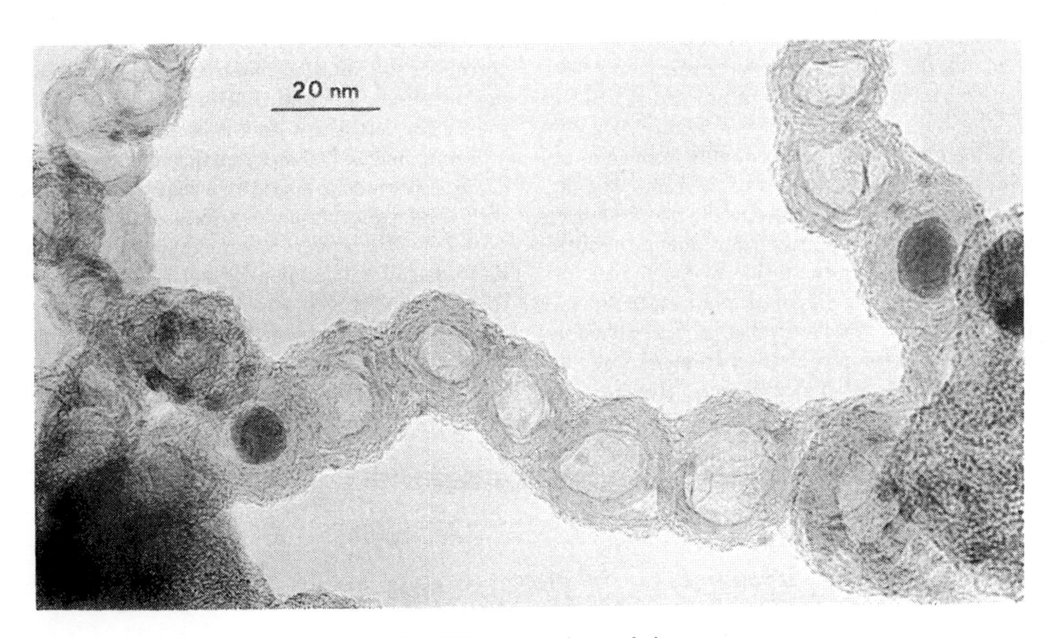

Fig. 11. TEM picture of nanochains.

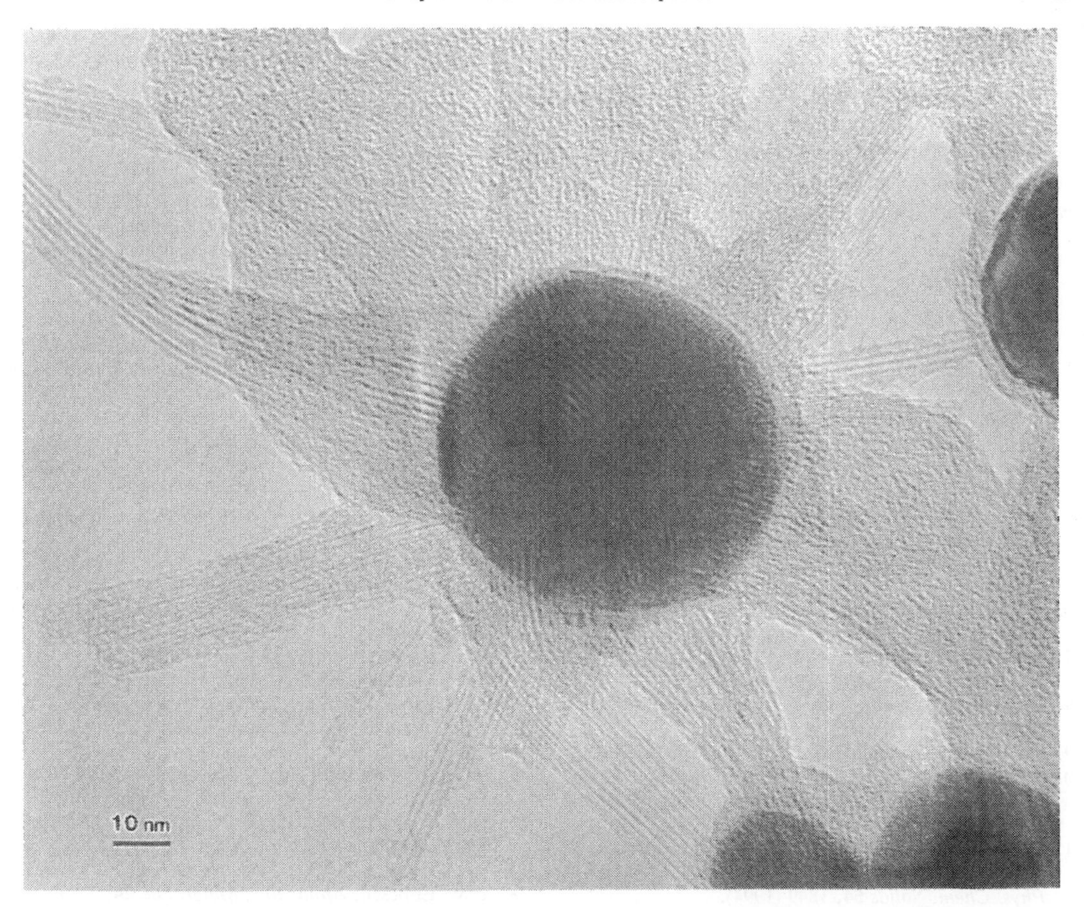

Fig. 12. TEM picture of single-wall tubes growing radially from a Ni-carbide particle.

as has been exemplified for Gd[42], Y[43], La[44] and Ce[45]. The morphology of the tubes that grow radially from these metal (or compound) particles and have "sea urchin"-like morphology is similar to that shown for Ni, but the length of tubes is shorter for the lanthanides (~100 nm long) than that of Ni (~μm long). The diameter of tubes produced from La is typically 1.8–2.1 nm, being larger (about twice) than that for Ni. It is reported that the addition of sulfur to Co catalyst promotes the formation of thicker SW tubes (in a range from 1 to 6 nm in diameter)[46]. The dependence of tube diameter on the catalysts employed suggests the possibility of producing SW tubes with any desirable diameter.

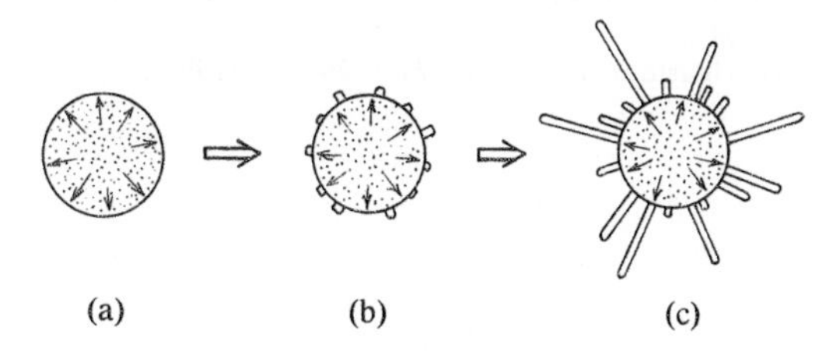

Fig. 13. Hypothetical growth process of SW tubes from a metal/carbon alloy particle: (a) segregation of carbon toward the surface, (b) nucleation of SW tubes on the particle surface and, (c) growth of the SW tubes.

A hypothetical growth process[40] of SW tubes from a core particle is illustrated in Fig. 13. When metal catalyst is evaporated together with carbon by arc discharge, carbon and metal atoms condense and form alloy (or binary mixed) particles. As the particles are cooled, carbon dissolved in the particles segregates onto the surface because the solubility of carbon decreases with the decrease of temperature. Some singular surface structures or compositions in an atomic scale may catalyze the formation of SW tubes. After the nuclei of SW tubes are formed, carbon may be supplied from the core particle to the roots of SW tubes, and the tubes grow longer, maintaining hollow capped tips. Addition of carbon atoms (and C_2) from the gas phase to the tips of the tubes may also help the growth of tubes.

4.3 *Anti-oxidation of wrapped iron nanocrystals*

The protective nature of graphitic carbon against oxidation of core nanocrystals was demonstrated by an environmental test (80°C, 85% relative humidity, 7 days)[44]. Even after this test, XRD profiles revealed that the capsulated iron particles were not oxidized at all, while naked iron particles with similar size, about 50 nm, were oxidized seriously to form oxides (rhombohedral-Fe_2O_3 and cubic-Fe_2O_3).

The magnetism of nanocrystallites contained in graphitic cages is intriguing, not only for scientific research, but also for applications such as magnetic fluids, magnetic ink, and so on.

REFERENCES

1. S. Iijima, *Nature* **354**, 56 (1991).
2. T. W. Ebbesen and P. M. Ajayan, *Nature* **358**, 220 (1992).
3. Y. Saito, T. Yoshikawa, M. Inagaki, M. Tomita, and T. Hayashi, *Chem. Phys. Lett.* **204**, 277 (1993).
4. Y. Chai, T. Guo, C. Jin, R. E. Haufler, L. P. F. Chibante, J. Fure, L. Wang, J. M. Alford, and R. E. Smalley, *J. Phys. Chem.* **95**, 7564 (1991).
5. H. Shinohara, H. Sato, M. Ohkohchi, Y. Ando, T. Kodama, T. Shida, T. Kato, and Y. Saito, *Nature* **357**, 52 (1992).
6. D. S. Bethune, R. D. Johnson, J. R. Salem, M. S. de Vries, and C. S. Yannoni, *Nature* **366**, 123 (1993).
7. R. S. Ruoff, D. C. Lorents, B. Chan, R. Malhotra, and S. Subramoney, *Science* **259**, 346 (1993).
8. M. Tomita, Y. Saito, and T. Hayashi, *Jpn. J. Appl. Phys.* **32**, L280 (1993).
9. S. Iijima and T. Ichihashi, *Nature* **363**, 603 (1993).
10. D. S. Bethune, C. H. Kiang, M. S. de Vries, G. Gorman, R. Savoy, J. Vazquez, and R. Beyers, *Nature* **363**, 605 (1993).
11. Y. Saito, T. Yoshikawa, M. Okuda, N. Fijimoto, K. Sumiyama, K. Suzuki, A. Kasuya, and Y. Nishina, *J. Phys. Chem. Solids* **54**, 1849 (1994).
12. Y. Saito, M. Okuda, T. Yoshikawa, A. Kasuya, and Y. Nishina, *J. Phys. Chem.* **98**, 6696 (1994).
13. M. Yoshida and E. Osawa, *Fullerene Sci. Tech.* **1**, 55 (1993).
14. Y. Saito, T. Yoshikawa, S. Bandow, M. Tomita, and T. Hayashi, *Phys. Rev. B* **48**, 1907 (1993).
15. M. S. Dresselhaus, G. Dresselhaus, K. Sugihara, I. L. Spain, and H. A. Goldberg, In *Graphite Fibers and Filaments*, p. 42. Springer-Verlag, Berlin (1988).
16. Y. Saito and M. Inagaki, *Jpn. J. Appl. Phys.* **32**, L954 (1993).
17. Y. Murooka and K. R. Hearne, *J. Appl. Phys.* **43**, 2656 (1972).
18. W. Finkelburg and S. M. Segal, *Phys. Rev.* **83**, 582 (1951).
19. D. R. McKenzie, D. Muller, and B. A. Pailthorpe, *Phys. Rev. Lett.* **67**, 773 (1991).
20. Pöttgen and W. Jeitschko, *Inorg. Chem.* **30**, 427 (1991).
21. Y. Saito, M. Okuda, T. Yoshikawa, S. Bandow, S. Yamamuro, K. Wakoh, K. Sumiyama, and K. Suzuki, *Jpn. J. Appl. Phys.* **33**, L186 (1994).
22. G. Adachi, N. Imanaka, and Z. Fuzhong, In *Handbook on the Physics & Chemistry of Rare Earths* (Edited by K. A. Gschneider, Jr. and L. Eyring) Vol. 15, Chap. 99, p. 61. Elsevier Science Publishers, Amsterdam (1991).
23. Y. Saito, T. Yoshikawa, M. Okuda, M. Ohkohchi, Y. Ando, A. Kasuya, and Y. Nishina, *Chem. Phys. Lett* **209**, 72 (1993).
24. R. E. Honig and D. A. Kramer, *RCA Rev.* **30**, 285 (1969).
25. H. Funasaka, K. Sugiyama, K. Yamamoto, and T. Takahashi, presented at 1993 Fall Meeting of Mat. Res. Soc., Boston, November 29 to December 3 (1993).
26. R. S. Ruoff, S. Subramoney, D. Lorents, and D. Keegan, presented at the 184th Meeting of the Electrochemical Society, New Orleans, October 10–15 (1993).
27. L. Moro, R. S. Ruoff, C. H. Becker, D. C. Lorents, and R. Malhotra, *J. Phys. Chem.* **97**, 6801 (1993).
28. Y. Saito, T. Yoshikawa, M. Okuda, N. Fijimoto, S. Yamamuro, K. Wakoh, K. Sumiyama, K. Suzuki,A. Kasuya, and Y. Nishina, *Chem. Phys. Lett.* **212**, 379 (1993).
29. T. Hihara, H. Onodera, K. Sumiyama, K. Suzuki, A. Kasuya, Y. Nishina, Y. Saito, T. Yoshikawa, and M. Okuda, *Jpn. J. Appl. Phys.* **33**, L24 (1994).
30. R. D. Heidenreich, W. M. Hess, and L. L. Ban, *J. Appl. Cryst.* **1**, 1 (1968).
31. T. Hihara, K. Sumiyama, K. Suzuki, to be submitted.
32. E. M. Brunsman, R. Sutton, E. Bortz, S. Kirkpatrick, K. Midelfort, J. Williams, P. Smith, M. E. McHenry, S. A. Majetich, J. O. Artman, M. De Graef, and S. W. Staley, *J. Appl. Phys.* **75**, 5882 (1994).
33. S. A. Majetich, J. O. Artman, M. E. McHenry, N. T. Nuhfe, and S. W. Staley, *Phys. Rev. B* **48**, 16845 (1993).
34. Y. Saito and T. Yoshikawa, *J. Cryst. Growth* **134**, 154 (1993).
35. A. Oberlin, M. Endo, and T. Koyama, *J. Cryst. Growth* **32**, 335 (1976).
36. M. R. Pederson and J. Q. Broughton, *Phys. Rev. Lett.* **69**, 2689 (1992).
37. P. M. Ajayan and S. Iijima, *Nature* **361**, 333 (1993).
38. T. W. Ebbesen, *Annu. Rev. Mater. Sci.* **24**, 235 (1994).
39. S. Seraphin, S. Wang, D. Zhou, and J. Jiao, *Chem. Phys. Lett.* **228**, 506 (1994).
40. Y. Saito, M. Okuda, N. Fujimoto, T. Yoshikawa, M. Tomita, and T. Hayashi, *Jpn. J. Appl. Phys.* **33**, L526 (1994).
41. S. Seraphin and D. Zhou, *Appl. Phys. Lett.* **64**, 2087 (1994).
42. S. Subramoney, R. S. Ruoff, D. C. Lorents, and R. Malhotra, *Nature* **366**, 637 (1993).
43. D. Zhou, S. Seraphin, and S. Wang, *Appl. Phys. Lett.* **65**, 1593 (1994).
44. Y. Saito, In *Recent Advances in the Chemistry and Physics of Fullerenes and Related Materials* (Edited by K. M. Kadish and R. S. Ruoff) p. 1419. Electrochemical Society, Pennington, NJ (1994).
45. Y. Saito, M. Okuda, and T. Koyama, *Surface Rev. Lett.*, to be published.
46. C. H. Kiang, W. A. Goddard, III, R. Beyers, J. R. Salem, and D. S. Bethune, *J. Phys. Chem.* **98**, 6612 (1994).
47. S. Bandow and Y. Saito, *Jpn. J. Appl. Phys.* **32**, L1677 (1993).
48. Y. Murakami, T. Shibata, T. Okuyama, T. Arai, H. Suematsu, and Y. Yoshida, *J. Phys. Chem. Solids* **54**, 1861 (1994).
49. D. Ugarte, *Chem. Phys. Lett.* **209**, 99 (1993).

ONION-LIKE GRAPHITIC PARTICLES

D. UGARTE
Laboratório National de Luz Síncrotron (CNPq/MCT), Cx. Postal 6192,
13081-970 Campinas SP, Brazil;
Institut de Physique Expérimentale, Ecole Polytechnique Fédérale de Lausanne,
1015 Lausanne, Switzerland

(*Received* 18 *July* 1994; *accepted* 10 *February* 1995)

Abstract—Nanometric graphitic structures (fullerenes, nanotubes, bucky-onions, etc.) form in different harsh environments (electric arc, electron irradiation, plasma torch). In particular, the onion-like graphitic particles may display a wide range of structures, going from polyhedral to nearly spherical. High-resolution electron microscopy is the primary tool for studying these systems. On the basis of HREM observations, we discuss the energetics and possible formation mechanism of these multi-shell fullerenes. The better understanding of the underlying processes would allow the development of an efficient production method.

Key Words—Graphite, fullerenes, HREM, nanostructures, electron irradiation.

1. INTRODUCTION

The high melting temperature of carbon materials (≈4000°K) has made difficult the preparation and study of carbon clusters. In recent years, the field of nanometric carbon particles has found an unexpected and overwhelming development. In a first step, the sophisticated laser evaporation source revealed the existence of fullerenes, but with the limitation that their study could only be performed in cluster-beam experiments[1]. Second, in 1990, the current and simple electric arc-discharge[2] allowed the synthesis and study of these molecules by a large number of laboratories, generating a burst of revolutionary discoveries. A wide family of nanometric graphitic systems may be synthesized by making slight modifications to the electric arc experiment (nanotubes[3,4], nanoparticles[5–7], metal-filled nanoparticles[8–10], etc.).

High-resolution transmission electron microscopy (HREM) is the technique best suited for the structural characterization of nanometer-sized graphitic particles. In-situ processing of fullerene-related structures may be performed, and it has been shown that carbonaceous materials transform themselves into quasi-spherical onion-like graphitic particles under the effect of intense electron irradiation[11].

In this paper, we analyze the methods of synthesizing multi-shell fullerene structures and try to gather some information about their formation mechanism. We also discuss some particularities of the energetics of onion-like graphitic particles. The understanding of the parameters involved would allow the development of efficient production procedures.

2. SYNTHESIS OF MULTI-SHELL FULLERENES

The electric arc is the easiest and most frequently used experiment to produce onion-like particles. A dc arc-discharge is used to generate a carbon deposit on the negative electrode following the procedure for large-scale synthesis of nanotubes[4]. The central part of the deposit consists of a black powder containing a mixture of graphitic nanotubes and nanoparticles. Although this procedure allows the generation of these particles in macroscopic quantities, the purification of the different soot components (e.g., nanotubes) has not been easily performed. Centrifugation and filtration methods have been unsuccessful, and a rather inefficient oxidation procedure (99% of the material is lost) has allowed the production of purified nanotubes samples[12].

The electric arc is a transient phenomenon, where the region of the electrode producing the arc changes permanently all over the surface of contact. The generated temperature gradients induce an important range of conditions for the formation of graphitic nanoparticles; this fact leads to wide size and shape distributions (see Fig. 1a). The particles usually display a clear polyhedral morphology, and a large inner empty space (3–10 nm in diameter). Macroscopic quantities of nanometric graphitic particles may be obtained by a thermal treatment of "fullerene black"[13,14]; this method yields particles with similar structure to those ones generated in the arc, but with a narrower size distribution, in particular located in the ≤10 nm range.

The high-energy electron irradiation of carbonaceous materials produces remarkably symmetrical and spherical onion-like particles[11] (see Fig. 1b, 2). These particles are very stable under electron bombardment, even when formed by a small number of shells (2–4) [15]. The generation of these quasi-spherical graphitic systems (nicknamed bucky-onions) is realized *in situ* in an electron microscope. However, the particles are only formed in minute quantities on an electron microscopy grid, and their study may only be performed by transmission electron microscopy-associated techniques. Two major structural aspects differentiate them from the graphitic particles discussed previously: (a) the shape of the concentrical arrangement of graphitic layers is a nearly perfect sphere, (b) the innermost shell

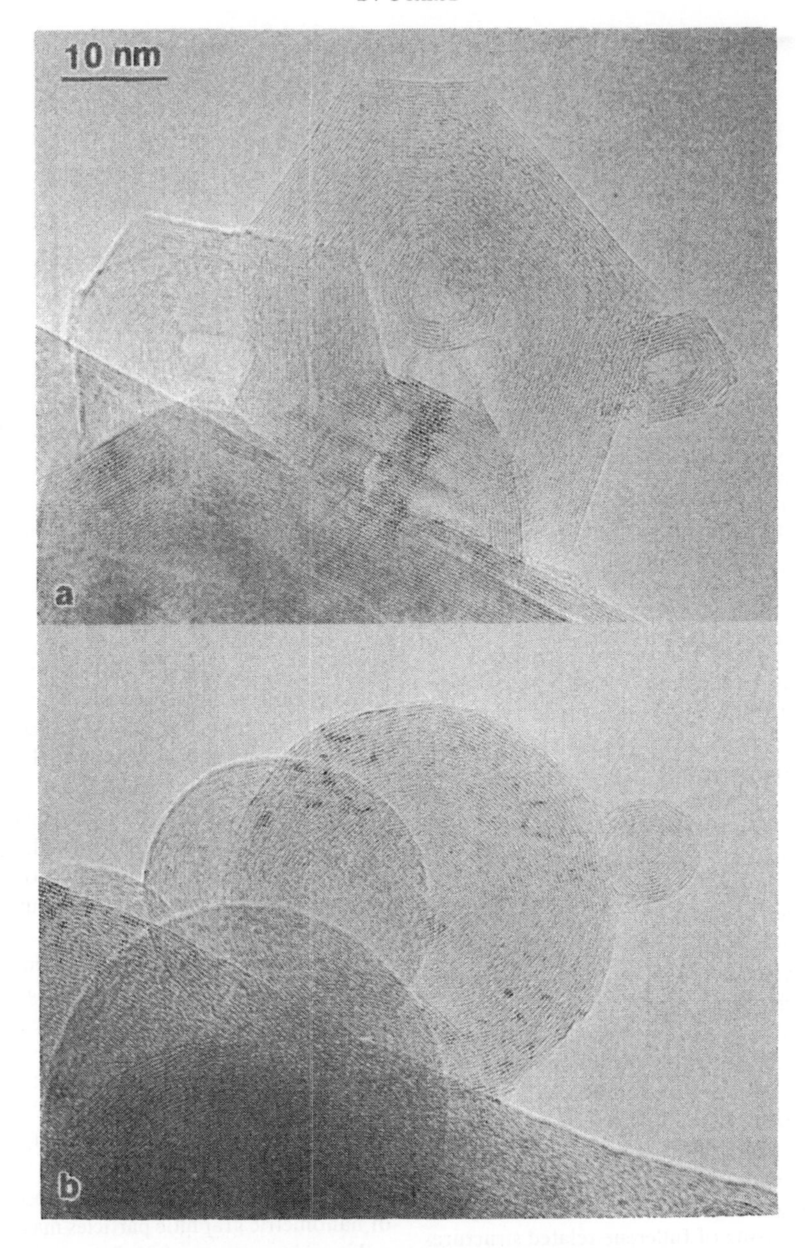

Fig. 1. High-resolution electron micrographs of graphitic particles: (a) as obtained from the electric arc deposit, they display a well-defined faceted structure and a large inner hollow space, (b) the same particles after being subjected to intense electron irradiation (note the remarkable spherical shape and the disappearance of the central empty space); dark lines represent graphitic layers.

is always very small, displaying a size very close to C_{60} ($\approx$0.7–1 nm) (see Fig. 2).

Onion-like graphitic clusters have also been generated by other methods: (a) shock-wave treatment of carbon soot[16]; (b) carbon deposits generated in a plasma torch[17], (c) laser melting of carbon within a high-pressure cell (50–300 kbar)[18]. For these three cases, the reported graphitic particles display a spheroidal shape.

In particular, the laser melting experiment produced two well-differentiated populations of carbon clusters: (a) spheroidal diamond particles with a radial texture in the 0.2–1 μm range, (b) small particles (<0.2 μm) also spherical in shape, but formed by concentric graphitic layers. Unfortunately, no detailed HREM study of these particles was undertaken, but an apparent size effect connecting diamond and graphitic particles was insinuated. Recently, an interesting experiment also revealed a relation between ultradispersed diamonds (3–6 nm in diameter) and onion-like graphitic particles. The nanometer diamond sample was obtained by a chemical purification of detonation soot. The annealing treatment (1100–1500°C) of nanodiamonds generates onion-like graphitic particles with a remark-

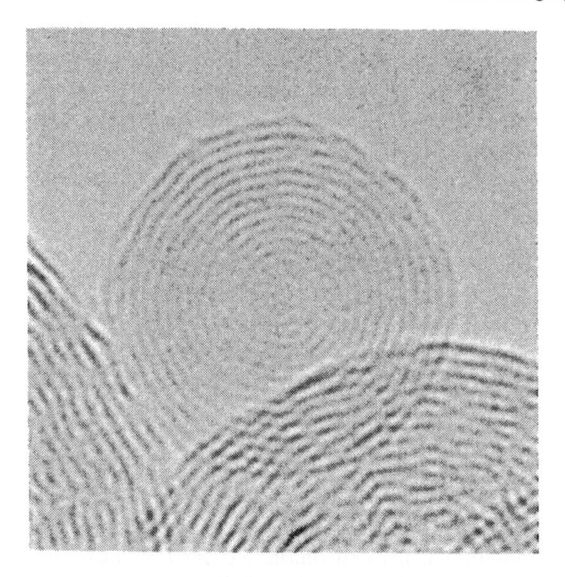

Fig. 2. HREM image of a quasi-spherical onion-like graphitic particles generated by electron irradiation (dark lines represent graphitic shells, and distance between layers is 0.34 nm).

able spheroidal shape[19], and a compact structure very similar to the bucky-onions generated by electron irradiation[11].

3. FORMATION MECHANISM

In the preceding section, we have described different experiments generating graphitic nanoparticles. As for the case of fullerene synthesis, the procedures are rather violent (electric arc, plasma torch, shock waves, high-temperature treatment, electron irradiation, etc.) and clearly display the present incapacity of generating nanometric curved and closed graphitic systems by standard chemical techniques. For the synthesis of C_{60}, it has been found that a temperature on the order of or higher than 1200°C are necessary to anneal the carbon clusters in the gas phase and efficiently form C_{60} molecules[20]. In the case of the graphitization process, the dewrinkling and elimination of defects in graphitic layers begins at 2000°C[21]. The extremely stable carbon-carbon bonds are responsible for the high-energetic process necessary to anneal graphitic structures.

The formation mechanism of fullerenes and related structures is not well understood. The fascinating high aspect ratio of nanotubes is associated to the electric field of the arc[7,22], but this fact has not yet been confirmed and/or been applied to control their growth. Moreover, the arc process generates simultaneously a large quantity of polyhedral graphic particles. The formation of multi-shell graphitic particles from the gas phase by a spiral growth mechanism has also been suggested[23,24], but no convincing experimental data of a spiral structure has been reported.

The thermal treatment of fullerene black generates nanometric polyhedral particles[13]. This experiment is an example of the graphitization of a "hard carbon" (when subjected to heat treatment, it yields irregularly shaped pores or particles instead of extended flat graphitic planes)[14,21].

In the case of carbon melting experiments[16] and electric arc[5,6], it has been suggested that the onion-like particles are generated by the graphitization of a liquid carbon drop. The growth of graphite layers is supposed to begin at the surface and progress toward the center (see Fig. 3a–d). Saito *et al.*[7] has suggested a similar mechanism but, instead of liquid carbon, they considered a certain carbon volume on the electrode surface, which possesses a high degree of structural fluidity due to the He ion bombardment.

The progressive ordering from the surface to the center has been experimentally observed in the case of the electron irradiation-induced formation of the quasi-spherical onion-like particles[25]. In this case, the large inner hollow space is unstable under electron bombardment, and a compact particle (innermost shell $\approx C_{60}$) is the final result of the graphitization of the carbon volume (see Fig. 3e–h).

The large inner hollow space observed in polyhedral particles is supposed to be due to the fact that the initial density of the carbon volume (drop) is lower than graphite[7]. Then, in order to prepare more compact graphitic particles (smaller inner shell), the starting carbon phase should have a density closer to graphite (2.25 gr/cm^2). This basic hypothesis has been confirmed by subjecting nanodiamonds to a high-temperature treatment (diamond is much denser than graphite, 3.56 gr/cm^2)[19]. Experimentally, it has been observed that the formation of graphitic layers begins at the (111) diamond facets, then generates closed-surface graphitic layers, and subsequently follows the formation of concentric shells epitaxially towards the center. At an intermediate stage, the onion-like graphitic particles contain a tiny diamond core (see Fig. 2 in ref. [19]). This process yields carbon on-

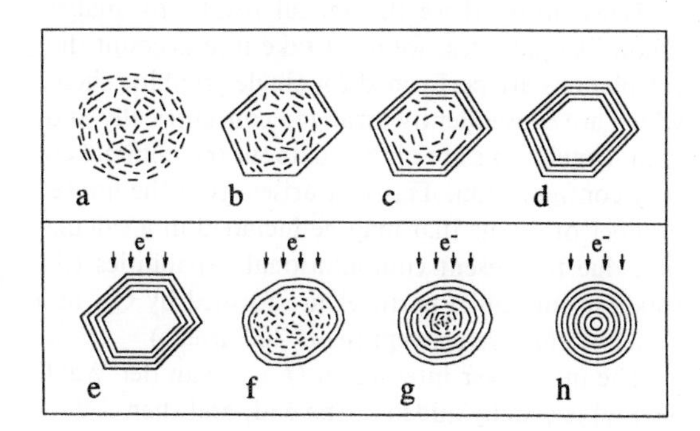

Fig. 3. Schematic illustration of the growth process of a graphitic particle: (a)–(d) polyhedral particle formed on the electric arc; (d)–(h) transformation of a polyhedral particle into a quasi-spherical onion-like particle under the effect of high-energy electron irradiation; in (f) the particle collapses and eliminates the inner empty space[25]. In both schemes, the formation of graphite layers begins at the surface and progresses towards the center.

ions with a very small inner empty space, which contrasts with the polyhedral particles prepared by heat treatment of fullerene black[13].

4. ENERGETICS

The elimination of the energetic dangling bonds present at the edges of a tiny graphite sheet is supposed to be the driving force to induce curvature and closure in fullerenes; this phenomenon is also associated with the formation of larger systems, such as nanotubes and graphitic particles.

The remarkable stability of onion-like particles[15] suggests that single-shell graphitic molecules (giant fullerenes) containing thousands of atoms are unstable and would collapse to form multi-layer particles; in this way the system is stabilized by the energy gain from the van der Waals interaction between shells [15,26,27].

Graphite is the most stable form of carbon at ambient conditions, and it is formed by the stacking of planar layers. The extreme robustness of the concentric arrangement of spherical fullerenes led to the hypothesis that the quasi-spherical onion-like graphitic particles are the most stable form of carbon particles[15]. This controversy between planar configuration of the sp^2 bonding in macroscopic graphite and the apparent spherical shape in nanometric system, has attracted a great deal of interest[28]. Calculations of the structure of giant fullerenes are not able to give a definitive answer: (a) models based on elastic or empirical potentials predict that the minimal energy structure is a slightly relaxed icosahedron with planar facets, the curvature being concentrated at the corner of the polyhedron (pentagons)[26,29,30]; (b) *ab initio* calculations predict two local energy minima for C_{240}. One represents a faceted icosahedron, and the second, a nearly spherical structure distributing the strain over all atoms, is slightly more stable (binding energy per atom -7.00 and -7.07 eV for polyhedral and spherical fullerenes)[31].

When using these theoretical results to analyze onion-like particles, we must take into account that calculations are performed for single graphitic shells, which are subsequently arranged concentrically and, then, conclusions are obtained about the minimal energy configuration. This fact arises from the limited number of atoms that may be included in a calculation due to present computational capabilities (the smallest onion-like particles are formed by C_{60} in a C_{240} and this system represents 300 atoms).

The inter-layer interaction (E_{vdW}, van der Waals energy) is usually added at the end, and then it does not participate in the energy minimization. Evaluating inter-shell interaction by a simple Lennard-Jones pair potential shows that concentrical spherical shells are more stable than icosahedral ones (E_{vdW} between C_{240} and C_{540} is -17.7 and -12.3 eV for spherical and icosahedral shells, respectively). An interesting comparison may be performed by considering that corannulene $C_{20}H_{10}$, which is a pentagon surrounded by 5 hexagons (non-planar structure with a bowl shape) is a very floppy structure. This molecule is the basic unit of C_{60}, and also the atomic arrangement present at the corners of faceted giant fullerenes. It has been observed by NMR that, at room temperature, it spontaneously presents a bowl inversion transition, with an energy barrier of only 440 meV[32]. Then, we may use this value to get a bare estimation of the energy necessary to crash the 12 corners of a polyhedral graphitic cage (containing the corannulene configuration) into a spherical one, and it turns out to be of the same order of magnitude that the gain in van der Waals energy mentioned previously (≈ 5 eV). This fact indicates that a global evaluation of multi-shell structures should be performed to answer the question of the minimal energy structure of onion-like graphitic particles.

From a different point of view, the sphericity of the irradiation generated onion-like particles have also been attributed to imperfect shells with a large number of defects[33].

Concerning the possible arrangement of concentric defect-free graphitic cages, a polyhedral graphitic particle has all the shells in the same orientation, so that all the corners (pentagons) are perfectly superimposed (see Fig. 4a). If the pentagons of the concentric shells are not aligned, the final shape of graphitic particles should be much closer to a sphere. In the spherical particles, an interesting issue is the fact that the shells may be rotating relative to each other[15]. This behavior has been predicted for C_{60} in C_{240}[30] and for multi-shell tubes[34].

5. SUMMARY AND PERSPECTIVES

The multi-shell fullerenes constitute the transition from fullerenes to macroscopic graphite. They present both the closed graphitic surface of fullerenes and the stacked layers interacting by van der Waals forces, as in graphite.

One of the main scientific issues of the discovery of the bucky-onions is the unresolved question of minimal energy configuration of carbon clusters (onion-

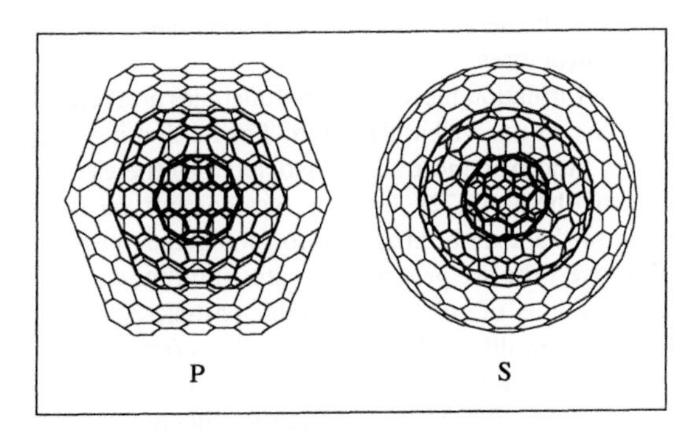

Fig. 4. Onion-like graphitic particles formed by three concentric layers (C_{60}, C_{240}, C_{540}): polyhedral (marked P) and spherical (marked S) structures. For clarity, only a half part of each shell is shown.

like particle: spherical or polyhedral) and the size where the transition is between these closed surface particles and the macroscopic planar graphite. There are still several difficulties in answering this question: (a) from the theoretical point of view, the large number of atoms to be considered renders the computational evaluation very difficult; (b) from the experimental point of view, the quasi-spherical structures have only been synthesised *in situ* in an electron microscope.

At present, major efforts are being done to develop the production and purification of macroscopic quantities of the quasi-spherical onion-like particles. The understanding of the formation mechanism and energetics involved would allow the development of efficient production methods. The annealing on nanometer-sized diamonds reported by Kuznetsov *et al.*[19] presents a promising way to generate compact graphitic particles, if we are able to overcome the difficulty of producing large quantities of nanodiamonds.

We hope that macroscopic samples of quasi-spherical onion-like particles will soon become available, and then we will be able to characterize these systems in detail. Probably a new generation of carbon materials can be generated by the three-dimensional packing of quasi-spherical multi-shell fullerenes.

Acknowledgements—The author is most grateful to W. de Heer for invaluable discussions and advice. We are indebted to R. Monot and A. Chatelain for several useful remarks. We thank the Brazilian National Council of Science and Technology (CNPq) and Swiss National Science Foundation for financial support.

REFERENCES

1. H. W. Kroto, J. R. Heath, S. C. O'Brien, R. F. Curl, and R. E. Smalley, *Nature* **318**, 162 (1985).
2. W. Krätschmer, L. D. Lamb, K. Foristopoulos, and D. R. Huffman, *Nature* **347**, 354 (1990).
3. S. Iijima, *Nature* **354**, 56 (1991).
4. T. W. Ebbesen and P. M. Ajayan, *Nature* **358**, 220 (1992).
5. S. Iijima, *J. Cryst. Growth* **50**, 675 (1980).
6. D. Ugarte, *Chem. Phys. Lett.* **198**, 596 (1992).
7. Y. Saito, T. Yoshikawa, M. Inagaki, M. Tomita, and T. Hayashi, *Chem. Phys. Lett.* **204**, 277 (1993).
8. R. S. Ruoff, D. C. Lorents, B. Chan, R. Malhotra, and S. Subramoney, *Science* **259**, 346 (1993).
9. M. Tomita, Y. Saito, and T. Hayashi, *Jpn. J. Appl. Phys.* **32**, L280 (1993).
10. D. Ugarte, *Chem. Phys. Lett.* **209**, 99 (1993).
11. D. Ugarte, *Nature* **359**, 707 (1992).
12. T. W. Ebbesen, P. M. Ajayan, H. Hiura, and K. Tanigaki, *Nature* **367**, 519 (1994).
13. W. A. de Heer and D. Ugarte, *Chem. Phys. Lett.* **207**, 480 (1993).
14. D. Ugarte, *Carbon* **32**, 1245 (1994).
15. D. Ugarte, *Europhys. Lett.* **22**, 45 (1993).
16. K. Yamada, H. Kunishige, and A. B. Sawaoka, *Naturwissenschaften* **78**, 450 (1991).
17. N. Hatta and K. Murata, *Chem. Phys. Lett.* **217**, 398 (1994).
18. L. S. Weathers and W. A. Basset, *Phys. Chem. Minerals* **15**, 105 (1987).
19. V. L. Kuznetsov, A. L. Chuvilin, Y. V. Butenko, I. Y. Mal'kov, and V. M. Titov, *Chem. Phys. Lett.* **222**, 343 (1994).
20. R. E. Smalley, *Acc. Chem. Res.* **25**, 98 (1992).
21. A. Oberlin, *Carbon* **22**, 521 (1984).
22. R. E. Smalley, *Mater. Sci. Eng. B* **19**, 1 (1992).
23. Q. L. Zhang, S. C. O'Brien J. R. Heath, Y. Liu, R. F. Curl, H. W. Kroto, and R. E. Smalley, *J. Phys. Chem.* **90**, 525 (1986).
24. H. W. Kroto and K. McKay, *Nature* **331**, 328 (1988).
25. D. Ugarte, *Chem. Phys. Lett.* **207**, 473 (1993).
26. A. Maiti, C. J. Bravbec, and J. Bernholc, *Phys. Rev. Lett.* **70**, 3023 (1993).
27. D. Tománek, W. Zhong, and E. Krastev, *Phys. Rev. B* **48**, 15461 (1993).
28. H. W. Kroto, *Nature* **359**, 670 (1992).
29. K. G. McKay, H. W. Kroto, and D. J. Wales, *J. Chem. Soc. Faraday Trans.* **88**, 2815 (1992).
30. M. Yoshida and E. Osawa, *Ful. Sci. Tech.* **1**, 55 (1993).
31. D. York, J. P. Lu, and W. Yang, *Phys. Rev. B* **49**, 8526 (1994).
32. L. T. Scott, M. M. Hashemi, and M. S. Bratcher, *J. Am. Chem. Soc.* **114**, 1920 (1992).
33. A. Maiti, C. J. Brabec, and J. Bernholc, *Modern Phys. Rev. Lett. B* **7**, 1883 (1993).
34. J.-C. Charlier and J.-P. Michenaud, *Phys. Rev. Lett.* **70**, 1858 (1993).

METAL-COATED FULLERENES

U. Zimmermann, N. Malinowski,* A. Burkhardt, and T. P. Martin
Max-Planck-Institut für Festkörperforschung, Heisenbergstr. 1, 70569 Stuttgart, Germany

(*Received* 24 *October* 1994; *accepted* 10 *February* 1995)

Abstract—Clusters of C_{60} and C_{70} coated with alkali or alkaline earth metals are investigated using photoionization time-of-flight mass spectrometry. Intensity anomalies in the mass spectra of clusters with composition $C_{60}M_x$ and $C_{70}M_x$ ($x = 0\ldots500$; $M \in \{Ca, Sr, Ba\}$) seem to be caused by the completion of distinct metal layers around a central fullerene molecule. The first layer around C_{60} or C_{70} contains 32 or 37 atoms, respectively, equal to the number of carbon rings constituting the fullerene cage. Unlike the alkaline earth metal-coated fullerenes, the electronic rather than the geometric configuration seems to be the factor determining the stability of clusters with composition $(C_{60})_nM_x$ and $(C_{70})_nM_x$, $M \in \{Li, Na, K, Rb, Cs\}$. The units $C_{60}M_6$ and $C_{70}M_6$ are found to be particularly stable building blocks of the clusters. At higher alkali metal coverage, metal-metal bonding and an electronic shell structure appear. An exception was found for $C_{60}Li_{12}$, which is very stable independently of charge. Semiempirical quantum chemical calculations support that the geometric arrangement of atoms is responsible for the stability in this case.

Key Words—Fullerenes, mass spectrometry, clusters, electronic shells, icosahedral layers.

1. INTRODUCTION

In their bulk intercalation phase compounds of C_{60} and alkali or alkaline earth metals have been studied intensively, spurred particularly by the discovery of superconductivity in several of these metal fullerides, such as $C_{60}K_3$, $C_{60}Rb_3$, $C_{60}Ca_5$, etc.[1-5]. However, despite the wealth of information that could yet be extracted from these fullerene compounds, we would still like to return briefly to looking at some interesting experiments that can be done by bringing just one single fullerene molecule in contact with atoms of the metals commonly used for the doping bulk fullerite. The properties of these very small metal-fullerene systems then termed clusters, can be studied quite nicely in the gas phase[6]. We observed that, in the gas phase, such a single fullerene molecule can be coated with layers of various alkali and alkaline earth metals[7,8]. In this contribution, we will focus primarily on the structure, both electronic and geometric, of this metal coating of the fullerenes C_{60} and C_{70}.

The method we use to study these metal-fullerene clusters is photoionization time-of-flight mass spectrometry (section 2). The clusters are produced by coevaporation of fullerenes and metal in a gas aggregation cell. By ionizing and, in some cases, heating the clusters with a pulsed laser, various features appear in the mass spectra that contain the information necessary to suggest a geometric or electronic configuration for the cluster investigated.

When building clusters by coating the fullerenes with metal, features similar to the electronic and geometric shells found in pure metal clusters[9] are observed in the mass spectra. In the case of fullerene molecules coated with alkaline earth metals (section 3), we find that a particularly stable structure is formed each time a new layer of metal atoms has been completed around a central fullerene molecule, the stability of these clusters seeming to be purely geometric in origin. The first layer contains exactly the same number of metal atoms as there are rings in the fullerene cage. In growing additional layers, the metal might be expected to prefer the icosahedral shell structure observed in pure alkaline earth clusters[10,11]. However, our measurements suggest a different growth pattern.

Coating the fullerenes with alkali metals (section 4), the resulting structures seem to be primarily governed by the electronic configuration. For example, the charge transfer of up to 6 electrons to the lowest unoccupied molecular orbital (LUMO) of C_{60} observed in bulk alkali fullerides[5] is also observed in our experiments, leading to the very stable building block $C_{60}M_6$ for clusters, where M is any alkali metal. An exception to this is the cluster $C_{60}Li_{12}$. Supported by semiempirical quantum chemical calculations, we find the high stability of $C_{60}Li_{12}$ to be caused by the geometric arrangement of the metal atoms rather than by the electronic configuration[12]. As predicted by *ab initio* calculations, this arrangement most likely has perfect icosahedral symmetry[13]. At higher alkali metal coverage, the coating becomes increasingly metallic and an oscillating structure caused by the successive filling of electronic shells shows up in the mass spectra, if photon energies near the ionization threshold are used.

Note that we always speak of C_{60} and C_{70} as a molecule and not a cluster. We reserve the word 'cluster' to refer to units composed of several fullerenes and metal atoms.

2. EXPERIMENTAL

Figure 1 shows a schematic representation of the experimental setup used to study the metal-fullerene

*Permanent address: Central Laboratory of Photoprocesses, Bulgarian Academy of Sciences, 1040 Sofia, Bulgaria.

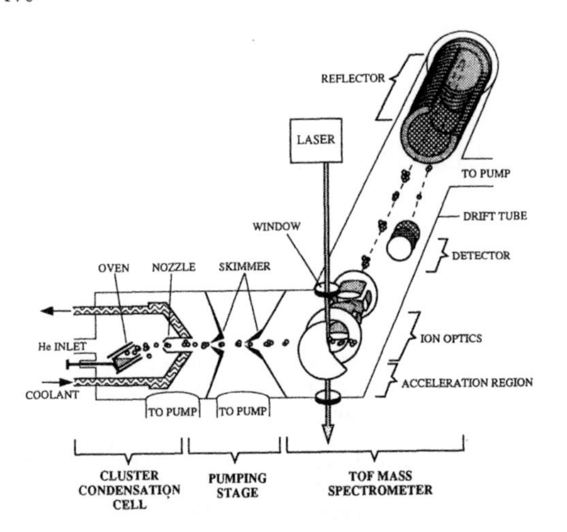

Fig. 1. Experimental setup: the clusters are emitted from the cluster condensation cell, passing as a particle beam through a differential pumping stage into the focus of a time-of-flight mass spectrometer, where they are ionized by a laser pulse.

clusters. The cluster source on the left is a low-pressure, inert gas condensation cell filled with approximately 1 torr He gas and cooled by liquid nitrogen flowing through the outer walls of the cell. Inside the cell, two electrically heated ovens, one containing a fullerene and one containing a metal, produce interpenetrating vapor clouds of the two materials. This mixture is cooled by collisions with the He gas, thereby supersaturating the vapor and causing clusters to condense. The size distribution of the clusters thus produced is rather broad, but the mean composition of the clusters depends on the relative density of the vapor components and can be adjusted by the temperatures of the ovens. However, the range of cluster compositions that can be studied using mass spectrometry is limited, despite the high resolution of the mass spectrometer employed: Due to the various natural isotopes exhibited by most of the metals studied, the mass spectra become increasingly confused with rising metal content, making exact identification of the peaks impossible. For metals with more than one significant isotope we can, therefore, only study clusters with either small metal and high fullerene content or high metal content and just one fullerene per cluster. The formation of pure metal clusters has to be avoided for the same reason. By keeping the temperature of the metal oven below the threshold for formation of pure metal clusters and introducing only small amounts of fullerenes as condensation seeds into the metal vapor, it is possible to generate cluster distributions consisting almost completely of compositions $C_{60}M_x$ or $C_{70}M_x$ with $M \in \{Li, Na, K, Rb, Cs, Ca, Sr, Ba\}$ and $x = 0 \ldots 500$.

After condensation, the clusters are transported by the He-flow through a nozzle and a differential pumping stage into a high vacuum chamber. For ionization of the clusters, we used excimer and dye laser pulses at various wavelengths. The ions were then mass analyzed by a time-of-flight mass spectrometer, having a two-stage reflector and a mass resolution of better than 20,000.

The size distribution of the clusters produced in the cluster source is quite smooth, containing no information about the clusters except their composition. To obtain information about, for example, the relative stability of clusters, it is often useful to heat the clusters. Hot clusters will evaporate atoms and molecules, preferably until a more stable cluster composition is reached that resists further evaporation. This causes an increase in abundance of the particularly stable species (i.e., enhancing the corresponding peak in the mass spectrum, then commonly termed 'fragmentation spectrum'). Using sufficiently high laser fluences ($\simeq 50\ \mu J/mm^2$), the clusters can be heated and ionized simultaneously with one laser pulse.

3. COATING WITH ALKALINE EARTH METALS

In this section, we will investigate the structure of clusters produced when the metal oven is filled with one of the alkaline earth metals Ca, Sr, or Ba.

A mass spectrum of $C_{60}Ba_x$ is shown in Fig. 2. The mass peaks corresponding to singly ionized clusters have been joined by a connecting line. Note that the series of singly ionized clusters shows a very prominent peak at the mass corresponding to $C_{60}Ba_{32}$, implying that this cluster is particularly stable. In searching for an explanation for the high stability of this cluster, we can obtain a first hint from looking at the doubly ionized clusters also visible in Fig. 2 (the peaks not connected by the line correspond to doubly ionized clusters). Again, the peak at $x = 32$ is particularly strong. This seems to indicate that the stability of $C_{60}Ba_{32}$ is not caused by a closed-shell electronic configuration. Instead, the high stability is expected to be of geometric origin. Remembering that the total number of faces or rings constituting the cagelike structure of the C_{60} molecule is 32 and, thus, equal to the number of Ba-atoms required to form this highly

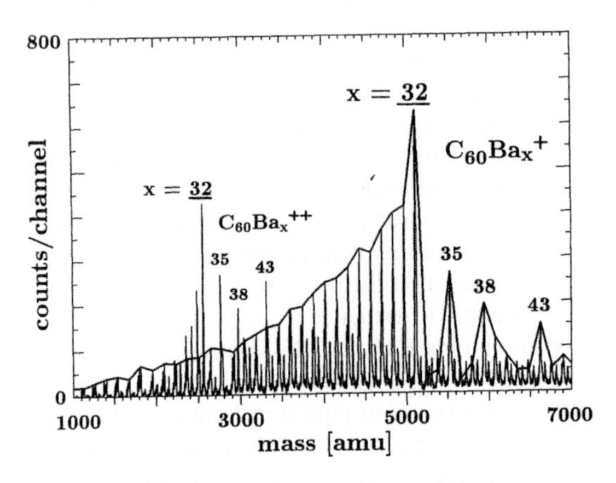

Fig. 2. Mass spectrum of photoionized $C_{60}Ba_x$ clusters containing both singly and doubly ionized species: the solid line connects peaks of singly ionized clusters. The sharp edge occurs at 32 metal atoms, equal to the total number of hexagonal and pentagonal rings of the C_{60} molecule.

stable cluster, the arrangement of metal atoms in this cluster becomes obvious. By placing one Ba atom onto each of the 12 pentagons and 20 hexagons of the C_{60} molecule, a structure with full icosahedral symmetry (point group I_h) is obtained that can be visualized as an almost close-packed layer of 32 Ba-atoms coating the C_{60} molecule. It seems reasonable that this structure exhibits an unusually high stability, somewhat similar to the geometric shells observed in pure alkaline earth clusters[10,11]. Any additional metal atoms situated on this first metal layer are likely to be only weakly bound to the layer underneath and, thus, evaporate easily, causing the mass peaks of $C_{60}Ba_x$ with x greater than 32 to disappear almost completely. The small peaks at $x = 35$, 38, and 43 might signal the completion of small stable metal islands on the first metal layer. We can, however, presently only speculate on the nature of these minor structures.

For a rough estimate of the packing density of this first metal layer, assume the atoms to be hard spheres having the covalent radii of the respective atoms (0.77 Å for C; 1.98 Å for Ba[14]). Placing the carbon spheres at the appropriate sites of the C_{60} structure with bond lengths 1.40 Å and 1.45 Å[15] and letting the Ba spheres rest on the rings formed by the carbon atoms, the Ba spheres placed on neighboring hexagons will almost touch, spheres on neighboring pentagons and hexagons will overlap by a few tenths of an Ångstrøm. The distance of the metal atoms to the center of the molecule is almost equal for atoms on hexagonal and pentagonal faces. In this simple picture, the packing of the metal layer is almost perfectly dense, the Ba atoms having an appropriate size. Incidentally, this argument also holds in a similar manner for Sr- and Ca-atoms.

Of course, this simple picture constitutes only a crude approximation and should be valued only for showing that the completion of a metal layer around C_{60} with 32 Ba-atoms is, indeed, plausible. More precise predictions would have to rely on *ab initio* calculations, including a possible change in bond lengths of C_{60}, such as an expansion of the double bonds of C_{60} due to electron transfer to the antibonding LUMO (as was found in the case of $C_{60}Li_{12}$[12,13]).

The significance of the magic number 32 found in the experiment may also be stated in a different manner. If a cluster containing Ba and a fullerene molecule will be stable and, thus, result in a clearly discernible structure in the mass spectra every time there is exactly one Ba-atom situated on each of the rings of the fullerene molecule, this property might be used to 'count the rings' of a fullerene. Of course, such a proposal has to be verified using other fullerenes, for example, C_{70} which is available in sufficient quantity and purity for such an experiment.

In investigating the metal coating of C_{70}, we will also replace Ba by Ca in the data presented. The coating of the fullerenes with the latter material is basically identical but exhibits additional interesting features that will be discussed below. Figure 3 shows two mass spectra, the upper one of $C_{60}Ca_x^+$, the lower of $C_{70}Ca_x^+$, both obtained under similar conditions as the spectrum in Fig. 2 but with a higher metal vapor density. A slight background caused by fragmentation of clusters inside the drift tube of the mass spectrometer has been subtracted. The lower axis is labeled with the number of metal atoms on the respective fullerene.

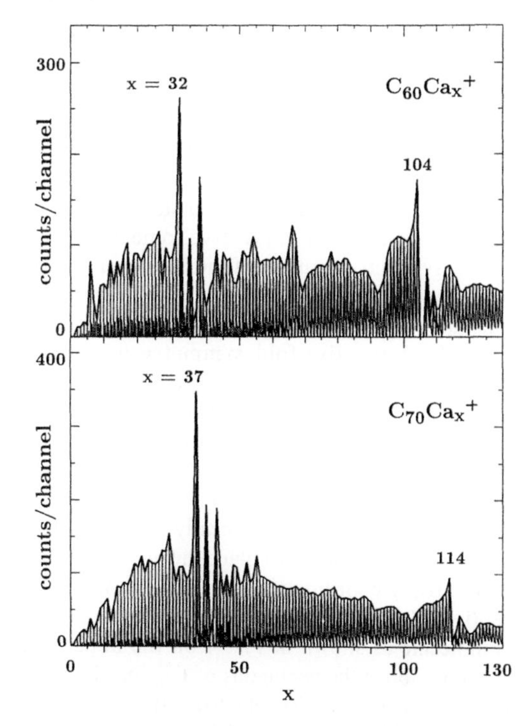

Fig. 3. Mass spectra of photoionized $C_{60}Ca_x^+$ (top) and $C_{70}Ca_x^+$ (bottom): the lower axis is labeled by the number of metal atoms on the fullerene molecule. The peaks at $x = 32$ for $C_{60}Ca_x$ and $x = 37$ for $C_{70}Ca_x$ correspond to a first metal layer around the fullerenes with one atom located at each of the rings. The edges at $x = 104$ and $x = 114$, respectively, signal the completion of a second metal layer.

Again, the coverage of C_{60} with 32 Ca atoms leads to a pronounced peak in the fragmentation mass spectrum. In the spectrum containing C_{70}, a very strong peak at $C_{70}Ca_{37}^+$ is observed. Note that C_{70}, just as C_{60}, has 12 pentagons but 5 additional hexagons on the equator around the remaining fivefold axis, totaling 37 rings. The 'ring-counting' thus seems to work for C_{70} also. However, the applicability of this 'counting method' to even higher fullerenes has to be verified as these become available in sufficient quantities for performing such an experiment.

If it is possible to put one layer of metal around a fullerene molecule, it is tempting to look for the completion of additional layers also. In the spectra in Fig. 3, the sharp edges at $C_{60}Ca_{104}^+$ and $C_{70}Ca_{114}^+$ would be likely candidates for signaling the completion of a second layer. As we will see below, there is, in fact, a very reasonable way of constructing such a second layer with precisely the number of metal atoms observed in the spectrum.

In proposing an arrangement of the atoms in the second layer, we will focus first on the metal coating of C_{60}. Note that we speak of layers, not shells. The term 'shell' implies self-similarity which, as we will see later, does not apply in our case. In the following paragraphs we will often specify the positions of the metal atoms relative to the central C_{60} molecule. This is done for clarity and is not meant to imply any direct interaction between the C_{60} and the atoms of the second layer.

In constructing the second layer, it seems reasonable to expect this layer to preserve some of the characteristic symmetry elements of the first layer (i.e., the fivefold axes). The second layer on C_{60} contains 72 atoms, a number being indivisible by 5. This requires that each of the five-fold symmetry axes passes through two metal atoms. Consequently, in the second layer there must be one metal atom situated above each of the 12 pentagonal faces of C_{60}. Let us first assume that the second layer has the full icosahedral symmetry I_h of the first layer. The remaining 60 atoms may then be arranged basically in two different ways. The first would be to place the atoms such that they are triply coordinated to the atoms of the first layer (i.e., placing them above the carbon atoms of the C_{60} molecule as shown in Fig. 4 on the upper left). The atoms above the pentagons of C_{60} (black) constitute the vertices of an icosahedron, the other atoms (white) resemble the C_{60}-cage. This structure can also be visualized as twelve caps, each consisting of a 5-atom ring around an elevated central atom, placed at the vertices of an icosahedron. This structure, however, does not result in an even coverage: there are 20 large openings above the hexagonal faces of C_{60} while neighboring caps overlap above the double bonds of C_{60}. Pictured on the upper right in Fig. 4 is a second way to arrange the 60 atoms with I_h symmetry, obtained by rotating each of the caps described above by one-tenth of a turn (36°) around the 5-fold axis through its center. The coordination to the atoms of the first layer will then be only two-fold, but the coverage will be quite even, making the latter of these two structures the more probable one.

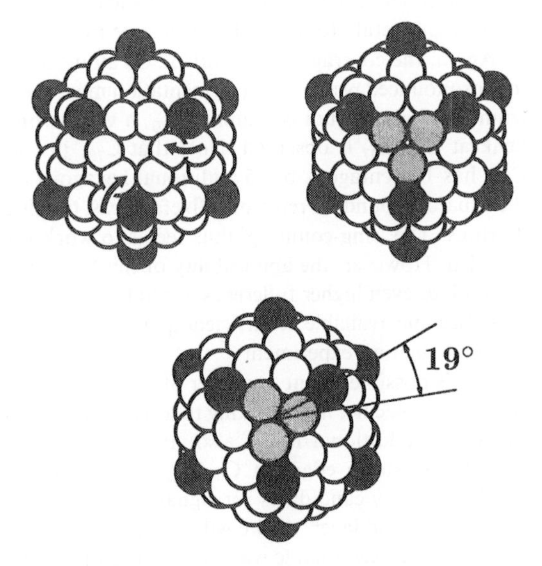

Fig. 4. Three possible geometries for arranging the 72 atoms of the second layer: the atoms above the pentagons of C_{60} are shaded. The structure on the upper left can be transformed into the more evenly distributed arrangement of atoms on the upper right by 36° turns of the caps around the five-fold axes. From this, the structure on the bottom can be obtained by rotating each triangular face of atoms by 19°.

The latter structure could be described as an 'edge-truncated icosahedron' with 20 triangular faces, each face consisting of the three atoms at the icosahedral vertices with a smaller, almost densely packed triangle of three atoms set in between (exemplarily, one of these triangles has been shaded). Note that this layer, having no atoms right on the edges, is not identical to a Mackay icosahedron[16] which is formed by pure alkaline earth metal clusters[10,11]. However, in this structure the two rows of atoms forming the truncated edges are not close-packed within the layer. This might be a hint that with the structure depicted on the upper right in Fig. 4 we have not yet found the most stable configuration of the second layer.

Up to this point, we have assumed that the second layer of atoms preserves the full symmetry (I_h) of the fullerene inside. Let us now allow the second layer to lower its symmetry. This can be done in a simple way: model the interaction between metal atoms by a short-range pair potential with an appropriate equilibrium distance and let the atoms of the second layer move freely within this potential on top of the first layer. This allows the atoms to move to more highly coordinated positions. Starting with atoms in the arrangement with I_h-symmetry, the layer will relax spontaneously by rotating all 20 triangular faces of atoms around their three-fold axes by approximately 19°. The resulting structure is shown at the bottom of Fig. 4. One of the rotated triangles has been shaded and the angle of rotation marked. In a projection on a plane perpendicular to the threefold axis, each pair of atoms at the edges of the triangle lie on a straight line with one of the three atoms on the surrounding icosahedral vertices. The two rows of atoms along the former truncated edges have now shifted by the radius of one atom relative to each other in direction of the edge, leading to close packing at the edges. Of course, the triangles could have been rotated counterclockwise by the same angle, resulting in the stereoisomer of the structure described above. This structure no longer has I_h-symmetry. There are no reflection planes and no inversion symmetry. Only the two-, three-, and five-fold axes remain. The structure belongs to the point group I (order 60). I is the largest subgroup of I_h. The layer has, thus, undergone the minimum reduction in symmetry.

Of the three arrangements of atoms in the second layer shown in Fig. 4, we find the one on the bottom (symmetry I) the most probable. It optimizes the coordination of neighboring atoms within the layer and, as we will see further down, this arrangement can also be well extended to C_{70} coated with metal.

Of course, after having observed two complete layers of metal around a fullerene, we searched for evidence for the formation of additional layers. However, before looking at experimental data, let us try to con-

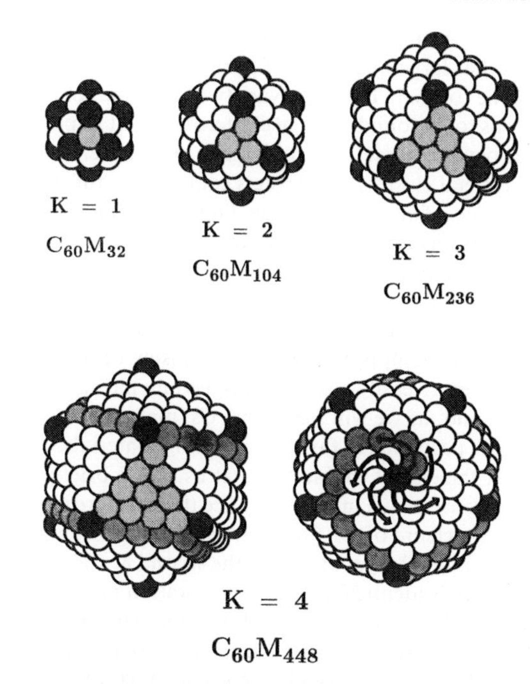

Fig. 5. Proposed arrangements of the atoms in the first four layers of an alkaline earth metal around a C_{60} molecule: the atoms at the icosahedral vertices are drawn in black and one of the triangular faces of atoms has been shaded in each layer. Note the spiral of atoms (dark grey) in the fourth layer.

struct the third and fourth layers around C_{60} in a manner similar to the second layer with I-symmetry: place one atom above each of the icosahedral vertices; for each additional layer, increase the length of the edges of the triangles between the vertices by one atom with respect to the underlying layer; rotate the triangles so that each edge points toward a different icosahedral vertex. For the second layer, this angle of rotation is 19°. For the third and fourth layer it is approximately 14° and 11°, respectively. These angles are measured relative to the position with full I_h-symmetry. The atoms stacked as triangular faces above the hexagonal rings of C_{60} resemble a tetrahedron with one tip pointing towards the center of the cluster, having a slight twist due to the difference in orientation of a few degrees between consecutive layers. The resulting structures of the first four layers are depicted in Fig. 5. For clarity, one of the triangular faces has been shaded. The atoms at the icosahedral vertices are drawn black. The number of atoms required to complete the third and fourth layer in this manner are 236 and 448.

At the bottom of Fig. 5, the fourth shell is shown from two directions. Note the spiral of atoms that are emphasized by a dark grey. This spiral can be wound around any of the five-fold axes from tip to tip. Similar spirals exist in the other layers, too. Each layer can be envisioned to consist of five such spirals of atoms. For each layer, there is also the stereoisomer with the opposite sense of chirality.

To express the number of atoms needed to complete such layers mathematically, let us introduce a layer index K. Define K as the number of atoms along the edge of a triangular face without including the atoms on the vertices above the C_{60}-pentagons. The first layer then has $K = 1$, the second $K = 2$. The number of atoms in the Kth layer can easily be calculated to

$$10K^2 + 10K + 12. \tag{1}$$

The total number of atoms $N(K)$ in a cluster composed of K complete layers around C_{60} becomes

$$N(K) = \tfrac{1}{3}(10K^3 + 30K^2 + 56K). \tag{2}$$

Note that the coefficient of the leading order in K, determining the shell spacing on an $N^{1/3}$ scale, is equal to that of an icosahedral cluster of the Mackay type[17]. Inserting $K = 1 \ldots 4$ into eqn (2), we find $N(1) = 32$, $N(2) = 104$, $N(3) = 236$, $N(4) = 448$, $N(5) = 760$, etc.

Did we predict the number of atoms required to complete additional layers around the metal-coated C_{60} correctly? Figure 6 shows a spectrum of C_{60} covered with the largest amount of Ca experimentally possible (note the logarithmic scale). Aside from the edges of $x = 32$ and $x = 104$ which we have already discussed, there are additional clear edges at $x = 236$ and $x = 448$ (completion of a third layer was also observed at $C_{60}Sr_{236}$). Note that these values are identical to the ones just predicted above for the completion of the third and fourth layer of metal atoms. We, therefore, feel confident that the alkaline earth metals studied do, in fact, form the distinct layers around a central C_{60} molecule with the structures depicted in Fig. 5.

It should be pointed out again that these layers would, of course, contain identical numbers of atoms if the triangular faces had not been rotated and, thus, the I_h-symmetry had been preserved[7]. The reason for preferring the arrangement with I-symmetry (which can still be called icosahedral) is that it leads to higher coordination of the atoms at the borders between the triangular faces.

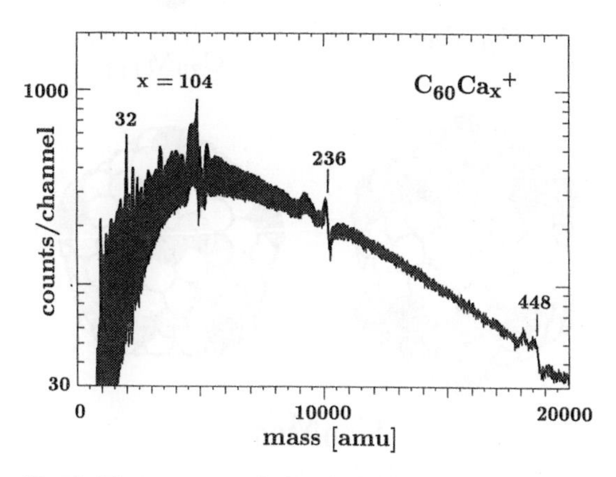

Fig. 6. Mass spectrum of photoionized $C_{60}Ca_x$ clusters with high metal content: additional edges, interpreted as completion of a third and fourth layer, are observed at $x = 236$ and $x = 448$.

Note that the structures depicted in Fig. 5 are not self-similar because the angle of rotation of the faces differs for each layer. The layers should, therefore, not be called 'shells' as they are called in the case of pure alkaline earth-metal clusters. With increasing size, the shape of the cluster will converge asymptotically to that of a perfect icosahedron.

With C_{70} at the center of the cluster, we observed the completion of layers at $x = 37$, 114, and 251. For completion of the observed three layers around C_{70}, each layer requires 5 atoms more than the corresponding layer around C_{60}. The arrangement of atoms in the first layer is again obvious: place one atom above each of the 37 rings of the fullerene.

Attempting to preserve the D_{5h}-symmetry of C_{70} molecule and of the first layer when constructing the second and third layer, results in some ambiguity of placing the atoms on the equator around the five-fold axis. Also, we found no structure that was sufficiently close packed to be convincing. Lowering the demand on symmetry by removing the symmetry elements containing a reflection (as was done in the case of the coated C_{60}) leads to the point group D_5. Similar to C_{60}, close-packed layers can be obtained by rotating the 10 remaining triangular faces around their normal by 19°. The remaining atoms can be placed in a close-packed arrangement on the remaining faces on the equator. Fig. 7 shows these first three layers. For the third layer, shown from two different directions, one spiral of atoms is indicated by a dark grey shading. Again, the layers can be envisioned to consist of five spirals of atoms around the five-fold axis.

Very high metal vapor pressures are required to produce the multilayered clusters discussed above, so high that large quantities of pure metal clusters may also be formed. The great variety of isotopic compositions to be found in large clusters makes it impossible, beyond some size, to distinguish between these pure metal clusters and clusters containing a fullerene molecule. This complication limits the amount of metal atoms that can be placed on one fullerene and, thus, the number of layers observable. This maximum amount differs for each alkaline earth metal and is lowest in the case of Ba coating. For this reason, it is desirable to suppress pure metal-cluster formation. This is more easily achieved with certain metals, such as Ca and, as we will see below, Cs, making these elements particularly favorable coating materials.

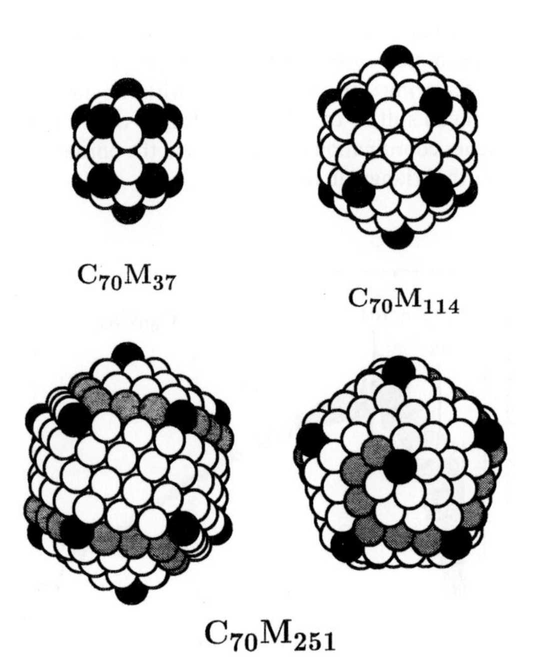

Fig. 7. Proposed arrangements of the atoms in the first three layers of an alkaline earth metal around a C_{70} molecule: the atoms at the icosahedral vertices are drawn in black. Note the spiral of atoms shaded in the third layer.

At the end of this section, let us return briefly to the spectra shown in Fig. 3. Notice the structure in the mass spectrum of $C_{60}Ca_x$ between the completion of the first metal layer at 32 and the second at 104. This structure is identical in the fragmentation mass spectra of fullerenes covered with Ca and with Sr. It is reminiscent of the subshell structure of pure Ca clusters. The subshells could be correlated with the formation of stable islands during the growth of the individual shells[10,11]. The 'sublayer' structure we observe here may also give some clue to the building process of these layers. However, the data is presently insufficient to allow stable islands to be identified with certainty.

4. COATING WITH ALKALI METALS

The structures observed in the mass spectra of fullerene molecules covered with alkaline earth metals, as described in the previous section, all seem to have a geometric origin, resulting in particularly stable cluster configurations every time a highly symmetrical layer of metal atoms around a central fullerene molecule was completed. When replacing the alkaline earth metals by an alkali metal (i.e., Li, Na, K, Rb, or Cs), a quite different situation arises.

Let us begin with clusters having a low metal content but containing several fullerene molecules. Figure 8 shows a fragmentation mass spectrum of $(C_{60})_nRb_x$ (a weak background has been subtracted). Mass peaks belonging to groups of singly ionized clusters with the same number of fullerenes have been joined by a connection line to facilitate assigning the various peaks. This spectrum is clearly dominated by the peaks corresponding to $(C_{60}Rb_6)_nRb^+$. Of the peaks corresponding to doubly ionized clusters, also visible in Fig. 8, the highest peak of each group $(C_{60}Rb_6)_nRb_2^{2+}$ with odd n, has been labeled '++' (note that every other peak of doubly ionized clusters with an even number of fullerenes coincides with a singly ionized peak). Writing the chemical formula of these particularly stable clusters in this way makes the systematics behind these magic peaks immediately clear: one or two Rb atoms are needed to provide the electrons for the charged state of the cluster, the remaining cluster consists of apparently exceptionally stable building blocks

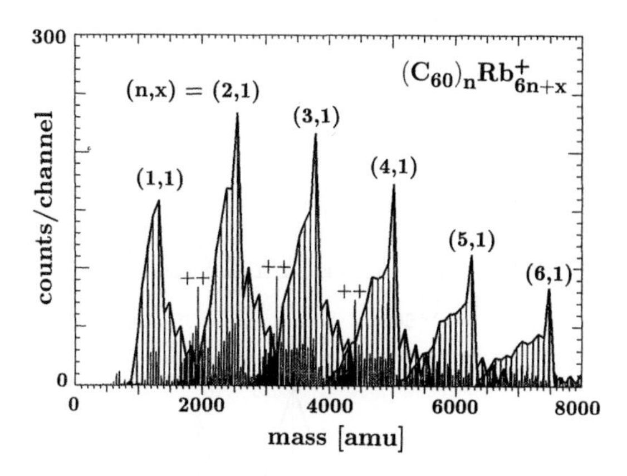

Fig. 8. Mass spectrum, with background subtracted, of photoionized $(C_{60})_n Rb_x$ clusters containing both singly and doubly ionized species: the solid line connects peaks belonging to groups of singly ionized clusters with a fixed value of n. Note the dominant peaks corresponding to $(C_{60}Rb_6)_n Rb^+$ and $(C_{60}Rb_6)_n Rb_2^{2+}$ (marked "++").

$C_{60}Rb_6$. The corresponding building block can be found in the mass spectra of clusters containing any alkali metal and C_{60}. Only Na is a minor exception to the extent that the clusters $(C_{60}Na_6)_n Na^+$ do not show up as especially strong peaks in the fragmentation mass spectra. They do, however, mark a sharp falling edge and a distinct change in the character of the spectra, as we will see later.

It seems quite obvious that the origin of the stability of these building blocks is not geometric. More likely, the electronic configuration of this unit is responsible for the stability, the six valence electrons of the metal transferred to the six-fold degenerate t_{1u} LUMO of the C_{60} molecule. Such a transfer of six electrons to the LUMO of C_{60} has also been observed in the bulk intercalation phases of $C_{60}M_6$ with $M \in \{K, Rb, Cs\}$[5]. These alkali metal fullerides become insulators due to the complete filling of the t_{1u} derived band (which was found to be only slightly disturbed by the presence of the alkali ions[5]). The appearance of such a building block is not limited to clusters containing C_{60}. Mass spectra of $(C_{70})_n M_x$ show exactly the same intensity anomalies at $(C_{70}M_6)_n M^+$ and $(C_{70}M_6)_n M_2^{2+}$. An explanation similar to the one given for C_{60} regarding the stability of the building block observed holds for C_{70}[18].

Adhering to this interpretation, the bonding of the first six or seven alkali metal atoms will be primarily ionic in nature. How will additional atoms attach to the C_{60} molecule? Will they continue transferring their valence electrons to the next unoccupied orbital of C_{60}, again showing high stability when this six-fold degenerate t_{1g} orbital becomes filled? Looking for information supporting this hypothesis, we will begin with an investigation of clusters having the composition $C_{60}Li_x$. Based on *ab initio* calculations, it has been suggested that the cluster $C_{60}Li_{12}$ should be stable with the valence electrons from the Li atoms filling both the t_{1u} and the t_{1g} orbitals[13].

Figure 9 shows fragmentation mass spectra of sin-

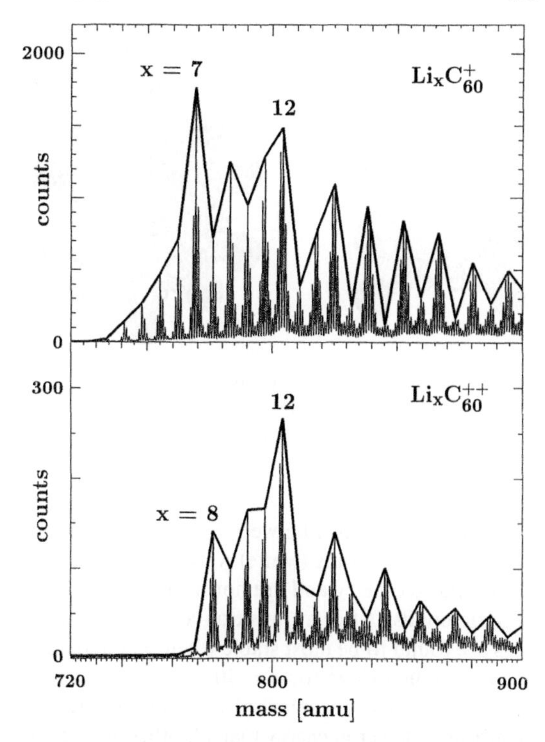

Fig. 9. Mass spectra of singly (top) and doubly (bottom) ionized $C_{60}Li_x$ clusters: note the prominent features at $x = 7$ for singly ionized and $x = 8$ for doubly ionized clusters and at $x = 12$ in both spectra.

gly and doubly ionized $C_{60}Li_x$ clusters. Mass peaks are, again, joined by a connecting line. The fine structure of the peaks is caused by the two natural isotopes of Li. Again, we find prominent peaks at $x = 7$ for singly ionized and $x = 8$ for doubly ionized clusters. Additionally, there are prominent peaks at $x = 12$ in both spectra. Twelve is exactly the number of electrons needed to fill the t_{1u} and t_{1g} orbitals, so it seems, at first, that we have found what we were looking for. However, remember that these clusters are charged, so the t_{1g} orbital obviously cannot be filled completely. Since the appearance of the magic number 12 is independent of charge, it seems more promising to try a geometric interpretation. *Ab initio* calculation shows that the twelve Li atoms have their equilibrium position above each of the twelve pentagonal faces and, thus, retain the icosahedral symmetry[13]. It seems likely that this highly symmetrical arrangement of atoms is responsible for the high stability of $C_{60}Li_{12}$, independent of the state of charge, rather than a complete occupation of vacant molecular orbitals.

To support this interpretation, we performed semi-empirical quantum chemical calculations using the modified-neglect-of-diatomic-overlap (MNDO) method[19,20]. For $x = 1 \ldots 14$, we searched for the most stable ground state geometries of $C_{60}Li_x$. We found that for $x = 1 \ldots 8$ for Li atoms preferred to be centered above the hexagonal faces of C_{60}[12]. Exemplarily, the geometry of $C_{60}Li_8$ is shown in Fig. 10 on the left. The eight Li atoms are situated at the corners

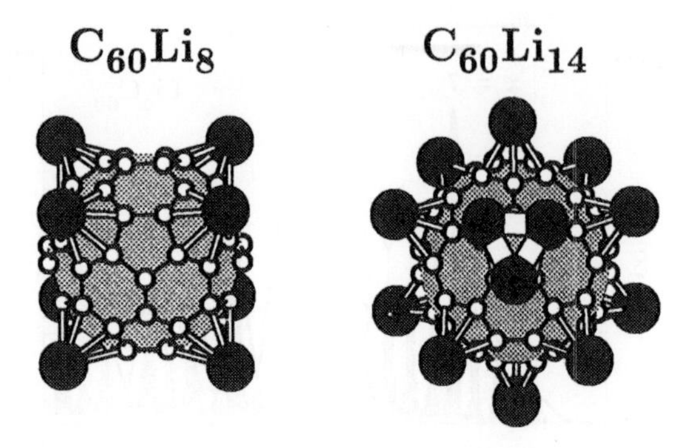

Fig. 10. Most stable ground-state geometries found for $C_{60}Li_8$ and $C_{60}Li_{14}$ by the MNDO calculations: the Li atoms are represented by the filled black circles.

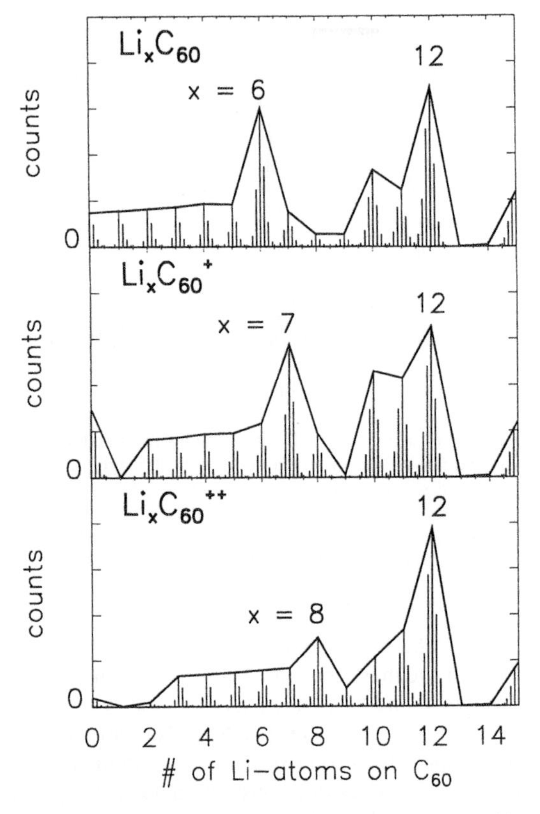

Fig. 11. Abundance mass spectra of differently charged hot $C_{60}Li_x$ clusters evaporating atoms calculated with a Monte-Carlo simulation (the Li and C_{60} isotope distributions are included). Energies required to remove Li atoms were calculated using the MNDO method. The peaks at $x = 12$ and at $x = 6 + n$ (where n is the cluster charge) observed in experiment (Fig. 9) are well reproduced.

of a cube. The bonds between the Li atoms (black) and the carbon atoms (white) were drawn merely to clarify the geometry and are not meant to imply any specific bonds. After a transition at $x = 9$, all Li atoms are found to be most stable when centered above the pentagonal rings for $x = 10 \ldots 12$. For $C_{60}Li_{12}$, the icosahedral arrangement of Li atoms proved to be significantly lower in energy than all other isomers, independent of the charge of the cluster, while for clusters with x around 7, the number of electrons in the cluster dominated over the geometry in determining the total binding energy of the cluster. Interpreting the magic numbers $x = 7$ and $x = 8$ to be of electronic and $x = 12$ to be of geometric origin thus seems reasonable.

For $C_{60}Li_{13}$, the most stable geometry has 12 Li atoms above the pentagons and one above a hexagon. If a fourteenth atom is placed near the Li atom above a hexagon, the arrangement of Li atoms becomes unstable. The two Li atoms initially not above a pentagon of C_{60} will then slide on top of a pentagon. The resulting most stable geometry of $C_{60}Li_{14}$ has one equilateral Li trimer (Li—Li bond length of 2.23 Å) lying flat above a pentagon and 11 Li atoms centered above the remaining pentagons of C_{60} as shown in Fig. 10 on the right. For comparison: MNDO calculates a bond length of 2.45 Å for the isolated Li_3^+ (equilateral triangle) and 2.19 Å for the two short bonds of neutral Li_3.

From the binding energies calculated for the different cluster compositions, we determined abundance mass spectra for heated $C_{60}Li_x$ clusters from a simple Monte Carlo simulation. Figure 11 shows the simulated mass spectra resulting from these calculations, including the Li and C_{60} isotope distributions. The peaks at $x = 12$ and at $x = 6 + n$ (where n is the cluster charge) observed in the experiment (Fig. 9) are well reproduced. For more details, see ref. [12].

For values of x greater than 14, a strong even-odd alternation becomes visible in the spectra shown in Fig. 9, peaks corresponding to clusters with an even number of available metal valence electrons being stronger. We suggest that this even-odd alternation, similarly observed in pure alkali metal clusters, signals the onset of metal-metal bonding of the metal atoms on the surface of C_{60} (remember that the MNDO calculations already show the formation of a metal trimer for $x = 14$). The electronic configuration of the clusters would, then, again determine their relative stability just as it does for pure alkali metal clusters. Consistent with this 'electronic' interpretation, the even-odd alternation displayed by the doubly ionized clusters is shifted by one atom with respect to the singly ionized clusters, an additional Li ion required to supply the charge of the cluster.

Such an even-odd alternation is observed to a different degree for all alkali metals covering fullerene molecules (see also Fig. 8). It is especially strong for Na. Fig. 12 shows a fragmentation mass spectrum of singly charged $C_{60}Na_x$. A strong even-odd alternation starts above $x = 7$, the point at which we suggested the metal-metal bonding to begin, and extends up to approximately $x = 66$. Note that $x = 12$ does not appear as a magic number in these spectra. In fact, Li is the only metal for which this magic number is observed. One possible explanation as to why Li behaves differently is the ability of Li atoms to form covalent bonds with carbon because the Li 2s orbital is close enough in energy to the carbon valence orbitals. Other than Li, the higher alkali metals form essentially ion pairs

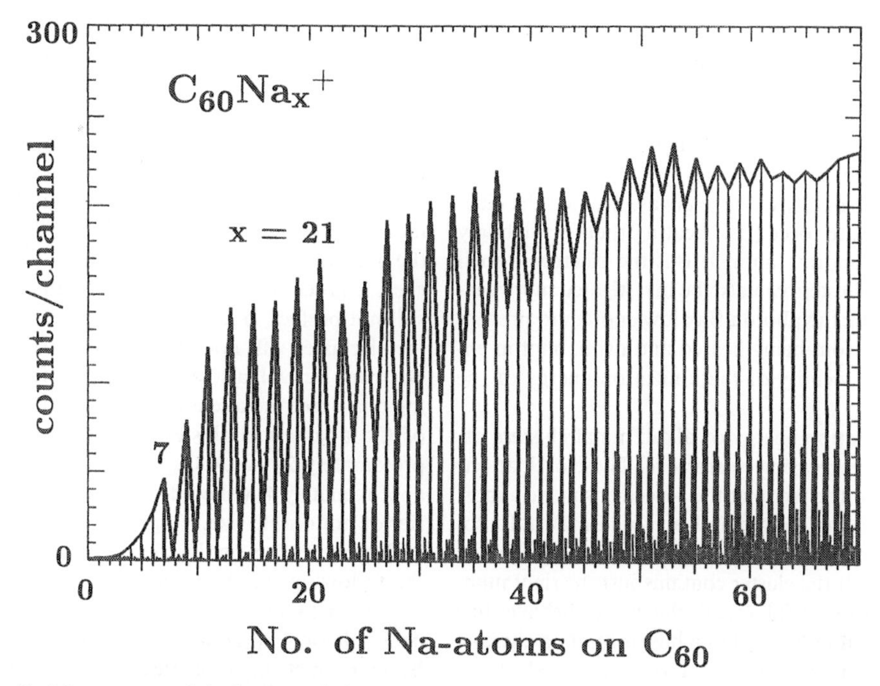

Fig. 12. Mass spectra of singly charged clusters composed of a single C_{60} molecule coated with a large amount of Na (background subtracted). The even-odd alternation extends up to approximately $x = 66$. Note that $x = 12$ does not appear as a magic number in these spectra.

in the gas phase (a Li^+ ion is exceptionally small and has, therefore, an exceptionally high charge-radius ratio, comparable to that of Mg^{2+}). A neighboring negatively charged fullerene would be polarized to such an extent that the description as ion pair would not be justified. The configuration of Li atoms around C_{60} might, therefore, be influenced more strongly by the structure of the fullerene molecule than is the case for other alkali metals, resulting in the unique configuration and stability of $C_{60}Li_{12}$.

Unfortunately, in the case of fullerenes covered with alkali metals, clear evidence is lacking regarding the geometry of the clusters. We can, therefore, only present speculation that may appear plausible but cannot be proven presently. The first seven Na ions of the $C_{60}Na_x^+$ clusters arrange themselves as far from each other as possible to minimize coulomb repulsion while adhering to the C_{60} molecule. Additional Na atoms might successively attach to these 7 ions in pairs of two, forming Na_3^+ trimers similar to the one calculated for $C_{60}Li_{14}$. Every time such a stable trimer, each containing two metal valence electrons, is completed, a strong peak is observed in the spectrum, resulting in an even-odd alternation. The abrupt change in the strength of this alternation at $x = 21 = 3 \times 7$ Na atoms fits this speculation.

When coating fullerenes with larger alkali metal atoms, the even-odd alternation is interrupted before reaching $x = 21$, so the structural sequence must be different for these. Nevertheless, we do suggest that the first 6 alkali metal atoms, having transferred their valence electron to the fullerene molecule, will remain distributed over the surface of the fullerene, gathering additional metal atoms around them as the cluster increases its metal content. This would result in at least one metallic layer coating the molecule (so speaking of metal-coated fullerenes seems justified). However, we do not have any evidence from the spectra indicating when this layer will be completed (a rough estimate shows that a first metal layer, for example of Cs, would require around 30 atoms for completion). As we have already mentioned, the stability of the alkali-fullerene clusters seems to be primarily determined by the electronic configuration. Therefore, it is not too surprising that completion of a layer of atoms, which would be a geometrically favorable structure, does not lead to any pronounced features in the mass spectra. Furthermore, it should be emphasized that to obtain these fragmentation spectra, the clusters have been heated up to a temperature at which they evaporate atoms on a μsec time scale. This corresponds to a temperature at which bulk alkali metals are molten. Incidentally, a similar behavior is observed in pure metal clusters: small alkali clusters (less than 1500 atoms) show electronic shells and alkaline earth clusters show geometric shells[9,10].

When the cluster, containing one fullerene, continues to grow by adding more metal, it will probably assume the more or less spherical shape observed for pure alkali metal clusters. It could, then, be viewed as a metal cluster with a large 'impurity': the fullerene. Alkali metal clusters containing small impurities, such as $(SO_2)_n$ or O_n, have already been studied[21,22], showing that the main influence of the impurity is to shift the number of atoms at which electronic shell closings are observed upwards by $2n$, 2 being the number of electrons bonded by SO_2 and O. What effect does C_{60} as an impurity have on the electronic shell structure? Will it merely shift the shell closings by 6 (the number of electrons possibly transferred to the C_{60} molecule)? We will investigate this in the following paragraphs.

Up to this point, we have always studied the clusters using brute force (i.e., heating them so strongly that they evaporate atoms). But the electronic shell structure of clusters can also be investigated more gently by keeping the photon flux low enough to prevent the clusters from being heated and using photon energies in the vicinity of the ionization energy of the clusters.

The ionization energy of alkali metal clusters oscillates with increasing cluster size. These oscillations are caused by the fact that the s-electrons move almost freely inside the cluster and are organized into so-called shells. In this respect, the clusters behave like giant atoms. If the cluster contains just the right number of electrons to fill a shell, the cluster behaves like an inert gas atom (i.e., it has a high ionization energy). However, by adding just one more atom (and, therefore, an additional s-electron), a new electronic shell must be opened, causing a sharp drop in the ionization energy. It is a tedious task to measure the ionization energy of each of hundreds of differently sized clusters. Fortunately, shell oscillations in the ionization energy can be observed in a much simpler experiment. By choosing the wavelength of the ionizing light so that the photon energy is not sufficient to ionize closed-shell clusters, but is high enough to ionize open-shell clusters, shell oscillations can be observed in a single mass spectrum. Just as in the periodic table of elements, the sharpest change in the ionization energy occurs between a completely filled shell and a shell containing just one electron. In a threshold-ionization mass spectrum this will be reflected as a mass peak of zero intensity (closed shell) followed by a mass peak at high intensity (one electron in a new shell). This behavior is often seen. However, it is not unusual to find that this step in the mass spectrum is 'washed out' for large clusters due to the fact that the ionization threshold of a single cluster is not perfectly sharp.

Figure 13 shows a set of spectra of $C_{60}Cs_x$ clusters for three different wavelengths of the ionizing laser. Note the strong oscillations in the spectra. Plotted on a $n^{1/3}$ scale, these oscillations occur with an equal spacing. This is a first hint that we are dealing with a shell structure. Because this spacing is almost identical to the one observed in pure alkali metal clusters, these oscillations are most certainly due to electronic rather than geometric shells. The number of atoms at which the shell closings occur are labeled in Fig. 13 and listed in Table 1. Note that these values do not correspond to the minima in the spectra as long as these have not reached zero signal.

Also listed in Table 1 are the shell closings observed in pure alkali metal clusters[9,21,23]. These values and the ones observed for the Cs-covered C_{60} have been arranged in the table in such a way as to show that there is some correlation between the two sets of numbers, but no exact agreement. If we make the simplifying assumption that six Cs atoms transfer their valence electrons to the C_{60} molecule and that these electrons will no longer contribute to the sea of quasi-free electrons within the metal portion of the cluster, the number 6 should be subtracted from the shell closings observed for metal-coated C_{60}. This improves the agreement between the two sets of shells. However, it is really not surprising that the agreement is still not perfect, because a C_{60} molecule present in a metal cluster will not only bond a fixed number of electrons but will also act as a barrier for the remaining quasi-free metal electrons. Using the bulk density of Cs, a spherical cluster Cs_{500} has a radius of approximately 24 Å. A C_{60} molecule with a radius of approximately 4 Å should, therefore, constitute a barrier of noticeable size. To get some idea of the effect such a barrier has on the shell closings, let us consider the following simple model.

The metal cluster will be modeled as an infinitely deep spherical potential well with the C_{60} represented by an infinitely high spherical barrier. Let us place this barrier in the center of the spherical cluster to simplify the calculations. The simple Schrödinger equation, containing only the interaction of the electrons with the static potential and the kinetic energy term and neglecting any electron-electron interaction, can then be solved analytically, the solutions for the radial wave functions being linear combinations of spherical Bessel and Neumann functions.

Such a simple model, without the barrier due to the C_{60} at the center, has been used to calculate the electronic shell structure of pure alkali metal clusters[9].

Table 1. Comparison of experimentally observed electronic shell closings with model calculations*

Experiment		Potential well	
$C_{60}Cs_x$	M_x [21,23]	With barrier	Without barrier
12 ± 0	8	8	8
27 ± 1	20	20	20
33 ± 1	34	32	34
44 ± 0	40		40
61 ± 1	58	50	58
		80	
98 ± 1	92	90	92
146 ± 2	138	130	138
		178	186
198 ± 0	198 ± 2		196
255 ± 5	263 ± 5	252	254
352 ± 10	341 ± 5	330	338
445 ± 10	443 ± 5	428	440

*See text. The first two columns give the numbers of metal atoms at which electronic shell closings have been observed in experiment for Cs-covered C_{60} and for pure alkali metal clusters, respectively. The columns on the right list the number of electrons required for shell closings in an infinitely deep potential well with and without a central barrier. The numbers in the different columns are mainly arranged in a manner to show correlations.

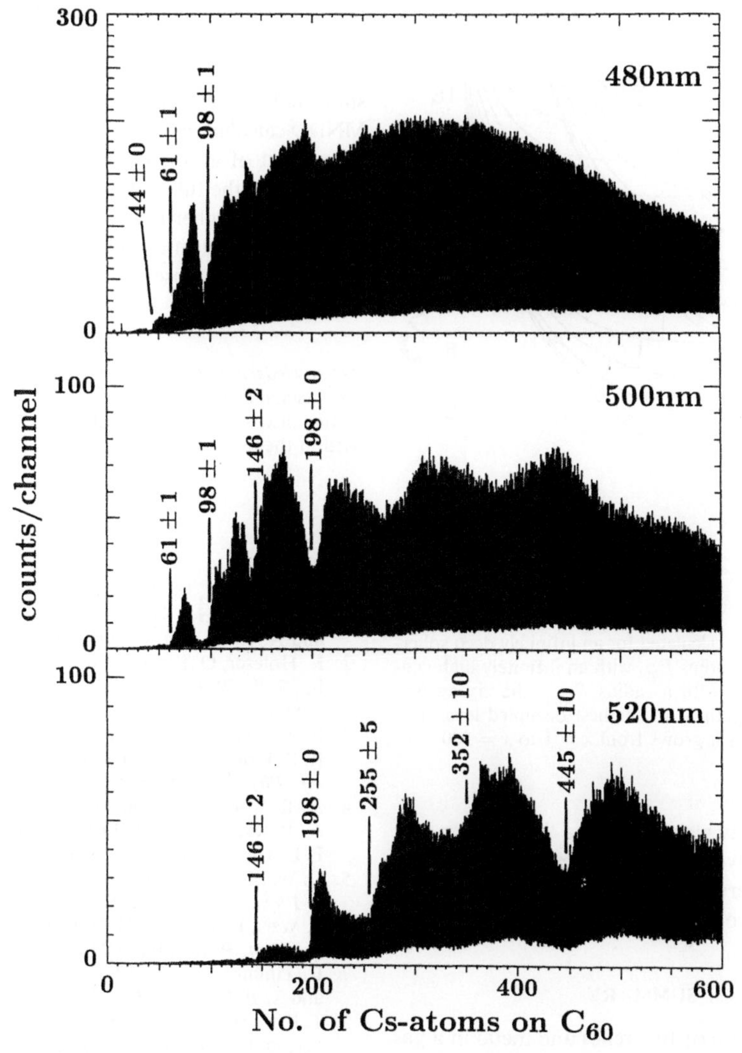

Fig. 13. Mass spectra of $C_{60}Cs_x$ clusters ionized at different photon energies near the ionization threshold; the values of x corresponding to the closing of electronic shells are indicated.

The electronic shell closings obtained from this model are listed in Table 1. Note that the agreement with the shells found experimentally in pure metal clusters is quite good. We should mention, at this point, that an intensity anomaly is not observed in the mass spectrum each time a new energy level (subshell) is filled. For large clusters only a 'bunching' of the subshells on the energy scale leads to a pronounced shell structure (it is plausible that, for example, the filling of a two-fold degenerate s-state will have little effect on a system containing hundreds of electrons).

Consider now the solutions of the spherical potential well with a barrier at the center. Figure 14 shows how the energies of the subshells vary as a function of the ratio between the radius of the C_{60} barrier $R_{C_{60}}$ and the outer radius of the metal layer R_{out}. The subshells are labeled with n and l, where n is the principal quantum number used in nuclear physics denoting the number of extrema in the radial wave function, and l is the angular momentum quantum number.

The energy E of the levels is more conveniently represented on a momentum scale. The sequence of levels at the left vertical axis corresponds to the infinitely deep well without the central barrier. The presence of the barrier primarily affects energy levels with low angular momentum because only these have a high probability density near the center of the well. Also drawn in Fig. 14 is the zigzagging 'path' of the highest occupied level of a C_{60} Cs_x cluster taking on various values of $R_{C_{60}}/R_{out}$ as it grows from $x = 1$ to $x = 500$. To determine this path, we used $R_{C_{60}} = 4$ Å and the Cs-density bulk value of 0.009 atoms per Å^3. The (sub-)shells resulting from this path are listed in Table 1. Obviously, the agreement with the experimentally observed shell closings has not been improved by including C_{60} as an impenetrable barrier at the center of the metal cluster. Varying $R_{C_{60}}$ and the Cs-density within reasonable bounds does not significantly improve the situation. On the other hand, this simple model shows that the shell structure of a metal sphere does not

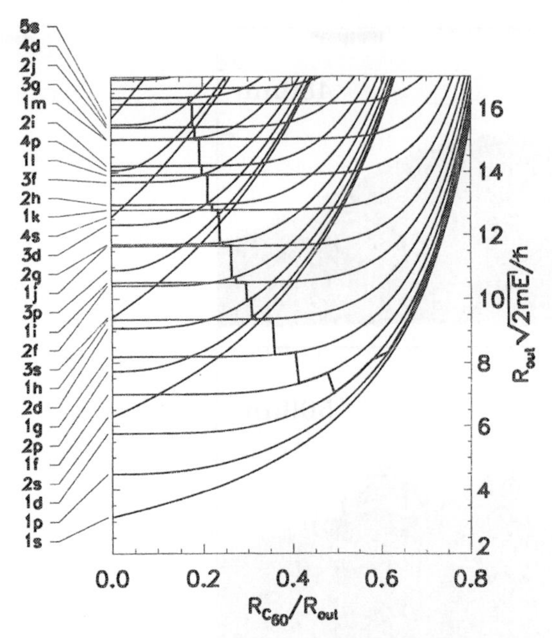

Fig. 14. Energy levels calculated for an infinitely deep spherical potential well of radius R_{out} with an infinitely high central potential barrier with a radius $R_{C_{60}}$; the zigzag line corresponds to the path of the highest occupied level of a $C_{60}Cs_x$ cluster as it grows from $x = 1$ to $x = 500$.

change significantly when placing a 'hole' in its center. This qualitative result is in agreement with the experimental observation. Similar results can be obtained from self-consistent jellium calculations[24].

5. SUMMARY

By coevaporation of fullerenes and metal in a gas aggregation cell, metal-fullerene clusters having a variety of compositions can be produced. Investigating such clusters using time-of-flight mass spectrometry, we found that alkaline earth metals will coat single fullerene molecules with up to four distinct layers of metal atoms. Clusters with complete metal layers proved to be particularly stable and appeared with enhanced intensity in the mass spectra. The number of atoms required to complete such a layer is identical for each alkaline earth metal. A geometrical arrangement of atoms, having I-symmetry in the case of coated C_{60} and D_5-symmetry in the case of coated C_{70}, was proposed for each layer. The number of alkaline earth atoms in the first layer of metal on C_{60} or C_{70} is identical to the number of carbon rings on the surface of the fullerene coated, so it seems possible to 'count' these rings.

In coating fullerenes with alkali metals, the stability of the cluster seemed to be determined primarily by the electronic configuration. The units $C_{60}M_6$ and $C_{70}M_6$, where M is any alkali metal, proved to be exceptionally stable cluster building blocks. Coating a fullerene with more than 7 alkali metal atoms led to an even-odd alternation in the mass spectra, interpreted to signal the onset of metal-metal bonding. An exception to the electronically determined cluster stability is $C_{60}Li_{12}$, which was observed to be particularly stable independent of the cluster charge. Supported by MNDO calculations, we found that the geometrical arrangement of atoms in this cluster, one above each pentagon of the fullerene, was most important for the stability. At higher alkali metal coverage of the fullerene, an electronic shell structure similar to pure metal clusters is observed in the ionization threshold of the clusters.

Acknowledgements—We would like to thank H. Schaber for his outstanding technical assistance, U. Näher for many stimulating discussions, and A. Mittelbach for providing the C_{70} used in the experiments.

REFERENCES

1. A. F. Hebard, M. J. Rosseinsky, R. C. Haddon, D W. Murphy, S. H. Glarum, T. T. Palstra, A. P. Ramirez, and A. R. Kortan, *Nature* **350**, 600 (1991).
2. K. Holczer, O. Klein, S.-M. Huang, R. B. Kaner, K.-J. Fu, R. L. Whetten, and F. Diederich, *Science* **252**, 1154 (1991).
3. P. W. Stephens, L. Mihaly, J. B. Wiley, S.-M. Huang, R. B. Kaner, F. Diederich, R. L. Whetten, and K. Holczer, *Phys. Rev. B* **45**, 543 (1992).
4. A. R. Kortan, N. Kopylov, S. Glarum, E. M. Gyogy, A. P. Ramirez, R. M. Fleming, O. Zhou, F. A. Thiel, P. L. Trevor, and R. C. Haddon, *Nature* **360**, 566 (1992).
5. D. W. Murphy, M. Z. Rosseinsky, R. M. Fleming, *et al.*, *J. Phys. Chem. Solids* **53**, 1321 (1992).
6. P. Weis, R. D. Beck, G. Bräuchle, and M. M. Kappes, *J. Chem. Phys.* **100**, 5684 (1994).
7. U. Zimmermann, N. Malinowski, U. Näher, S. Frank, and T. P. Martin, *Phys. Rev. Lett.* **72**, 3542 (1994).
8. T. P. Martin, N. Malinowski, U. Zimmermann, U. Näher, and H. Schaber, *J. Chem. Phys.* **99**, 4210 (1993).
9. T. P. Martin, T. Bergmann, H. Göhlich, and T. Lange, *J. Phys. Chem.* **95**, 6421 (1991).
10. T. P. Martin, U. Näher, T. Bergmann, H. Göhlich, and T. Lange, *Chem. Phys. Lett.* **183**, 119 (1991).
11. T. P. Martin, T. Bergmann, H. Göhlich, and T. Lange, *Chem. Phys. Lett.* **176**, 343 (1991).
12. U. Zimmermann, A. Burkhardt, N. Malinowski, U. Näher, and T. P. Martin, *J. Chem. Phys.* **101**, 2244 (1994).
13. J. Kohanoff, W. Andreoni, and M. Parinello, *Chem. Phys. Lett.* **198**, 472 (1992).
14. L. Pauling, *J. Am. Chem. Soc.* **69**, 542 (1947).
15. C. S. Yannoni, P. P. Bernier, D. S. Bethune, G. Meijer, and J. K. Salem, *J. Am. Chem. Soc.* **113**, 3190 (1991).
16. A. L. Mackay, *Acta Crystallogr.* **15**, 916 (1962).
17. U. Näher, U. Zimmermann, and T. P. Martin, *J. Chem. Phys.* **99**, 2256 (1993).
18. J. H. Weaver, *J. Phys. Chem. Solids* **53**, 1433 (1992).
19. M. J. S. Dewar and W. Thiel, *J. Am. Chem. Soc.* **99**, 4899 (1977).
20. M. J. S. Dewar and W. Thiel, *J. Am. Chem. Soc.* **99**, 4907 (1977).
21. H. Göhlich, T. Lange, T. Bergmann, and T. P. Martin, *Phys. Rev. Lett.* **65**, 748 (1990).
22. H. Göhlich, T. Lange, T. Bergmann, and T. P. Martin, *Z. Phys D* **19**, 117 (1991).
23. W. D. Knight, K. Clemenger, W. A. de Heer, W. A. Saunders, M. Y. Chou, and M. L. Cohen, *Phys. Rev. Lett.* **52**, 2141 (1984).
24. S. Satpathy and M. Springborg, private communications.

SUBJECT INDEX

AUTHOR INDEX

Printed and bound by CPI Group (UK) Ltd, Croydon, CR0 4YY

Printed and bound by CPI Group (UK) Ltd, Croydon, CR0 4YY
13/05/2025
01869820-0001